Cities and Nature

Cities and Nature fosters high-quality multi-disciplinary research addressing the interface between cities and the natural environment. It provides a valuable source of relevant knowledge for researchers, planners and policy-makers. The series welcomes empirically based, cutting-edge and theoretical research in urban geography, urban planning, environmental planning, urban ecology, regional science and economics. It publishes peer-reviewed edited and authored volumes on topics dealing with the urban and the environment nexus, including: spatial dynamics of urban built areas, urban and peri-urban agriculture, urban greening and green infrastructure, environmental planning, urban forests, urban ecology, regional dynamics and landscape fragmentation.

More information about this series at https://link.springer.com/bookseries/10068

Carla Brisotto · Fabiano Lemes de Oliveira
Editors

Re-Imagining Resilient Productive Landscapes

Perspectives from Planning History

Springer

Editors
Carla Brisotto
Florida Institute for Built Environment Resilience
University of Florida
Gainesville, FL, USA

Fabiano Lemes de Oliveira
Department of Architecture and Urban Studies (DASTU)
Politecnico di Milano
Milano, Italy

ISSN 2520-8306 ISSN 2520-8314 (electronic)
Cities and Nature
ISBN 978-3-030-90444-9 ISBN 978-3-030-90445-6 (eBook)
https://doi.org/10.1007/978-3-030-90445-6

This Springer imprint is published by the registered company Springer Nature Switzerland AG
The registered company address is: Gewerbestrasse 11, 6330 Cham, Switzerland

Contents

Productive Urban Landscapes

Contributors

Izaskun Aseguinolaza-Braga Departamento de Arquitectura, Profesora Adjunta, Universidad del País Vasco (UPV/EHU), Leioa, Spain

Paola Branduini ABC Department, Politecnico di Milano, Milano, Italy

Carla Brisotto Florida Institute for Built Environment Resilience (FIBER), University of Florida, Gainesville, FL, USA

Silvio Cristiano Department of Environmental Sciences, Informatics and Statistics, Università Ca' Foscari Venezia, Venice, Italy;
Department of Civil, Environmental and Architectural Engineering, Università degli Studi di Padova, Padua, Italy;
Interuniversity Department of Regional and Urban Studies and Planning, Politecnico di Torino, Turin, Italy

Robert Freestone University of New South Wales (UNSW), Sydney, Australia

Kristin Jones Illinois Institute of Technology, Illinois, USA

Fedor Kudryavtsev Moscow Institute of Architecture (state academy), Department of Urban Planning, Moscow, Russia

Luca Lazzarini Department of Architecture and Urban Studies (DAStU), Politecnico di Milano, Milan, Italy

Fabiano Lemes de Oliveira Department of Architecture and Urban Studies (DASTU), Politecnico di Milano, Milan, Italy

Teresa Marat-Mendes Iscte – Instituto Universitário de Lisboa, DINÂMIA'CET, Lisbon, Portugal

Cecilia Muzika-Minteer The University of Texas at Austin, Texas, TX, USA

Andrea Pane Department of Architecture, University of Naples Federico II, Napoli, Italy

Denise Piccinini Delft University of Technology, Rotterdam, The Netherlands

Dirk Schubert HafenCity University Hamburg, Hamburg, Germany

Kathryn R. Terzano Arizona State University, Arizona, USA

Herman van Bergeijk Delft University of Technology, Rotterdam, The Netherlands

Joshua Zeunert University of New South Wales (UNSW), Sydney, Australia

Rethinking the Urban–Rural Relationships and Productive Urban Landscapes

Fabiano Lemes de Oliveira and Carla Brisotto

Abstract This chapter reflects upon how lessons from past urban planning experiences can inform current debates on the relationships between the city and the countryside and on productive urban landscapes. Today, the extent of urbanisation has put in check categorical definitions of the urban and the rural. Besides, productive urban landscapes have been posited as instruments for the positive transformation of the agricultural sector and to approach territorial fragility while promoting social cohesion, food security, and broader environmental and economic benefits. The chapter initially discusses key questions related to urban–rural planning and food production in cities. This discussion is followed by the analysis of cross-cutting themes: domains of human activity, the question of reconnection, transcalarity, and resilience and sustainability of food systems, and food security. Findings suggest that an efficient food system is both sustainable and resilient. Connecting the different scales of the territory from regional to local, focusing on local sources, and creating a flexible and adaptable system would allow productive urban landscapes to adjust any environmental, social, or cultural condition guaranteeing both the rights of the individuals as much as of the public at large.

Keywords Urban–rural · Food · Productive urban landscapes · Planning · History · Resilience

1 Introduction

The continuous process of urbanisation of the globe, which will potentially lead to over 70% of the world's population living in cities by 2050, poses a range of questions on regional planning and territorial management. A recent report by the United

F. Lemes de Oliveira (✉)
Department of Architecture and Urban Studies (DASTU), Politecnico di Milano, Milan, Italy
e-mail: fabiano.lemes@polimi.it

C. Brisotto
Florida Institute for the Built Environment Resilience (FIBER), University of Florida, Gainesville, FL, USA
e-mail: c.brisotto@ufl.edu

C. Brisotto and F. Lemes de Oliveira (eds.), *Re-Imagining Resilient Productive Landscapes*, Cities and Nature, https://doi.org/10.1007/978-3-030-90445-6_1

Nations (2020) on the progress of the Sustainable Development goals showed that over the period between 1990 to 2015, most urban areas recorded a general increase in the amount of built-up area per person and that the physical expansion of cities was faster than their rates of population growth. This phenomenon suggests that the disintegration of the categories of the 'urban' and the 'rural' into a multifaceted and polysemic network of conceptual domains will continue. The 'purity' of both the urban and the rural, if that ever existed, now presents a range of hybrid characterisations that challenge the city's traditional urbanity and the countryside's ruralness (Sieverts 2003; Balducci et al. 2017). Besides, productive urban landscapes have been posited as instruments for the positive transformation of the agricultural sector and to approach territorial fragility while promoting social cohesion, food security, and broader environmental and economic benefits (Horst et al. 2017; Lohrberg et al. 2016; Warren et al. 2015).

This book explores how lessons from past urban planning experiences can inform current debates on the relationships between the city and the countryside and on urban agriculture, informing practitioners, students, researchers, policymakers, and food no-profits to reimagine urban–rural relationships, intra-urban productive landscapes, and urban agricultural systems today.

It asks two inter-related research questions: how were views on the urban–rural domains conceptualised and operationalised in significant planning experiences during the nineteenth and twentieth century? And, how did such cases of the past approach the question of productive urban landscapes within their visions? The focus is on planning ideas, either implemented or not, and the roles of key individual architects and planners. *Re-imaging Resilient Productive Landscapes: Perspectives from Planning History* aims to: (1) provide an exploration of critical reflections on the relationships between the city and the countryside in planning history; (2) bring to light significant approaches to urban agriculture from the perspective of architects or planners; (3) offer a new interpretation of historical city plans by analysing them through the urban–rural lens and from the perspective of productive landscapes; and (4) enlarge the understanding of the phenomenon of urban agriculture by comparing different global approaches.

Fourteen other chapters addressed two central themes: urban–rural relationships and productive urban landscapes. In many cases, chapters have addressed them concomitantly, albeit with different emphases. In the following sections, we present crucial current questions concerning urban–rural planning and food production, followed by a critical reflection on the two major themes mentioned above and then an analysis of the identified cross-cutting themes.

2 Challenges and Perspectives for Changing Territories

It is essential to consider that with current governmental discourses on fighting climate change and meeting the Paris Agreement commitments, containing land take and revisiting land uses become crucial. The European Union recently announced its

Green Deal, which aims to cut its greenhouse gas emissions (GHG) to at least 55% by 2030, compared to 1990, and become carbon neutral by 2050 (European Commission 2020c). Associated strategies include the EU Biodiversity Strategy, which promises the planting of three billion threes by 2030, and the Farm to Form strategy, focused on making food systems more sustainable. Today agriculture is responsible for 10% of GHG emissions in the EU, but food systems worldwide account for nearly one-third of GHG emissions (European Commission 2020b). In turn, agriculture will be one of the most impacted sectors by climate change, given processes of desertification, soil erosion, water scarcity, among others, leading to significant challenges regarding food security (European Commission 2020a).

With the increased need for afforestation and reforestation to fight climate change, agricultural land is likely to shrink and become more hybridised with forestry. In the UK, for instance, agricultural areas have declined consistently over the past 20 years because of transport infrastructure, building and woodland expansion, and the growth of other non-agricultural uses. It is predicted that a further 21% of agricultural land will be needed for actions to reduce emissions by 2050 (CCC 2020). In the EU, a reform of the Common Agricultural Policy is under discussion to help achieve the objectives of the *Green Deal*. It includes proposals to bring at least 10% of agricultural areas back under high-diversity landscape features and to destine at least 40% of the rural development budget to climate targets (European Commission 2017).

This context puts forward the paradox of planning for the forecast that 70% more food would need to be produced in the following decades (EEA 2015) while eventually reducing the availability of land for agricultural production. A range of nature-based solutions has been put forward in urban areas to help with climate change adaptation and disaster risk reduction, including urban green and blue spaces and urban agriculture (EEA 2021). Finally, a fundamental reconsideration of the relationships between core urban areas, peri-urban areas, and landscapes beyond is due if these challenges are to be met. Such reassessment will need to involve land uses as well as a renewed call for hybridism and the multifunctionality of landscapes.

The United States has tried to create a New Green Deal through the Democratic party proposal presented at the US Congress in 2019. The deal's main goal was to tackle climate change, similarly to the European initiatives, although the emphasis was in gaining along the process more equitable social living conditions. Compared to its predecessor—the New Deal by Franklin Roosevelt in 1929—the proposal aimed to boost the economy via climate change initiatives with the clear objective of transforming these actions into opportunities for equitable development across demographics, especially people of colour and low-income individuals. Amongst the recommendations for reducing GHG emissions, building new infrastructures, and pursuing renewable energy, special attention was given to the issue of food and agriculture. First, the deal included healthy food (that is, the stress is on providing nutritional food, not any kind of food) in the list of basic necessities for current and future generations. Second, the deal described the agriculture sector as partially responsible for climate change and at the same time as a contributor to its solution. The proposal suggested that investing in the design of a better food system and supporting small farming are essential actions to pursue food sustenance, equitable food accessibility, and small farmers' economic strive (HRES 109 2019).

The New Green Deal has remained a proposal and can be seen as a naive attempt to address an extremely complex problem as climate change. Yet, it has the value of exposing the connections between climate change, agriculture (i.e. GHG emissions, soil degradation and loss of biodiversity), and issues of environmental justice (food insecurity). Furthermore, it emphasised how economic actions can become opportunities not only to achieve environmental sustainability but also equity. The proposed idea to support small farming businesses would especially aid black farmers in the US Southern States that account for 88% of the total black farmers in the US (NASS 2017). According to the United States Department of Agriculture, the majority of these farmers live in poverty (Hoppe 1986). Involving them in the redesigning of the food system would increase local food production, enhance farmers living conditions, and create job opportunities while supporting the wellbeing of the entire population.

A focus on food justice is also the central focal point of a recent initiative by the United States Department of Agriculture that provides grants for urban agriculture and innovative production to not-for-profit organisations and Native American tribal governments. As the program manager of the grant Leslie Glover II states, 'with 80 percent of the U.S. population living in or near urban centers, urban agriculture can make a significant positive impact on the health and well-being of many individuals' (USDA 2021). Established in 2020, this grant is an important acknowledgement of how urban agriculture is an instrument to create proximity between production and consumption, reaching out to American cities' most vulnerable populations that often live in food-desert areas.

This brief introduction to the challenges and perspectives of changing productive landscapes does not aim to be comprehensive. However, the main goal was to review a selection of initiatives that show examples of how the Western world—which accounts for the largest rate of global GHG emissions—is reflecting upon the urgency of rethinking the food system. Interestingly, two powers such as Europe and the United States pursue this objective by focusing on two different applications, the first on the governance of land, the latter on equitable outcomes. Yet, both look for innovative approaches to cultivation and systems. As designers, planners, and architects, the challenge will be to support these and other ambitions giving shape to new urban-rural relations and forms of urbanity.

3 Urban and Rural: Dichotomies, Relationships and the Overcoming of Categories

The urban–rural dichotomy, or the contrast between the city and the countryside, is posed as a consequence of the Industrial Revolution. The fundamental economic imbalance towards urban areas led to mass rural to urban migration with associated effects to both urban and rural areas. Additionally, the high overall population growth in many countries across the nineteenth century majorly contributed to urban sprawl and the worsening of living conditions in cities (Sutcliffe 1981). The timing of such processes varied across the world, with Britain and Germany forging ahead in Europe. For instance, in 1850, England and Wales had already crossed the 50% threshold of urban population (Lawton 1973), a point only crossed globally a few years ago. The population of London more than doubled between 1801 and 1851, reaching 2.4 million people and 6.5 million by the turn of the twentieth century (Jefferies 2005). Similarly, the population of Berlin increased nearly fivefold between 1850 and 1914 to reach over two million people (Lenger 2002). Implications to urban areas included poor public health, squalor, social inequality and sprawl (Hall 1988). The consequences to the rural areas were no less problematic. They comprised depopulation with associated impact on the stewardship of the countryside, the loss of agricultural land, the rural character of the countryside and the physical and psychological disconnection of people from the land.

The urban–rural dichotomy was further exacerbated given the unequal access to open spaces by urban residents. While the wealthier tended to be historically closer to high-quality green areas within cities and was also able to move out to newly created suburbs, the increasing lack of access to open spaces was felt the most by the poor. They were also ever more distant from the countryside, now too far to reach on foot. Such discrepancies in access to urban green spaces and the countryside generated various 'park movements' and societies for the protection of the rural land in countries such as the USA, Britain, and Germany, as shown by Dirk Schubert in chapter two.

The relationship between the city and the countryside was thus a fundamental aspect of the development of town planning. Leonardo Benevolo (1967) showed that planning emerged as a reactive discipline attempting to tackle the consequences of the Industrial Revolution in towns and cities. Solutions were proposed to transform existing areas and to devise new urban models. In this context, two approaches rivalled on how to deal with the urban expansion of large cities. The first considered that they should be allowed to continue to grow and that one of the planning's most important tasks was to find ways to best organise it, such as through linear city (Soria y Mata), grid, or radio-concentric models (Abercrombie 1933). Schubert, for example, showed how Fritz Schumacher was indeed in favour of finding solutions to the large city, which was seen as a natural consequence of industrialisation and processes of agglomeration. The second approach vouched for modes of growth that would be done through new garden cities, satellite towns (Mumford 1961), or through new patterns generated by the dissolution of the urban and rural domains. The remaining of this section will discuss how these two broader perspectives manifested themselves through more precise conceptualisations of the relationships sought after between the urban and the rural areas. These are termed *proximity and differentiation*, and *integration and dissolution.*

3.1 Proximity and Differentiation

Various terms were used by planners and in literature to characterise the relationship between the urban and the rural: dichotomy, opposition, contrast, complementarity, polarities, domains, dimensions, etc. (Davoudi and Stead 2002). Dichotomy and 'opposition' are the points of departure. A common thread is the appreciation of the upper hand of the urban world upon the rural, which was seen as a subjugation of the latter through the weight of socio-cultural and economic development in urban areas, the squeezing of the rural population out to cities and pouring of expansion areas into the countryside. Recognising such opposition was a fundamental first step to consider how it could be addressed. Transforming it into complementarity was the second.

In this line of thought, although it was clear that one could not exist without the other, one must not *be* like the other. For example, emphasising the two domains, as shown by Izaskun Aseguinolaza in chapter four, Thomas Sharp was adamant that the binomials town-urban / countryside-rural ought to be maintained since it was in the very urbanity of the town and the rurality of the countryside where their power rested. He directly challenged the garden-city-based view of a town-country hybrid, which he saw as an aberration. Such thoughts resonate with those by Patrick Abercrombie who saw that the overcoming of this opposition could be achieved through conceptualising the future of the urban and the rural relationships as distinct but complementary 'polarities'. Yet, differently from Sharp, Abercrombie assumed 'tinctures' of one into the other, a park system in the city and rural development in the countryside, as necessary conditions for a new balance to be reached.

On urban–rural discourses where differentiation was to be established or maintained, the very edge between them became an essential topic of reflection. While for some, such as Sharp, an obvious boundary should be established, with no room for transition between town and country, others did consider the need for some flexibility and gradation. Fabiano Lemes de Oliveira, in chapter five, shows how Abercrombie's wartime plans for London employed the green belt as an enforcer to restrict urban growth and protect the outer countryside from encroachment while remaining primarily within the sphere of influence of the urban domain, as an overspill of recreational areas beyond the suburban ring. In addition, the application of green wedges allowed for a radial transition from parkland to countryside within London. Both the green belt and green wedges were too employed by Étienne de Groër as part of his vision to transform Lisbon into a garden-city region, as shown by Teresa Marat-Mendes in chapter seven. Joshua Zeunert and Robert Freestone in chapter twelve, reflecting on the case of Sydney, reveal how the will to establish a town-country boundary led to the definition of a green belt in the 1948 plan. Their criticism of the simplistic nature of such a belt when agricultural land was to be safeguarded, while ignoring soil quality in favour of a formal approach, can be extended to similar instances worldwide. In the case of Italy, Luca Lazzarini in chapter six argues that the city-country dichotomy persisted in planning legislation and thought throughout the immediate post-war, being Luigi Piccinato, Piero Bottoni, and Aldo

Lucchi's plan for Siena from 1956 and Giovanni Astengo's plan for Assisi from 1958 worthy exceptions that did consider planning also from a rural perspective. He shows that Astengo's case evidences a significant step forward in considering urban and rural planning simultaneously, and from their spatial and functional interdependencies. Paola Branduini and Andrea Pane's chapter further elaborate on the shift towards relationships, instead of dual opposition, looking into the plan for Amalfi and Sorrento coast in Italy made by Roberto Pane and Luigi Piccinato. They indicate how Pane and Piccinato's vision centred on enhancing a system of assets across the urban–rural areas linked by landscape relations, which pioneered a systemic approach to regional planning.

Although 'protection' of the countryside emerged as a firm principle across this volume, several plans did not see it as freezing the countryside in time. If, on the one hand, its beauty, cultural significance and role in food production were to be maintained, on the other hand, a view of its 'development' was in line to ensure the countryside could remain relevant in an industrial and urbanised society. Chapter four showed how Sharp called not for the 'protection' of the countryside as it was, but for a new growth based on its cultural landscape that would keep its rural character and put it in a position of strength, which chimes with many views on rural development today. Abercombie too envisaged the need to consider the countryside as part of a larger planning exercise, that of the region, which was seen as crucial for a common healthy future to both urban and rural areas.

If protection of the countryside was not enough, the creation of a new countryside from scratch was also contemplated and realised. Herman van Bergeijk and Denise Piccinini in chapter three narrated how the Netherlands's tradition of land reclamation gave rise, with the creation of polders, to the possibility to construct, literally and symbolically, a newly imagined countryside which should be beautiful, with both cultural and productive meaning. The country is here not only seen as agricultural land but also comprehensively defined to include villages, infrastructure and buildings.

3.2 Integration and Dissolution

While many attempted to resolve the dichotomy between the city and the countryside through the strengthening of their identities and better relationships between them, both in terms of the physical structure and flows, others envisioned either an integration of the urban and the rural into a new blended landscape, or the complete extinction of these domains altogether.

The former can be seen in Alfred Caldwell's 'the city in the landscape', as presented by Kristin Jones in chapter eleven. He put forward a model for the integration of the built and open environments; decentralising growth and the stewardship of the land. Silvio Cristiano's chapter revealed how Hans Widmer' (P.M.) *bolo'bolo* would allow for a systemic restructuring of the cities from within, based on the implementation of units blending living and food production. P.M advocated that large cities would need to be depopulated to make space for such units, and that, in turn, would repopulate villages, generating a multi-scalar network across the landscape. Remarkably, this is a phenomenon seen during the COVID-19 pandemic, which generated centrifugal forces out of large cities giving the rural areas a new lease of interest from urban dwellers (Beria and Lunkar 2021; Mukhra et al. 2020).

The latter approach is characterised by radical visions of how to overcome these domains. This would only be possible given the then-current and expected technological advances in telecommunication, energy distribution, transportation, and food production. The erosion of the urbanity of towns and the rurality of the countryside was considered as a manner of conceiving the landscape as a totality, freeing people from the attachments of historically defined traditional functions of urban and rural areas. This view initially took significant prominence after the Russian Revolution in the discussions over the urban–rural dichotomy and how the socialist city should be developed. Engels stated that human emancipation would only be achieved should the antithesis between town countryside be eradicated (Kopp 1970). For the architect El Lissitsky, this would be a necessary step for 'social evolution' to take place. In addition, the very centre-periphery structure of the capitalist city needed to be revisited since it reiterated unequal modes of production and social segregation (Lemes de Oliveira 2017). Fedor Kudryavstev unfolded in chapter ten how, through the lenses of the so-called 'disurbanists', both dichotomies were to be sublimated into a unifying landscape. It was not to be understood as the urbanisation of the countryside or the ruralisation of the city, but a new condition altogether. Technology and effective planning would resolve the dependency of the city on resources from the countryside, and the countryside's dependency on cities as centres for socio-cultural and economic development. In this future where centralities and 'peripheralities' would no longer exist, an ever-expanding constellation of networks would become the dominating model. As Carla Brisotto's chapter unveiled, this idea resonates with Yona Friedman's *Continent City* proposal, which called for transforming Europe into a 'networked continent city' made possible by high-speed train connections and balanced distribution of urban areas and food production. Similarly, Frank Lloyd Wright's Broadacre City would propose a radical reorganisation of the relationships between people and the land. Cecilia Klein's chapter argues that Wright's proposal for the resettling of people on 1 × 1-acre plots was based on the principle of communal individuality—a return to more organic ways of living anchored on agrarianism philosophy. This approach would lead to decentralisation and eventual abandonment of cities.

4 Productive Urban Landscapes

The analysis of urban–rural dichotomies enlightened how the Industrial Revolution induced the separation of these two domains. One impact of this separation has been the accessibility to food due to the detachment between producers and consumers. To overcome this impact, the recent return to growing food within cities is often framed within the idea of *Right to Food*—the right to eat an adequate amount of food, having physical access to it and the financial means to obtain it (FAO 2020). Mentioning the physical access to food unveils the relevance of planning the city's form to guarantee food for all, overcoming inequalities. This is the case of the development of food deserts in urban areas, even in those classified as rural, that disproportionally affect the low-income classes (Abel and Faust 2020; Wright et al. 2016). Although the Food and Agriculture Organization of the United Nations (FAO) defines this right as a human right, we argue that the *Right to Food* is a civic right—a political action. Thinking of food as a political action means to reject the idea that eating is only a means to survive, to pursue a bare life (Agamben 1995: 131–211). Growing food in the city defends the right to contribute to a change, to participate in the decision-making, and to shape the form of a city to adjust it to this need, recalling Henry Lefebvre's theory of the Right to the City (1996). This trend has been revamping contemporary planning and architectural discourses of how to design the urban form to include agriculture in it, building upon the lessons learned during the twentieth century. Starting from the Industrial Revolution, history showed that growing food inside the city is a resilient act used to overcome urban crises due to environmental challenges, economic distress, and wars.

Ebenezer Howard was one of the first planners to think of the form of the city as an instrument to solve the environmental issues generated by the Industrial Revolution. He initiated the Garden City movement at the end of the nineteenth century to advocate for the creation of new towns in the countryside to reverse the excessive urbanisation of existing cities generated by the massive migration from the rural areas to the industrial centres. The new city in the country aimed to maintain the beauty and pleasure of the country life with the entertainment and resources that the city life could offer. His slogan 'the country must invade the city' synthesises his theory and thought (Howard 1902: 147). The garden city proposed a model where the relationship between rural and urban areas was redefined. The city was also planned to be self-sustaining, relying on a local food system that would feed all citizens (Lickwar and Thoren 2020: 94). Each garden city of about 32,000 inhabitants was circular with public buildings and public spaces at the centre surrounded by residential areas with access to common gardens, allotments, and agricultural fields on the outskirts. Farmers cultivated the crops, whereas the allotments were cultivated by individual or small groups (Howard 1902: 26). Howard's garden city exemplifies how a productive urban landscape was part of a bigger picture to address uncontrolled urbanisation and its related problems (Brisotto 2019).

The case of German cities during World War I shows how a local food system supported the containment of the housing crisis. Notably, several German cities had allotments provided with garden sheds. The war generated a massive lack of housing with the consequent increase in the homeless population. Those Germans that had one of these allotment gardens moved in their sheds to overcome the crisis (Henderson 2010, 1999: 319). The post-war German program *Siedlung* unveils how this long tradition of gardening the city was recreated in a series of new settlements. The one in Frankfurt am Main, designed by architect Ernest May in 1926, being one of the most well-known. This new housing scheme, besides aiming at providing the conditions for better living, also sought to provide each house with enough space for a garden, which could be cultivated (Miller 1895: 208). In this case, the productive urban landscapes, either traditional or through the *Siedlung* program, became opportunities for addressing both the housing and food crises.

Urban productive landscapes were also used to address both economic and war distress. The Detroit potato patches (nineteenth and twentieth centuries) and the North American community gardens of the 70s and 80s are all examples of how the cities' open spaces were adjusted to the food emergencies of their times (Lawson 2005). This was a tradition that started with the Victory War Gardens of World War II that used parks and private backyards to grow food during wartime (Imbert 2015; Lawson 2005). Schubert's, Kathryn's Terzano's, and Lemes de Oliveira's chapters also report examples of war gardens, respectively, in Stadtpark in Hamburg (Germany) during World War I, the allotment gardens in Denmark during both World Wars, and with the 'Dig for Victory' initiative in the United Kingdom starting in 1939. Zeunert and Freestone's chapter reports on a similar program in Sydney, Australia, where the local farming activity supported the military forces during World War II.

This overview explained how productive urban landscapes have been connected with the *Right to Food* responding to historical shocks and influencing and impacting the design and uses of urban patterns. In the following sections, we explore how this notion of right raises questions about the *dilemma of the commons*, the balance between individual rights and communal rights. We also analyse how the *Right to Food* is a design action.

4.1 The Dilemma of the Commons

The presence of urban productive landscapes inevitably raises questions of leveraging and balancing individual rights over the community (groups of people sharing the same values) or public rights (the rights of the entire population beyond cultural, demographics or ethnic differences) (Edmonds 2020). How does this balance reflect on the distribution of private and public spaces? How can we accommodate these diverse levels of engagement? Do we prioritize one over the other or do resilient productive urban landscapes shift from one level to the other? The dimension of communities in relation to available resources, the connection between individual and collective needs, and the private and public uses of space are all elements playing a role in the definition of the commons within cities.

Controlling the size of a community seems a strategic approach to achieve a balance between individual needs and collective ones. The critical group size developed by Friedman—as Brisotto reported in her chapter—described exactly this concept. Friedman suggested the size of the group has to be calculated case by case according to the territory and type of the problem to be solved. In this light, the more resources are available the bigger the community can be and vice versa. This idea is a flexible and malleable way of designing a community within a territory receding from ideologies of homogenisation that ignore specific needs and situations. In controlling the size of the group and assigning it to a specific territory, the designer simultaneously achieves individual and community needs, and the public interest to preserve the environment at large. The question of size is also treated by Cristiano. In his chapter, we read that P.M. thought about the dimension of the group as key to sustainability—as per Friedman—but also spatial justice. He indicated a size of 300–500 individuals which allows members to know each other. Within this type of group, individuals cultivated food, detaching themselves from the system, and becoming 'calories independent'. At the same time, P.M.'s individual's *Right to Food* was a means to contribute to the community (*bolo'bolo*) overall self-sustenance. P.M. pushed the argument even further when allocating the spaces in-between *bolos* to persons that do not fit in the community cultures and values. These outsiders embodied the extreme individual right of living completely off the system. An idea that recalls several current attempts of living exclusively by self-grown produce.

The idea that small enterprises can help the community is also present in Jones', Klein's, and Kudryavstev's chapters. In them, this idea overlapped with different notions of private (Jones and Klein) and public (Kudryavtsev) properties. Interestingly, both approaches are united by philosophies on individualism. Jones explained that Caldwell envisioned a decentralized urban landscape for Chicago. The plan proposed family farms as the landscape cells to heal nation poverty, unemployment, and inflation. In this light, the farm's private properties are used to achieve the greater good. This conclusion reflects Wright's ideas as shown by Klein. The core of Broadacre's plan is made of individually owned farms belonging to a greater community. In this model, the citizen-farmer can thrive only if the neighbours thrive as well, rejecting attitudes like the 'not in my backyard' movements. Caldwell's landscape cells remind of Mikhail Okhitovich and the 'disurbanists' in Kudryavtsev's chapter. Here, the cells are areas delimited by public infrastructures arranged in a network system. This distributed urbanisation de-facto eliminated the concept of centre-periphery. The residents of the cells had equitable access to the given resources (food, industry, energy, etc.) while collectively contributing to the functionality of the system. Kudryavtsev argued that the uniformity of the network did not impede the individual rights of each resident. On the contrary, the network supported the integration of different scales of needs.

Schubert's chapter elaborated on the notion of private and public space in terms of food production. He argued that Migge's demand for more city allotment gardens was a call to help citizens' self-support rather than a call to beautify the city. Hence, according to Migge, planners had to include this activity in city planning. Schumacher seemed to embrace this invitation at different scales for the city of Hamburg in the early 1920s. He designed a series of terraced houses that had gardens coupled in pairs, two in the front of two houses and two in the back. The arrangement served to avoid narrow gardens by placing them orthogonally to the buildings. This project supported the individual *Right to Food* to each dweller by orienting the garden plot in such a position. If the individual access to food—in this case achieved through the family self-provision of vegetables and herbs—was achieved at the scale of housing, the public access to food was pursued through Schumacher's green corridors across the city. The corridors were necessary to let the city breathe, to regain access to green areas, and therefore to cultivable land on public areas. The latter was an attempt to provide land to cultivate to the low-income classes. Lemes de Oliveira's chapter provides another perspective of the balance between private and public space in the pursuit of the *Right to Food*. Abercrombie proposed for London allotments gardens near residential spaces. Although Abercrombie's maps vaguely survey private and public land for this purpose, what is important is the reference to collective productive spaces. This notion would counter-argue the current tendency of seeing the privatisation of public spaces that limits the access to vital spaces for low-income residents (Bodnar 2015; Narváez-Muelas 2019).

4.2 Right to Food as a Design Action

The *Right to Food* induces design actions since it impacts the shape of the urban settlements. Not surprisingly, planners and architects included the design of urban productive landscapes in their city plans throughout history. Henri Colboc and Georges Philippe designed the *Regis Market* in early twentieth-century Paris showing the attention of modern architecture to this function of the city (Tenhoor 2015). Migge designed private and public vegetable gardens in May's Siedlung in Frankfurt in Germany in the 1930s (Brisotto 2019; Haney 2015; Henderson 2010). Le Corbusier included cultivated fields assigned to each of his *The City of To-morrow* apartment buildings (Le Corbusier 1929). More recently Andrea Branzi envisioned a city—*Agronica*—that reproduced the flexibility of agricultural cycles to adapt to social and environmental changes (Branzi 2006; Brisotto 2019).[1]

The chapters of this book demonstrate how planners and architects draw their attention to the design of productive urban landscapes shifting the paradigm from a space designed by people to a space designed by professionals. As noted in Terzano's chapter, the Danish allotment gardens became a domain of landscape architecture design during World War II. Another appealing case of how urban productive landscapes are a realm of design came from van Bergeijk and Piccinini's description of the Dutch polders. These agricultural patches are reclaimed from water delivering utterly planned and constructed land. Analogously, Klein's report of Wright's one acre per citizen/farmer policy also defined the idea of productive land as an act of design, an operation of planning and distributing resources to achieve sustainability. The professional approaches to designing productive urban landscapes could be divided into two categories. Firstly, there is the tendency of integrating these spaces within the urban patterns. Secondly, there are attempts to use design to reduce the consumption of land for agricultural uses.

In the first case, Abercrombie drew the production of food directly inside the city of London by proposing green wedges that transition the rural into the urban domains (as in this book's chapter by Lemes de Oliveira) and by distributing allotments within residential areas. This integration of rural and urban is also visible in Cadwell's plan (as in this book's chapter by Jones). In his plan for Chicago, the green space is seen

[1] From a conversation between Andrea Branzi and Carla Brisotto, May 2016.

as a connective tissue of the urban pattern and therefore wrapped around houses and settlements. Both ideas carefully craft green productive space in a way that can be accessible to people. Accessibility is the main aim of Okhitovich in Russia as described in Kudryavtsev's chapter. Here, agricultural land is designed in triangular shapes surrounded by infrastructures (transportation, energy lines, industry) and in close proximity to residences.

The second approach of designing productive urban landscapes shows how the reuse of the existing built environment can reduce the consumption of land for agricultural purposes. This is seen in both Friedman (Brisotto) and P.M. (Cristiano) proposals. Friedman advocated for limiting the expansion of the cities by connecting existing settlements together, preserving the agricultural land in between (the *Continent City*). In turn, P.M. thought about his *bolo'bolo* as micro-systems that reused and converted existing structures. This approach would limit the consumption of land but also respect the cultural and geographical features and aspects of the local context. As Cristiano said, geography matters. Zeunert and Freestone's foodscapes of Sydney support this argument describing how the local foodshed of the Australian city developed out of geographical constraints that impede easy transitions between urban and rural areas.

5 Discussion

The two themes discussed in the previous sections led to a series of overarching topics across the chapters. Resilient productive urban landscapes involve reflections on human activities on the environment, reconnection between urban and rural domains and people and land, concepts of transcalarity, exploration of foodscape as systems, and the question of food security. The following sections will unfold each of these topics.

5.1 Domains of Human Activity on the Environment

The preoccupation of many planners with the overwhelming pressure of cities to expand into the countryside and consequently with the erosion of the categories of the urban and the rural proved to be sustained. After the war, this process intensified. The post-modern city saw the loss of the dominance of a single centrality towards much more polycentric urban regions and the weakening of the power of a single centre to hold urbanity together. In the United States, this was seen through processes of suburbanisation and post-suburbanisation (Fishman 1987; Garreau 1992). Europe, which already presented a more polycentric structure of towns and cities, saw urbanisation rates grow steadfastly (Bengs and Zonneveld 2002). Sieverts (2003) showed how the disaggregation of the urban through processes of suburbanisation, development in the peri-urban and non-urban areas and infrastructure building in Germany led to new conditions, which were not urban and not rural, but 'in between'. In the turn to the twenty-first century, much attention was given to the 'peripheries': seen as a new domain or territory of planning action between the former consolidated city and the stable countryside. It became a territory of potential, experimentation, and creativity (Holl 1996; Sola-Morales et al. 2008). Many definitions and terms have been employed ever since to characterise gradients between urban areas through to an archetypical rural area (Healey 2002). Eurostat, for example, defines 'rural areas' as 'all areas outside urban clusters', which in turn are clusters of contiguous grid cells of 1 km^2 with a density of at least 200 inhabitants per km^2 and a minimum population of 5,000 people. This is further broken down into 'predominantly urban', 'intermediate', and 'predominantly rural' (Eurostat 2021). Similarly, the US Census Bureau (N/A) defines the rural as any territory outside the urban area, distinguishing between urbanized areas (more than 50,000 residents) and urban clusters (between 2,500 and 50,000 residents). Land take remains a crucial problem with clear environmental implications. Perhaps well beyond characters such as Sharp, Abercrombie, or Schumacher might have envisioned, today the urban areas and the countryside are physically and semantically plural. Yet, a look back at such planning experiences shows us the importance of conceptualising the various domains of human activity in the environment, and considering the relationships between them as the starting point for effective and proactive planning. Besides, they show us that spatial planning and design matters also at such broad scales.

5.2 *The Question of Reconnection*

Reconnection emerged as two main topics in this book. Firstly, as a physical and psychological reconnection between people and the land; and secondly between the urban and rural domains. This not only took place in countries where the rural tradition was foremost in people's minds, such as in England; but across the spectrum of studies covered here. While Sharp and Abercrombie called for the possibility of access to the countryside, in its beauty and ruralness, others such as Caldwell claimed that further integration of people and the land was necessary on ecological grounds. Klein's exploration of the concept of Usonia is one of the stronger calls for a return to the land, here most of all in agrarian terms. The reconnection between the urban and the rural was proposed either from the point of departure of difference but complementarity, as well as from the perspective of integration and dissolution. Planners aligned with that first line of thought employed spatial planning strategies, such as green belts and wedges, in order to ensure that reconnection would not mean mutual destruction of polar identities. A transformation from the small scale of the plot, such as Wright's based on the land-owning citizen-farmer, in turn, was meant to confront the way in which such domains had become predominant. Those aligned with the second line of thought suggested how an overhaul of established categories would in fact eliminate the very need to 'reconnect' since the new domain (not urban and not rural) would overcome the need for such distinctions. Recently, with the growth of city-regions into networks, according to Balducci et al. (2017), a new urban condition altogether emerged: the post metropolitan territory. This conceptualisation brings to memory Friedman's and P.M.'s bold visions of highly integrated environments where established hierarchies are challenged, and new territories emerge.

5.3 *Transcalarity*

The theme of the multiple and inter-related scales in planning urban and rural areas was explored as a condition for achieving the above-discussed reconnection of people to the land. Where the urban and rural domains were to be maintained, this would take place through strategies put in place to mediate the relationships between them, such as dealing with the urban–rural edge and in providing a 'park system' articulating the urban open spaces from the plot scale to the open country, as seen in the case of Schumacher, Abercrombie or de Gröer. Identifying the importance of regional planning beyond the major cities' area of influence was another essential value discussed, in particular in the works by Sharp, Astengo, Pane, and Piccinato addressing the scales of towns and villages. Transcalar thinking would take place too in approaches more concerned with integration and dissolution, such as the continental networks proposed by Friedman, P.M., and the 'disurbanists'. Here, despite the will to show the scalability of their proposals, there is a keen eye to the importance of units and their connections across the proposed system.

5.4 *Resilience and Sustainability of Food Systems*

To understand food systems and how they could efficiently serve a territory, we must first define what we mean for sustainability and resilience and how they relate to each other. In 1987, the World Commission on Environment and Development defined sustainable development as the ability to meet present needs while preserving the ability of future generations to satisfy their own needs. This call advocated for a development that consciously uses the environment and its resources respecting cultures and social systems. However, as Cristiano noted, this definition of sustainability involves economic growth that might or might not achieve long-term sustainable goals. Resilience is defined as the capabilities of ecosystems to absorb shocks and restore the initial stability of ecological systems (Holling 1973). However, when applying resilience to the urban context, the definition becomes 'adaptation to new changes' (Caniglia et al. 2020; Walker and Salt 2006). Sustainability and resilience can jointly contribute to achieving a better food system to pursue long-term goals by focusing on short-term actions. At the same time, by adopting both definitions, the focus would shift from being purely economical to also being social and ecological.

As noted in Sydney's foodscape case by Zeunert and Freestone, the development of plans that favour urban development and economic growth in a neoliberal fashion led to the dismissal of intragenerational resilience. Flexibility, another key concept of resilience in ecosystem services (Holling 1973), could help to connect between generations. Friedman's (Brisotto) idea of infrastructures suspended over an untouched landscape would allow the freedom to use the space according to inhabitants' needs and uses, adapting to each situation, context, and condition. Another opportunity comes from the circular economy paradigm as noted in Kudryavtsev's exploration of Okhitovich and the 'disurbanist' movement. Okhitovich foresaw how agriculture could become circular by envisioning agricultural waste as an input to industrial cycles. In this light, the dispersed urban productive landscapes of the 'disurbanists' would feed the circular economy by always drawing from local sources. Local food systems are also at the centre of Klein's and Jones' arguments. In both chapters, the reliance on individual small farms distributed in the territory aims to achieve economic and environmental sustainability goals. This objective relies on the connection between landscape and residents through the figure of the citizen-farmer. The lessons learned suggest that sustainability and resilience can be simultaneously achieved by keeping the food system flexible and adaptable while relying on local sources.

5.5 *Food Security*

Food security is one of the UN sustainable goals and one of the main issues in post-disaster emergency plans often addressed with resilience-thinking approaches. This book's plans and concepts often refer to the need to provide land access to residents. The authors mentioned the struggles of the industrial workers in British cities and socialist settlements (Lemes de Oliveira, Aseguinolaza, Schubert and Kudryavtsev), the poor in developing countries (Brisotto), and the socially unfitted (Cristiano). Land ownership and land tenure are seen as crucial points to overcome such struggles. Both Caldwell and Wright saw private farming land as a means to a free society. However, this strategy does not guarantee equitable access to land, leaving individuals with fewer resources outside the system. As Klein argues, the inequality of Wright's system must turn into an opportunity of equality and accessibility—especially if we consider its application within the inflated current real estate market of large cities.

A self-sufficient city should produce enough food for its inhabitants, as Zeunert and Freestone noted, but also redistribute the surplus to citizens with no gardens, as seen in Lemes de Oliveira's and Cristiano's chapters. The latter action could be considered another aspect of a circular economy, as seen in the previous section, while also building an aid system. Friedman addressed this issue from the perspective of education, trying to improve the resilience of those with no resources by enhancing their food growing skills. After all, food independence should lead to freedom from exploitations, as suggested by P.M.

One risk and limitation of urban productive landscapes is the possible ignition of green gentrification processes. Green gentrification is the increase of real estate values in areas where there are new green interventions. Consequently, this process pushes low-income residents outside their own neighbourhoods (Anguelovski et al. 2019; Rigolon et al. 2020). A possible reason is shown by Zeunert and Freestone while observing that the 1948 County Cumberland Council (CCC) Planning Scheme failed because urban agriculture was valued more for its beautification significance rather than its foundational provision of food to the city residents. Terzano also explains how urban productive landscapes can lose their food security purpose to become a leisure space. The Brøndby Haveby gardens in Denmark designed to be allotment gardens for Danish individuals with no access to land are currently used almost exclusively as recreational summer houses.

6 Final Remarks

The analysis of the above-mentioned cross-cutting themes—domains of human activity, the question of reconnection, transcalarity, resilience and sustainability of food systems, and food security—demonstrates how the relationships between the urban areas and the countryside, and urban productive landscapes are complex territories to address environmental, social, and cultural issues. Already a key preoccupation of the first planners in the turn of the twentieth century, and reiterated in the post-war period, the intensive expansion of urbanisation in most parts of the world continuously poses new challenges to the definitions of categories of urban/rural and to food production, divided between disappearing rural land and contented urban areas. Due to climate change, the question of nature is becoming too more and more prominent. The need to substantially increase afforestation and reforestation adds a third vector to the polarities of urban–rural. Such categories of urban–rural-nature and their manifold gradations will require further research and deeper conceptualisations if we are to build sustainable and resilient landscapes. This objective can only be achieved by enhancing our understanding of the nuances of the complexity of the interactions across such polarities and hybrid landscapes, and by connecting the different scales of the territory from regional to local. This transcalarity implies the focus on local sources while keeping the system connected to a larger network. Such an approach would allow for a flexible and adaptable system capable of adjusting to singular conditions or emergency situations ensuring both the rights of the individuals and the public at large.

References

Abel KC, Faust KM (2020) Modeling complex human systems: An adaptable framework of urban food deserts. Sustainable cities and society, 52. https://doi.org/10.1016/j.scs.2019.101795

Abercrombie P (1933) Town and country planning, vol 163. Oxford University Press, London

Beria P, Lunkar V (2021) Presence and mobility of the population during the first wave of Covid-19 outbreak and lockdown in Italy Sustainable Cities and Society 65:102616 doi:https://doi.org/10.1016/j.scs.2020.102616

Agamben G (1995) Homo Sacer: Il Potere Sovrano e La Nuda Vita. Einaudi, Torino
Aktuna B and Brisotto C (2018) Agripoetic resistance in urban architecture and planning in the European world." In: Sadri H. (ed.) Neo-Liberalism and the End of the Profession of Architecture, pp. 207/228
Anguelovski I, Connolly JJ, Garcia-Lamarca M, Cole H, Pearsall H (2019) New scholarly pathways on green gentrification: What does the urban "green turn" mean and where is it going? Prog Hum Geogr 43(6):1064–1086. https://doi.org/10.1177/0309132518803799
Balducci A, Fedeli V, Curci F (eds) (2017) Post-metropolitan territories: Looking for a new urbanity. Routledge, London
Benevolo L (1967) The origins of modern town planning. Routledge and Kegan Paul, London
Bengs C, Zonneveld WIL (2002) The European discourse on Urban-Rural relationships: A new policy and research agenda built environment (1978-) 28:278–289
Bodnar J (2015) Reclaiming Public Space: Editorial. Urban Studies, 52(12), 2090–2104. http://usj.sagepub.com/content/by/year
Branzi A (2006) Weak and diffuse modernity: The world of projects at the beginning of the 21st century. Skira, Milan
Brisotto C (2019) City of agro-fragments: How ecological thinking influences architectural processes in the era of post-globalization. Dissertation. University of Florida, Gainesville
Caniglia B S, Frank B, Knott J J L, Sagendorf K S, and Wilkerson E A (2020) Regenerative urban development, climate change and the common good. Routledge Advances in Climate Change Research. Routledge, Abingdon, Oxon
CCC (2020) The sixth carbon Budget: The UK's path to NetZero. Committee on Climate Change, London
Davoudi S, Stead D (2002) Urban-Rural Relationships: An Introduction and Brief History Built Environment (1978-) 28:268–277
Edmonds A (2020) Connecting People. Intellect Books, Bristol, Place and Design
EEA (2015) European Environment—state and outlook. European Environment Agency, Copenhagen
EEA (2021) Nature-based solutions in Europe: Policy, knowledge and practice for climate change adaptation and disaster risk reduction. European Environment Agency, Luxembourg
European Commission (2017) Modernising & simplifying the CAP: Background Document. Climate and Environmental challenges facing EU agriculture and rural areas. EU Directorate-General for Agriculture and Rural Development, Brussels
European Commission (2020a) Analysis of climate change impacts on EU agriculture by 2050. Luxembourg
European Commission (2020b) Farm to fork strategy: for a fair, healthy and environmentally-friendly food system. Luxembourg
European Commission (2020c) Kick-starting the journey: towards a climate-neutral Europe by 2050. EU Climate Action Progress Report. European Commission, Luxembourg
Eurostat (2021) Applying the degree of urbanisation: a methodological manual to define cities, towns and rural areas for international comparisons. Publications Office of the European Union, Luxembourg
FAO (2020) Right on food. Food and gricultural Organization of the United Nations, accessed August 5, 2020, http://www.fao.org/right-to-food/en/
Fishman R (1987) Bourgeois utopias: The rise and fall of suburbia. Basic books, New York
Garreau J (1992) Edge City: life on the new frontier. Anchor Books, New York
H.RES. 109 (2019) Resolution: recognizing the duty of the Federal Government to create a Green New Deal. 116th Congress 1st session. Washington.
Hall P (1988) Cities of tomorrow: an intellectual history of urban planning and design in the twentieth century. Basil Blackwell, Oxford
Haney DH (2015) Small-scale agriculture as a solution to urban impoverishment in Britain and Germany, 1880–1933. In: Imbert D (ed) Food and the city: Histories of culture and cultivation. Dumbarton Oaks Research Library and Collection, Washington, D.C., pp 17–53

Henderson S R (2010) Romerstadt: The modern garden city. Planning Perspectives 25, no. 3 (2010): 323–346.
Henderson SR (1999) Self-Help Housing in the Weimar Republic: The Work of Ernst May. Hous Stud 14(3):311–328
Healey P (2002) Urban-Rural relationships, spatial strategies and territorial development built environment (1978-) 28:331–339
Holl S (1996) Edge of a city. Princeton Architectural Press, New York, Pamphlet Architecture
Holling CS (1973) Resilience and Stability of Ecological Systems. Annu Rev Ecol Syst 4:1–23
Hoppe R A, and Others, and Department of Agriculture W D (1986) Social and Economic Environment of Black Farmers. Rural Development Research Report Number 61
Horst M, McClintock N, Hoey L (2017) The Intersection of Planning, Urban Agriculture, and Food Justice: A Review of the Literature. J Am Plann Assoc 83:277–295. https://doi.org/10.1080/01944363.2017.1322914
Howard E (1902) Gardens Cities of Tomorrow. Swan Sonnenschein & Co., London
Imbert D (2015) Food and the city: Histories of culture and cultivation. Dumbarton Oaks Research Library and Collection, Washington, D.C.
Jefferies J (2005) The UK population: past, present and future. In: Chappell R (ed) Focus on People and Migration. Palgrave Macmillan UK, London, pp 1–17. doi:https://doi.org/10.1007/978-1-349-75096-2_1Lawson, L (2005) City bountiful: A century of community gardening in America. University of California Press, Berkeley
Lawton R (1973) Rural Depopulation in nineteenth century England. In: Mills DR (ed) English rural communities: The impact of a specialised economy. Macmillan Education UK, London, pp 195–219. doi:https://doi.org/10.1007/978-1-349-15516-3_10
Le Corbusier and Frederick Etchells (1929) The City of to-Morrow and Its Planning. J. Rodker, London
Le Corbusier (1967) 3rd Part: The New Age. In The Radiant City; Elements of a Doctrine of Urbanism to Be Used as the Basis of Our Machine-Age Civilization, edited by Le Corbusier, 64–87. Orion Press, New York
Lefebvre H (2003) The urban revolution. University of Minnesota Press, Minneapolis
Lefebvre H (1996) *Writing on Cities*. trans Eleonore Kofman and Elizabeth Lebas. Blackwell Publishers, Cambridge, MA
Lemes de Oliveira F (2015) Abercrombie's green-wedge vision for London: the County of London Plan 1943 and the Greater London Plan 1944 Town Planning Review 86:495–518. https://doi.org/10.3828/tpr.2015.30
Lemes de Oliveira F (2017) Green wedge urbanism: History, theory and contemporary Practice. Bloomsbury, London/New York
Lenger F (2002) Towards an urban nation: Germany Since 1780. Bloomsbury, London
Lickwar P, Thoren R (2020) Farmscape: The design of productive landscapes. Routledge, London and New York
Lissitzky E (1970) Russia: An architecture for world revolution. MIT Press, Cambridge
Lohrberg F, Licka L, Scazzosi L, Timpe A (eds) (2016) Urban Agriculture Europe. Jovis, Berlin
Narváez-Muelas AD (2019) Privatización del espacio público verde en la era neoliberal. Bitácora Urbano/territorial 29(1):121–130
Miller BL (1985) Architecture and Politics in Germany, 1918–1945. Harvard University Press, Cambridge, Mass
Mumford L (1961) The city in history: its origins, its transformations, and its prospects. Secker & Warburg, London
Mukhra R, Krishan K, Kanchan T (2020) COVID-19 Sets off Mass Migration in India Archives of Medical Research 51:736–738 doi:https://doi.org/10.1016/j.arcmed.2020.06.003
Rigolon A, Keith SJ, Harris B, Mullenbach LE, Larson LR, Rushing J (2020) More than "just green enough": Helping park professionals achieve equitable greening and limit environmental gentrification. J Park Recreat Adm 38(3):29–54

Sieverts T (2003) Cities without cities: An interpretation of the Zwischenstadt: Between Place and World, Space and Time, Town and Country. Routledge, London
Sola-Morales M, Frampton K, Ibelings H (2008) A matter of things. NAI, Rotterdam
Sutcliffe A (1981) Towards the planned city: Germany, Britain, the United States and France 1780–1914. Basil Blackwell, Oxford
Tenhoor M (2015) Markets and the food landscape in France, 1940–72. In: Imbert D (ed) Food and the city: Histories of culture and cultivation. Dumbarton Oaks Research Library and Collection, Washington, D.C., pp 333–351
U.S. Census Bureau (N/A) Urban and Rural Population: 1900–1990; Census 2000 Summary File 1 Table P002; 2010 Census Summary File 1 Table P2, A Century of Delineating a Changing Landscape: The Census Bureau's Urban and Rural Classification, 1910 to 2010. U.S. Census Bureau, N/A
USDA NASS (2017) Black producers. Census of Agriculture www.nass.usda.gov/AgCensus
USDA (2021) Urban Agriculture and Innovative Production (UAIP) Competitive Grants Program. https://www.grants.gov/web/grants/view-opportunity.html?oppId=333854
United Nations (2020) The Sustainable dDevelopment gGoals rReport 2020. Department of Economic and Social Affairs, New York
Warren E, Hawkesworth S, Knai C (2015) Investigating the association between urban agriculture and food security, dietary diversity, and nutritional status: A systematic literature review Food Policy 53:54–66 doi:https://doi.org/10.1016/j.foodpol.2015.03.004
Walker BH, Salt D (2006) Resilience thinking: sustaining ecosystems and people in a changing world. Island Press, Washington, D.C.
World Commission on Environment and Development (1987) Our common future. Oxford University Press, Oxford
Wright JD, Donley AM, Gualtieri MC, Strickhouser SM (2016) Food deserts: What is the problem? What is the solution? Society 53(2):171–181. https://doi.org/10.1007/s12115-016-9993-8

Urban-Rural Relationships

Paths to the Green City: On the Work of Fritz Schumacher

Dirk Schubert

Abstract This contribution traces the paradigm shift from decorative green spaces to "social green" spaces and the work of the urban planner Fritz Schumacher (1869–1947) from the beginning of the twentieth century. Initiatives to reform living conditions in cities are discussed, using the examples of Hamburg/Altona and Cologne and their concepts of decreasing urban densities, providing green infrastructure and designing urban parks until the Nazis came to power in Germany in 1933. The work of the cities' three directors of horticulture Otto Linne (1869–1937), Ferdinand Tutenberg (1874–1956) and Fritz Encke (1861–1931) is examined in relation to urban reform and town planning. The work of this generation of protagonists is placed in the context of upheavals the First World War brought with it. Further to examining the various models of urban green and open spaces from the time before and after the First World War, opportunities for their development, the adaptation to changing needs and the forms of appropriation are discussed.

Keywords Fritz Schumacher · Urban problems · Reforming the city · Hamburg · Cologne

1 Fritz Schumacher—A Life Working for a "Better City"

Fritz (Friedrich Wilhelm) Schumacher was born in Bremen. His father was Minister Resident in Bogota and New York, where Schumacher grew up. In 1883, the family returned to Bremen and Schumacher attended high school there. From 1889 to 1896, he studied at the Technical University of Munich, first mathematics and natural sciences and then architecture. After working at the municipal planning and building control office in Leipzig, he was appointed as professor at the Dresden University of Technology in 1901, was co-founder of the Deutscher Werkbund in 1907 and, in 1909 at the age of 40, became director of building construction in Hamburg.

D. Schubert (✉)
HafenCity University Hamburg, Felicitas-Kukuck-Strasse 14, 22765 Hamburg, Germany
e-mail: dirk.schubert@hcu-hamburg.de

C. Brisotto and F. Lemes de Oliveira (eds.), *Re-Imagining Resilient Productive Landscapes*, Cities and Nature, https://doi.org/10.1007/978-3-030-90445-6_2

In Hamburg, he was initially "only" director of building construction and held responsibility for the planning of all public buildings (Frank 1994). He soon sought to expand his field of activity by taking on urban planning tasks. Plans that were already underway—such as opening up a corridor for Mönckebergstrasse as part of major urban redevelopment in the aftermath of the 1892 cholera epidemic—were extended and redesigned. Schumacher wanted to implement his ideas not only in the design of buildings, but also in the layout of parks, such as the Stadtpark (city park). The First World War brought an extensive decline in construction activity, which gave Schumacher the opportunity to take part in a competition to redevelop the ring of fortresses in Cologne. He won the competition and, at the request of Mayor Konrad Adenauer, prepared a settlement masterplan for Cologne. In 1924, he returned to Hamburg with new powers—now as chief planning and building director.

In the meantime, the issue of "Greater Hamburg" with its associated area expansions and incorporations had become virulent and questions of regional and state-level planning were in urgent need of solutions. Above all, new low-rent apartments had to be planned and built. A number of residential areas were created under Schumacher's direction and implemented by local architects. His regional planning work was not taken beyond a preliminary stage and in 1933, after the Nazis came to power, he was forced to retire. He turned to writing and became an "apolitical" authority on planning issues, working in the background. Although he suffered from thrombosis and a disease of the spine which limited his mobility, he continued to work on book projects with great self-discipline.

After the end of the Second World War, he gave a landmark speech on reconstruction in October 1945. On 5 November 1947, a day after his 78th birthday, he died in Lüneburg, where he had moved after his home in Hamburg had been destroyed. Fritz Schumacher, of little international repute and nationally misunderstood as a "conservative reformer" caught between tradition and modernity, is to this day valued as an important architect and urban planner in Hamburg. He should be rediscovered, reassessed and re-evaluated again and again in order to embrace the complexity of his literary work, his architectural works, his planning concepts and the significant influence he had on the theory and practice of our building culture and urban design. This chapter aims to place the focus on Schumacher's contribution to urban reform and the significance of open space planning.[1]

2 Transformation of the Existing Urban Fabric: Reforming the City

Before the First World War, cities were generally viewed critically for different reasons. Rapid urbanisation had given rise to efforts such as the garden city and the land and housing reform movements. Green and open spaces gained in importance

[1] For a systematic work overview of publications and plans in a chronologically order check: https://fritzschumacher.de/gesellschaft/werkkatalog/.

as they set a counterweight to densely built-up areas. Camillo Sitte (first in 1889, 208) had already differentiated between "sanitary green" and "decorative green" and pointed out the health-promoting quality of green spaces. Martin Wagner (1915, 9) had further emphasised that it is not the mere existence of open spaces but what uses they offer that is relevant to the health of the urban population. Urban planners and medical practitioners made demands for different minimum standards. A battle for square metres ensued in which having too much green space was not a consideration. Wagner (1915, 34) demanded 19.5 square metres per person, differentiated into playgrounds and sports fields, promenades, parks and woodlands. But it was not only a matter of the quantity of open space requirements, but also a question of design and what uses they should provide.

Overcrowded apartments and increasing urban densities gave rise to the question of an inadequate supply of green spaces for urban dwellers, who had become alienated from nature. In the period of upheaval after the turn of the twentieth century, the call for more (usable) open space and a move away from stately gardens became louder. The reformers demanded that ornamental gardens, which resembled extended palace gardens, be replaced by recreational green spaces for the urban population. But Hamburg had no royal parks, as many other German cities did. Nineteenth-century garden art, with formal layouts, eclectic use of design elements, ornamental gardens and areas for "strolling" and idealizing nature should soon be replaced by functional space systems conceived for mass use in the form of public parks with play and sports facilities, which put the emphasis on public health.

In addition to reasons of public health, the poor physical fitness of recruits served as an argument in favour of more green spaces. It was not until the end of the nineteenth century that the bourgeoisie—women only accompanied by men—began to take walks, which was previously reserved for the nobility in its geometrically laid out parks. While most green spaces and parks were elaborately configured as "decorative green" for the upper and middle classes, the focus shifted to aspects of usability for all while incorporating educational and public health functions. "And for our terribly degenerating civilization, the garden will be a conserving counterweight and an inexhaustible source of recreative forces of nations". Leberecht Migge (1913, III) saw green and open space planning as a cultural task and Schumacher added that open spaces should also contain allotment gardens for the working population (Fig. 1).

The beginning of Fritz Schumacher's activity in Hamburg in 1909 must also be seen against this backdrop. After the First World War, he used Oswald Spengler's metaphor of "the fall of the West" with "that strange mixture of curiosity and resignation". For new "outlooks on life" and "living conditions" to take hold, it is important to develop a harmonious form that is derived "from the needs of the masses". He employed biologistical metaphors in order to establish a resilient base of reasoning for urban planning in the face of social upheaval. "Symptoms of disease" and "social suffering" would accumulate in cities. It is important to "heal" these, for which Schumacher attached considerable importance to the housing reform (Schubert 2021) and open spaces.

Abb. 48. Grünplan von Hamburg nach den reformierten Plänen.

Fig. 1 Green and open spaces in the Hamburg city area according to the reformed plans, signed by Fritz Schumacher, Erwin Ockert and Emil Metzel (around 1912). *Source*, Architekten und Ingenieurverein zu Haamburg, Hamburg und seine Bauten, Hamburg (1929, p. 40)

3 Hamburg's Stadtpark—A Park "To Live In"

In many German cities, as in Hamburg after the turn of the twentieth century, against the backdrop of insufficient green and open spaces—particularly for Hamburg's growing working population in the districts of Winterhude and Barmbek—a movement for public parks formed. Bourgeois circles had not initiated the construction of a city park until 1896, for which land had been bought in 1901. The need for a water tower on elevated terrain to serve new residential areas coincided with considerations for a park. Winterhude, a village far from the city centre which had been incorporated in 1894, was the site where a park was to be created "without any major sacrifices for the state treasury" (Bezirksamt Hamburg Nord 2014, 6). It was hoped that properties in the vicinity of the park would increase in value. Location, size, function and design were discussed controversially. Initial designs produced by Hamburg's chief engineer Eduard Vermehren in 1902 consisting of "pretzel paths" in line with a traditional understanding of garden art were discarded (Grunert 2014, 15).

The call for a park that would serve all strata of the population became loud, which—as Alfred Lichtwark put it with educational awareness of the mission—should be "accessible" and "playable". Elitist sports like tennis should be excluded. He suggested naming it "Alsterpark"", to underline its unique selling point. "The abundance of city parks robs the name of its individualising or suggestive power. And this effect is by no means indifferent". Lichtwark assumed that it was not advisable to lay down all the details for the park, but only its most important elements. "The detailed implementation may take decades" (Lichtwark 1917, 43, 69).

In 1905, a competition was held for the park in which only second and third prizes were awarded and some accreditations made. After a long quarrel over competence, the assertive Fritz Schumacher, who had only just been appointed to Hamburg, was to take over its planning in 1909, together in a difficult collaboration with the head of engineering Ferdinand Sperber (Frank 2014, 74). Schumacher criticised the fact that the nineteenth century had not contributed any solutions for public parks, but had only produced inflated private park concepts. It was to become a completely new task from then on. He had taken eight months' (unpaid) leave before assuming office to prepare numerous designs for the Stadtpark and some buildings as well as for other public edifices (Fig. 2).

The debate about the detailed designs for Hamburg's Stadtpark and the competition has already been sufficiently examined in the relevant literature (Frank 2014; Grunert (ed.) 2014, 2015). Here, the focus is on urban reform and redevelopment of the city. Schumacher's vision sought to put the prestigious elements behind social and usable elements. In 1912, he (and Karl Brunke, who was responsible for the construction of the Stadtpark) went on a trip to England in order to "let the masterly large parks take effect on me" and to be well equipped for the tasks he would face at the Hamburg Stadtpark. He distinguished between the "prestigious stylised grounds of French culture and the habitable stylised grounds of English culture", [...] the inhabited public park, the open house with many rooms that are accessible" (Schumacher 1936, 48, 40). The vibrancy of Hyde Park, the metaphor of the "inhabited people's

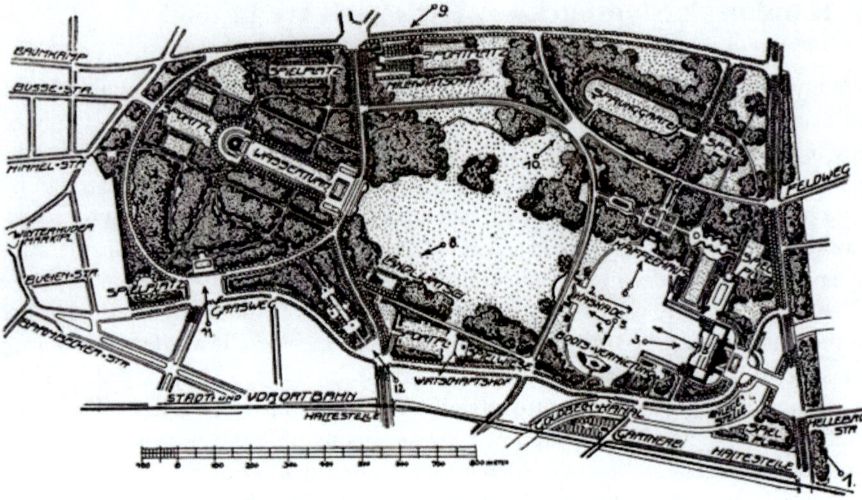

Fig. 2 Approved draft plan of the Hamburg Stadtpark from 1910 by Fritz Schumacher and Fritz Sperber. *Source* Bezirksamt Hamburg-Nord (2014) Hamburger Stadtpark 1914–2014, Hamburg, p. 14

park", was to serve as a model for the Stadtpark in Hamburg, while Hyde Park's artistic design was considered disappointing. "The prestigious park of earlier times has become the social park of our time, and the royal park has become the people's park" (Schumacher 1927, 8). More progressive urban planners and landscape architects were concerned with a paradigm shift from decorative design to social green spaces for specific purposes. Schumacher used the metaphor of a "house in a garden" to illustrate the symbiosis of the built environment and open spaces when occupying, using and appropriating them (Krings 2014, 50), which was later paraphrased by Hesse (2014, 220) as "garden monument to inhabit".

Leberecht Migge, who mutated from garden designer to settlement propagandist, had already formulated polemically in 1908: "In today's people's park, the people's natural desire to play is prevented" (Migge 1909, 7). He called for hundreds of allotments and "the real people's park should in future have a few square metres of land ready for anyone who does not own a garden and who wants one!" (Migge 1909, 12). A few weeks after the opening of the Hamburg Stadtpark on 1 July 1914, the First World War began, and from 1915 food had to be rationed. In these times of need, vegetables and turnips, kohlrabi and kale were grown in the Stadtpark, and the area around the water tower was transformed into a potato field. The produce was given away to the starving population in winter. In times of high unemployment and food shortages, parts of the Stadtpark were designated vegetable-growing areas (Fig. 3).

The completion of the buildings, for which Schumacher was responsible, and the laying out of the Stadtpark should drag on until the end of the 1920s. Schumacher

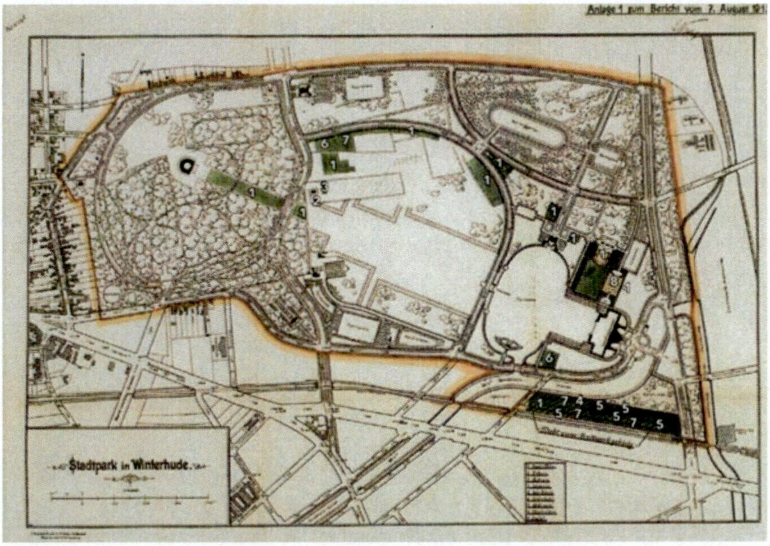

Fig. 3 Conversion of an area of the Stadtpark for growing vegetables during the First World War (potatoes, peas, beans, roots, etc.) marked with numbers. *Source* Grunert, Heino (ed.) (2014), Betreten erwünscht. Hundert Jahre Hamburger Stadtpark, Dölling and Galitz, Hamburg, p. 26

estimated its costs at 7 million marks, of which 1.5 million marks were for the buildings (Staatsbibliothek Hamburg, lecture, 8). The cascade, the city café, the drinking hall and the country house were built in brick, while the design for the half-timbered milk bar with a thatched roof was in the Northern German farmhouse tradition. The civic hall (built between 1911 and 1922 with interiors under the direction of the architect Hermann Höger) offered gastronomy with spaces for up to 10,000 patrons. The building was destroyed in air raids in 1943 and demolished in 1951. Otto Armand Linne, who was appointed director of horticulture to Hamburg in 1914, was responsible for landscape work in the park. He was "a man of action" and can be described as an "advocate of social green spaces". "Up until the war, the parks and green spaces were mostly decorative and ornamental, meticulously tidy, richly adorned with flower beds, but not open for children to play. After the war things changed", explained Linne (1922, 173).

Linne was also an advocate of allotment gardening. In 1922, around 33,000 Hamburg families cultivated a total of 1,350 hectares of garden land (Linne 1922, 175). By 1925, an allotment office had been set up to handle the parcels of garden land for 10,000 families and to meet the growing demand. Leberecht Migge had already proclaimed before the First World War that "the city needs gardens out of necessity" and "our urban misery, never totally comprehensible, finally creates gardens out of natural necessity" (Migge 1913, 6). His demand was not for more care, but for people to help themselves, and he wanted urban planners to show more consideration for the allotment garden movement in future. Thus, allotment gardens were assigned an almost "missionary" health, educational, moral, economic and aesthetic significance for the population. For the purpose of preventative health care, green spaces should compensate for unhealthy living conditions (Kuick 2014, 167, 174). "Care for the garden saves care for the sick" was Linne's mantra, who wanted to implement forward-looking green space policies in a constructive cooperation with Schumacher. The park with its buildings, green spaces, gardens and design elements could be rated as a total work of art. While the Hamburg Stadtpark was a concept designed at the drawing board, including spatial sequences and a monumental main axis, the Altona Volkspark in the neighbouring Prussian city, was a concept integrated into the landscape.

4 Excursus: Altona Volkspark—"From the People for the People"

After Breslau, Altona a separate city until 1937 directly adjacent to Hamburg) was the German city with the highest population density. While Germany had an average of 4,034 inhabitants per square kilometre, it was 8,352 in Altona (Nath-Esser 1997, 57). Densely populated Altona had many commercial and industrial businesses, a high proportion of workers and often problematic living conditions. In 1913, a year before Otto Linne, Johannes Ferdinand Tutenberg took over the office of Prussian director of horticulture in Altona, and in 1914 a horticultural exhibition was planned, but these preparations came to an end at the outbreak of war. However, the construction work in the park, initially called "Kaiser-Wilhelm-Park" and later "Altonaer Volkspark", was continued. In 1914, the work began with 1,000 unemployed labourers and was continued in 1918 in the second phase of the park also by setting the unemployed to work. "The Hamburg Stadtpark has therefore become a creation of garden art in which architecture, especially in the eastern part, plays a dominant role" was Altona's criticism. The park in Altona, on the other hand, is "a functional park through and through, which is intended to serve the population in every respect", as was polemicised (Hoffmann 1929, 153–154). Altona tried to outdo Hamburg by making the Volkspark larger with 205 hectares (with an additional 20 ha approx. for stadiums, sports and event areas) while the Hamburg Stadtpark, only measured 148 hectares. The park was to open up for people in need of recreation to camp and play in.

The pragmatist Tutenberg succeeded in converting the site and military training grounds, initially designed in 1914 for the 25th anniversary of his Majesty's service, into a functional garden for the population of Altona. The park was developed along old farm lanes with an organic network of paths offering myriad contrasts and exploiting the hilly topography. In 1923, work finally began on the third construction phase including large sports facilities, which were completed in 1925 with the building of a stadium and an inauguration attended by 50,000 spectators. The park was to provide a public health facility with air and light baths as a counterpoint to the overpopulated workers' living quarters. Model allotments with a small workshop were also created and instructions on practical horticulture were given in a teaching garden. The gardening school was worked by around 4,000 boys and girls in the 1920s (Fig. 4).

In a wooded area of the Volkspark, "woodland beauty" and idyllic woodlands were to be cultivated with sustainable forestry. A woodland park movement had emerged, which opposed deforestation and saw the maintenance of the beauty of woodlands as an important goal of protecting the homeland (Bey 2021, 142). In 1914, up to 1,000

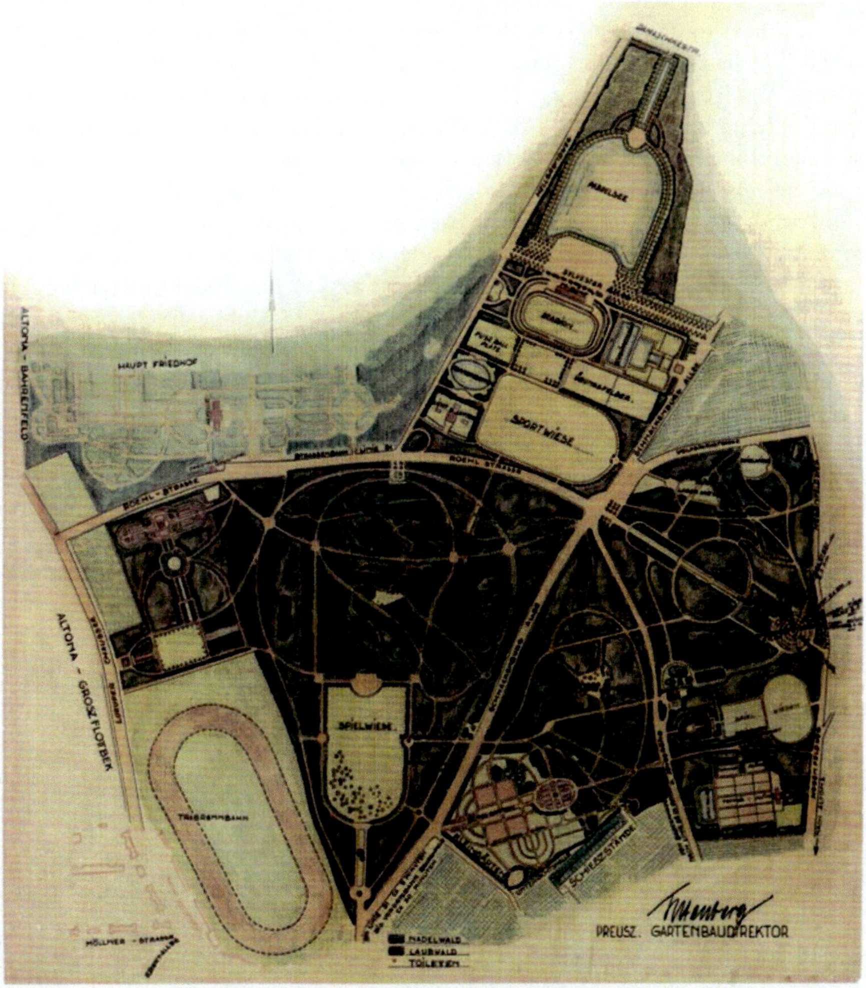

Schauplan des Volksparks um 1930

Fig. 4 Show plan of the Volkspark in Altona around 1930 signed by hortical director Ferdinand Tutenberg. *Source* (Bezirksamt Hamburg-Altona 2014) Altonaer Volkspark 1914–2014, Hamburg, p. 35

public relief workers were involved in the construction of paths in the new park, and in the late 1920s, the recruitment of unskilled workers was to help them earn a living while alleviating social tension (Bezirksamt Hamburg Altona 2014, 16).

Tutenberg pursued expediency as a design principle and a purpose rather than an "artistic revelation" (Nath-Esser 1997, 70). The entire urban population should have access to the park and be able to use it. The landscape architect Leberecht Migge, ambitious and later quite well-known, had formulated in 1913 that the "ideal people's park [...] simply has to give back to all strata of society everything that our urban life as such withholds from them" (Migge 1913). In 1923, the Altona main cemetery was built next to the Volkspark, further extending the large green area. Josef Brix and Gustav Oelsner had drawn up a settlement masterplan for the Prussian cities of Altona, Wandsbek and Harburg in 1923—without Hamburg's involvement, to the horror of Schumacher (Timm 1985, 78). The incorporation of areas into Altona and the establishment of the Volkspark made it possible to develop this plan into an open space concept based on three green belts. The first encompassed the core area of Altona and Ottensen, the second enclosed the suburbs with the Altona Volkspark forming its centrepiece and the third circular belt already incorporated parts of the surrounding areas. In contrast to Schumacher's axis or fan model for Hamburg, the Altona plan pursued a concept of rings that incorporated public green spaces.

Altona had pursued forward-looking land policies and owned about a third of the urban area at the time. Green and settlement policies were integral components in order to expose as many families as possible "to light, air and green" (Hoffmann 1929, 121). The Steenkamp estate, only a few minutes from the Altona Volkspark, documented this concern in an impressive way. "The settler has a home, garden and possibly stables for breeding small animals, but otherwise remains a consumer in the city and integrated into urban consumerism" (Hoffmann 1929, 122). Between 1914 and 1926, more than 700 residential units were built, mostly two-storey buildings with semi-detached and terraced houses and a garden. Tutenberg as "advocate of social green" was retired by the Nazis, and after 1933 the Altona Volkspark was increasingly used for forestry.

5 Garden City or Provisional Solution? The Langenhorn Small-Scale Housing Estate—Fritz Schumacher Estate

In Hamburg's neighbouring cities, Altona and Wandsbek, plans for garden cities had already been initiated before the First World War. The ideas of the reform movements became widely accepted after the end of the war. Schumacher was not uninvolved in the establishment of the first garden city in Germany (Dresden-Hellerau), but remained quite critical of garden cities and never joined the German Garden City Society. By 1912, he had already visited Port Sunlight near Liverpool, Hampstead Garden Suburb and Letchworth, "model cities of a new settlement idea" (Schumacher 1936, 57) and he had formulated his professional judgment. He was a member of the arts commission in Dresden-Hellerau, but did not see garden cities as a solution to urban problems. Ebenezer Howard's idea would leave cities untouched and ultimately plan their dissolution and redundancy by creating new garden cities. Schumacher vigorously opposed this approach, which would not solve the problems, but at best offer alternatives elsewhere. The city is a child of industrialisation and urbanisation, and the concentration of trade and industries are indispensable economic consequences. It is necessary to optimise these agglomeration advantages and control them better spatially.

Schumacher also had great sympathy for the land reform movement, as he was involved in reallocations and expropriation issues on a daily basis. Particularly when it came to retaining open spaces that could be used more profitably for development, municipal (pre-emption) rights had to be used. Here too, Schumacher proves to be a balanced, down-to-earth reformer, keeping an eye on long-term land law goals for urban planning and their implementation (Fig. 5).

After the First World War, physical exercise gained importance and allotment gardens for cultivating and improving the nutritional basis were intended to promote well-being and a healthy, natural way of life. In the time of post-war turmoil, under the direction of Schumacher and in difficult conditions, the Staatssiedlung—today Fritz Schumacher Siedlung—was realised in Hamburg's suburb of Langenhorn, which was not incorporated into Hamburg until 1912. The housing shortage, against which Schumacher wanted to appeal in the first edition of his volume on small apartments, had worsened (Schumacher 1917, 4; Clasen 1947). "Passionate desire" and the practical consideration born of necessity, of estates with small-scale housing and a garden would then be given great emphasis. As early as 1902, Hamburg had foresightedly bought up areas in Langenhorn and at the start of planning owned around a quarter of the development area of approximately 100 hectares. The housing commissioner had the authority to expropriate or buy land for the price of 2 marks per square metre (Wulff 1986, 15). In times of political turmoil, it was imperative for the senate to be seen to actively address housing policies.

Two schools, two shopping centres and a building for administration and events were initially envisaged. The plot sizes were designed for self-sufficient residents who should grow fruit and vegetables and could keep small animals, but also sheep, goats and pigs. It was not intended as a temporary solution, but many compromises had to be made as poor building materials and untried construction methods had to be used. To this day, a lively community has been consolidating over generations in the neighbourhood around the school (Volkshaus) in the centre of the estate. Workers who strive for a house of their own with their own garden must be the goal, according to Schumacher. The tightest of finances, unfavourable soil conditions with a high groundwater level and material shortages made implementation difficult. But, according to Schumacher, Langenhorn proved that "an estate of small-scale housing is not impossible in Hamburg". The provisional start of operation of the Langenhorn railway in 1918 was an important prerequisite for the construction of the estate.

Under the pressure of the housing shortage, the state took the initiative for the first time in Hamburg, a solution that had been unthinkable before 1918. Because of the scarcity of building materials, economical construction methods ("replacement

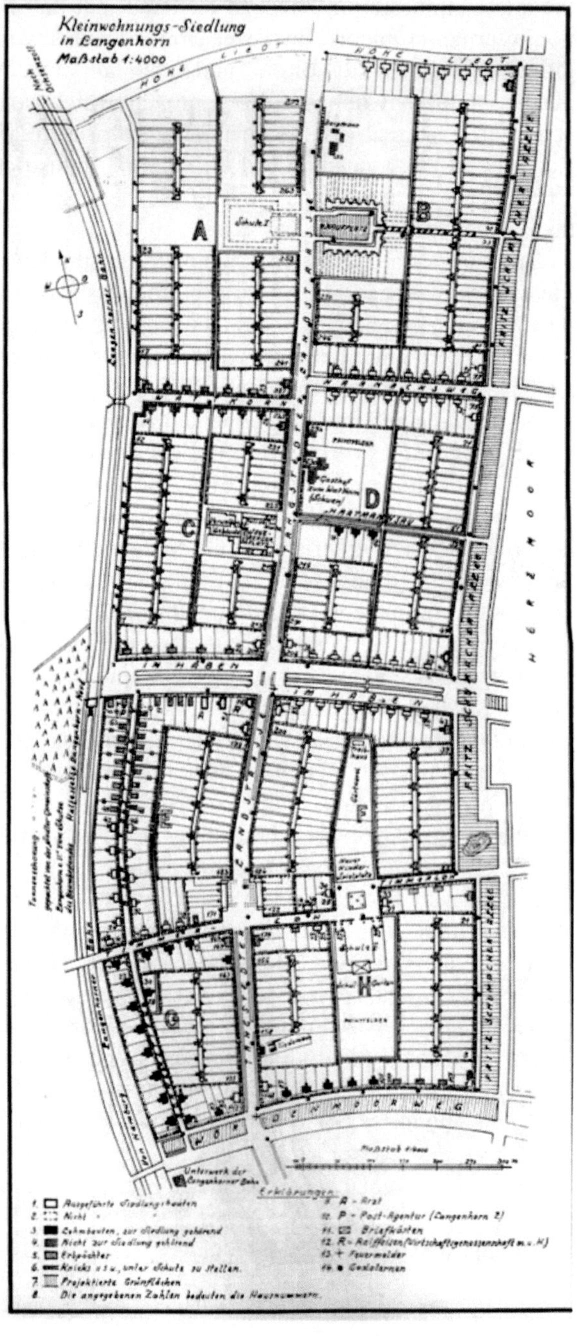

Fig. 5 Garden (suburb) Langenhorn small settlement (Later Fritz-Schumacher-Estate) with reserved areas for schools and shops. *Source* Wulff, Günter, Das Werden der Fritz-Schumacher-Siedlung 1919-1921, Erster staatlicher Wohnungsbau in Hamburg, Hamburg 1986

construction methods") had to be adopted and makeshift solutions like concrete and clay construction were used instead of the generally preferred brick. In 1920, the first houses were completed for occupancy. The Reich granted a "lost grant", which determined that mainly large families, families of war veterans and war invalids should be given preference in the allocation of housing. In 1920, the residents ("Börner") joined up to form an estate association (Wulff 1986, 84), which still is involved in all administrative issues to this day (except for the period 1933–1945).

The urban planning concept for the estate included: the creation of plots of sufficient size (750 m^2) with gardens large enough for cultivating and recycling household waste on the plot, minimisation of development and construction costs through a closed block development. There were no jobs on the estate; work had to be found in the city (Schumacher 1919, 208). In 1931, the school designed by Schumacher was completed. In order to avoid the usually narrow plots of land associated with terraced houses, Schumacher developed an idea of merging the gardens of two houses at the front and back, to make the vegetable gardens two houses wide. "A plot of land that can be used for gardening can only be achieved with an unusual trick, by making the garden the width of two terraced houses and arranging the garden in front and behind, so that in a group of four houses there are two gardens in front and two behind, which alternately span the front or rear of the neighbouring property" (Schumacher 1932a, 77). The topography further enhanced the idea developed on the drawing board. However, the estate "was built in the worst imaginable time after the war, when the most essential building materials were either not available at all or only in poor quality. [...] The workers achieved less and their work was worse. There was a lack of discipline on the construction site because the supervisors could not take action" (Jahresbericht (Annual Report) 1925, 126). Schumacher put the criticism into perspective:"I dared to build a permanent estate of 700 homes in Langenhorn, using newly invented construction methods instead of a provisional barrack town that was actually thought to be the only possibility," (Schumacher 1949, 422). However, this type of terraced house on the outskirts did not offer a long-term solution to the housing issue.

6 "Arteries" of Green Planning in and Around Cologne

Planning projects—not only in Hamburg—had to be postponed after the end of the First World War and languishing construction activities left Schumacher time for other things, which he used to take part in an urban planning competition in Cologne. He won the competition to redesign the "inner rayon" (ring of fortresses) in 1919 and was given a 3-year leave of absence from Hamburg at the urging of Cologne's Lord Mayor Konrad Adenauer (Schumacher/Arntz 1923). Adenauer, from 1906 to 1917 alderman and from 1918 to 1933 mayor of Cologne, justified the need for more green spaces for the population with public health arguments of "ventilation". Schumacher's work in Cologne fell into a difficult period of upheaval, as Cologne was occupied by allied forces until 1926 as part of the Rhineland occupation after the First World War, and its municipal independence was restricted.

The focus in Cologne was on the reorganisation of a ring-shaped zone around the old town and city centre, which is characterised by the ring of fortresses. The reallocation process was initiated by new legislation in 1911. The option of constructing taller buildings and achieving higher rents led to an increase in value and the owners had to contribute to the cost of creating new open spaces. This, in turn, increased the location advantages of the plots. “The mayor’s [Adenauer’s] ingenious idea” (quoted from Meynen 1979, 42), according to an assessment by the Nestor of German urban planning Josef Stübben, was not all that new. The planning of the Ringstrasse in Vienna and the ideas competition for Greater Berlin in 1909—won by Hermann Jansen—had already proposed and implemented similar green belts. At that time, this was one of the most important urban planning tasks in Germany.

The Berlin architect Hermann Jansen, Cologne’s urban planning inspector Alfred Stooß and Fritz Schumacher were invited to partake in the 1919 competition. Later Paul Bonatz also delivered a plan. Schumacher was described as one of “the most important urban planners in Germany” with a “characteristic style” (Meynen 1979, 43) and, not least thanks to his eloquence, the city council passed the resolution to implement the redevelopment according to his plan. Schumacher gave a speech to the city council in 1920 in which he highlighted the importance of planning “for all areas of life, for open space policies, for transport policies, for economic policies, for land policies, for housing policies […]” (Staats- und Universitätsbibliothek NFS IV C1, 1923). He designed the green belt with spaces for different functions, but without any horticultural details and he envisaged public buildings along the edges. The horticultural design was left to Fritz Encke, who was director of horticulture in Cologne from 1903 to 1929. He had been responsible for the design of many urban parks based on the idea of social green spaces with sports grounds and play areas.

An inner and an outer green belt formed the central elements of his plan, which was to be expanded from an initially limited sub-area to an urban development plan with regional connections. Large-scale recreation areas and decreasing densities towards the outskirts interspersed with allotment gardens characterised the plan. The green belt was to benefit as many Cologne citizens as possible, the design of which was to be co-financed by the landowners. The development was set around the green spaces, had no rear courtyards and integrated prestigious elements and monumentality. A model at the scale of 1: 500 was built to better describe the ideas. Exchange and purchase offers were made during the negotiations, but the complex reallocation process delayed implementation. From 1922 to 1924, 3,000 workers were deployed as part of public relief works on the inner belt, which was mostly completed by 1923 (Fig. 6).

The decrease in building densities towards the outskirts was to form a gradual transition from the city to the natural countryside and contribute to the recovery of cities. While the Hamburg Stadtpark was merely a green island in a sea of houses, the Altona Volkspark was already integrated into a system of three green belts. American park systems that stretched like networks across the entire city had been featured in German specialist journals. The insertion of radial and ring systems was considered

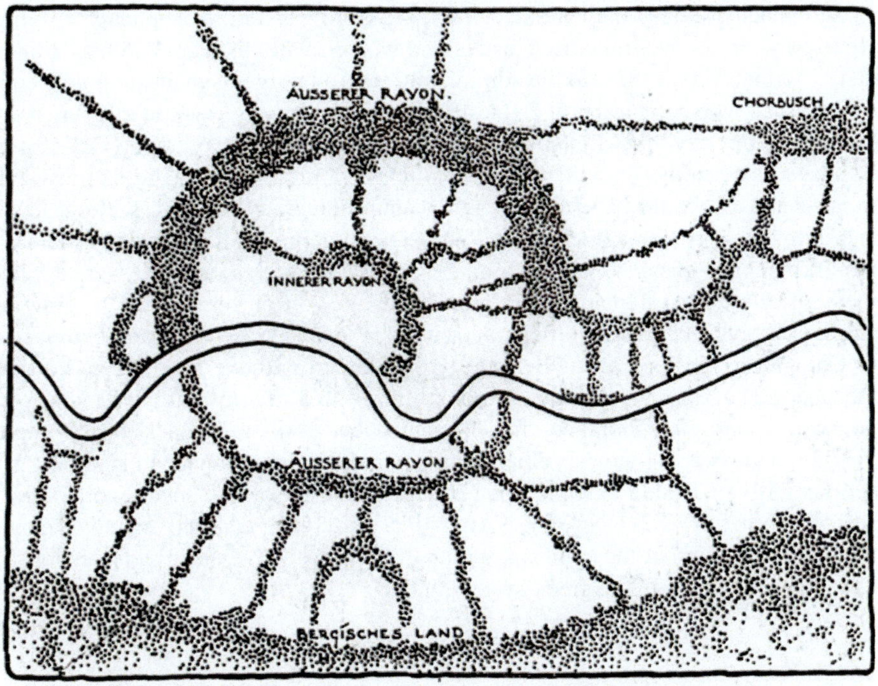

190 Das Kölner Grünsystem von F. Schumacher. Stand der Planungen 1923. Endgültiges Schema mit Durchdringung von Kreis- und Radialstreifen.

Fig. 6 Green belt around Cologne on former fortifications (1923) with outer and inner green belts and radial stripes. *Source* Schumacher, Fritz, Arnzt, Wilhelm, Köln. Entwicklungsfragen einer Großstadt, Köln 1923

to be the best solution. The green belt system in Cologne—best known was London's green belt (Thomas, 1970)—can be seen as a kind of synthesis of these urban spatial concepts; the green corridors had to be adapted to the existing topographical and structural conditions.

Schumacher used biological narratives and spoke of a "disastrous petrification" and that a "system of arteries" would rehabilitate the "urban organism" (Meynen 1979, 82). Based on his studies on the issue of small-scale housing in Hamburg, he analysed the type of apartment buildings found in Cologne and sought to improve natural daylight conditions and ventilation in the blocks by means of urban reforms and the inclusion of green spaces. Schumacher's systematic preparatory work, including a complex, problem-oriented survey and the consideration of influencing factors, the development of target perspectives for spatial development and an exemplary analysis of detailed problems should boost professionalism in urban design and urban planning, which was generally characterised by the everyday routine of drawing up partial development plans.

Schumacher did not plan public buildings in Cologne, despite Adenauer's urging. He lived with his two unmarried sisters and worked with his team and his "adjudicator" Wilhelm Arnzt next to the administration in a bourgeois mansion on the edge of the urban woods (Burkhardt 2014, 45). He did not pursue "drastic measures" for "smashing the city", but his aim was its reformability. The sprawling city was to be shaped by healthy growth. This should not be a prescribed image, but give rise to an ideal that would have to take into account integrative spatial, sectoral levels and individual problems (Genzmer 1924, 481). In doing so, he provided a state-of-the-art blueprint of what urban planning could achieve for a city, if it were applied methodically and systematically. He received offers from other cities to draw up similar surveys and carry out preparatory work. Adenauer tried to bind Schumacher to Cologne in the long term. He later wrote in his figurative style: "I was a man married to Hamburg who suddenly found himself in a new love affair. This was by no means a negligible moral conflict." (Schumacher 1949, 431). Back from Cologne in 1923, he was able—now as chief planning and building director in Hamburg but seriously ill—to obtain extended powers and a rise in salary (Staats- und Universitätsbibliothek NFS 731–8 A 769). After 1923, his colleague Arntz was to take over the implementation of the open space plan that was to "solve the urban question" in Cologne and complete the green system.

7 "A Belt Around Hamburg's Old Body"—New Residential Areas

In the period before the First World War, the need to create sufficient green and recreational areas was not recognised. "Today, serious accusations are rightly raised against the neglect of the simplest and now self-evident social tasks [...]" (Ockert / Linne 1929, 329). Schumacher had asked for around six square metres of open space per person; in 1904 it had only been 3.4 square metres (6 m^2 in Altona). Around 1929, the result for Hamburg was 4.38 square metres of open space and 1.07 square metres of play and sports grounds per capita of the population. More green and open spaces were an integral part of the campaign against the densely built-up tenement districts of the pre-war period.

In the new residential neighbourhoods built in the 1920s, courtyards were laid out as "play areas and romping areas for young people living in the blocks of flats" (Ockert/Linne 1929, 35). Smaller open spaces were divided up for target groups. Sandboxes, paddling pools, toddler rooms and nursing rooms for the youngest, lawns, play and gymnastics equipment for children, soccer fields, fistball and handball courts were accommodated in special areas and tranquil spaces, called "old people's gardens" for the elderly. The estates that were developed under Schumacher after the First World War, such as Jarrestadt, Barmbek-Nord and Veddel, had open spaces integrated into their new residential neighbourhoods (Schubert 2021).

Dulsberg, for example, with its traditional brick architecture and a development plan reformed by Schumacher, is crisscrossed by a green oasis. He had noted the existing development plan and ornamentally laid out square with a small playground and ample opportunity for densification with a "shudder". He presented a "reformed plan" in 1918, with a central green corridor around which the building blocks were grouped like the stays of a corset. The city itself was a client for the first buildings and Schumacher himself had designed five blocks with inner courtyards. The remaining development was left to private developers, housing cooperatives and well-known architects. The heights of the buildings were limited to three to five storeys (Fig. 7).

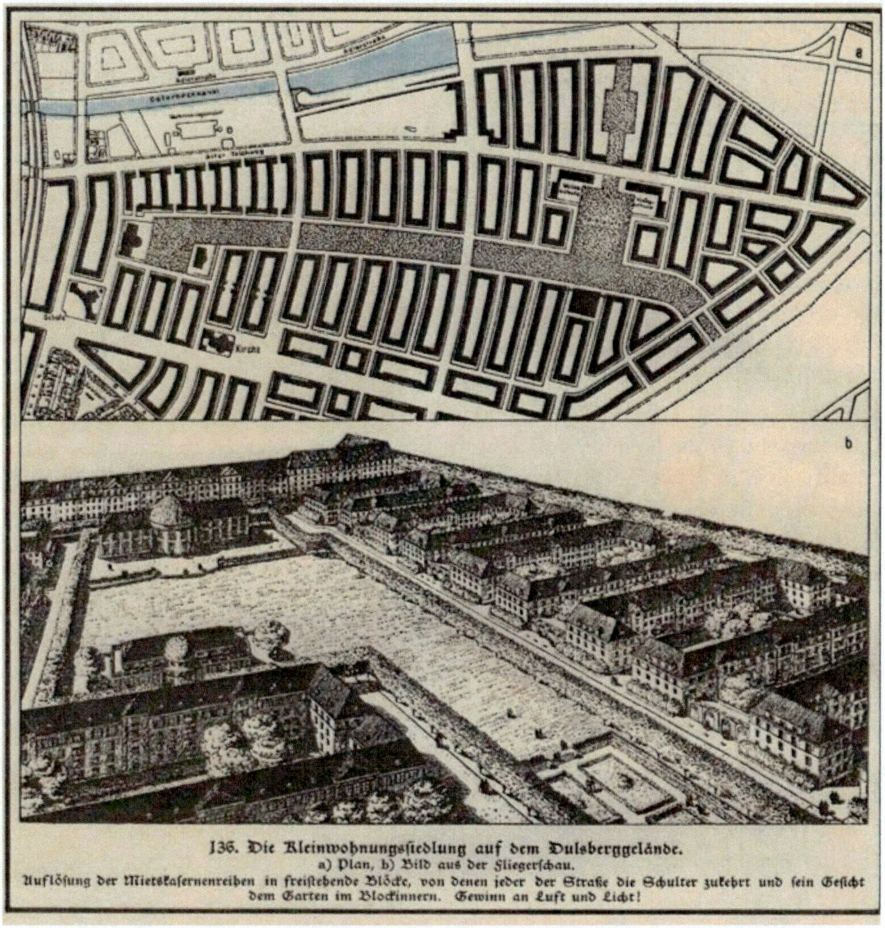

Fig. 7 Plan and perspective of Housing Estate Dulsberg with integrated green spaces and gardens inside the blocks. *Source* Wölfle, Karl, Hamburger Geschichtsatlas, Hamburg 1926, L. Friedrichsen and Co., p. 46

The Alster canal project had already been discussed in Hamburg before Schumacher took office. The office of engineering under Ferdinand Sperber developed a plan for a freight canal which was considered inadequate and other authorities such as Reinhold Baumeister, Hermann Jansen and Karl Henrici demanded that Schumacher be included in its planning. The problematic structure of responsibilities and the dominance of engineering became apparent, leading to local differences between engineering (Sperber) and building (Schumacher) practices. Schumacher sought to connect small green spaces into a network of green corridors and to give the different areas of building development distinct characters. From 1914, the position of horticultural director was established and filled by Otto Linne (1869–1937), who became involved in the horticultural design. The execution began in May 1914. However, green areas, in particular, should be affected by the aftermath of the war—they were leased as land for cultivation. Initially only intended as a temporary measure, the allotment gardeners and their strong lobby managed to hold onto the land.

When considering the canalisation of the Alster River, Schumacher's concern was not the construction of small-scale housing, but rather the question of how controlling and lowering the water table could benefit the development of attractive detached houses. The green corridor alongside the watercourse was soon lined with villas which were intended to reduce the emigration from Hamburg's territory (Beck / Voss 1999). Many wealthier citizens had settled in the periphery of Hamburg, where land was cheaper to buy and the landscape beautiful. "Numerous taxable citizens had moved to Prussian territory, where the feverish urban development had not yet had as devastating an impact as in flourishing Hamburg, and as a result the charms of nature were still unfolding their familiar benefits. This became dangerous for Hamburg and so the housing policy at the time is dominated by the idea of opening up attractive areas within Hamburg's state territory to keep the taxable population under the influence of its sovereignty" (Schumacher, 1932a, 67). Schumacher had outlined a problem that a century later should still be on the agenda for the city-state of Hamburg (Fig. 8).

Schumacher sought also to prevent private parks from being built over by cleverly arranging buildings along their periphery. This gave rise to high-quality residential developments at Hayns Park on the Alster in Eppendorf and around other parks, whose planning integrated public open spaces.

Fritz Schumacher's influence, communicated in his "polemics", was initially "only" concerned with the "inward reform of the city" and thus limited to the city centre of Hamburg. Against the backdrop of the supply crises and inflation brought about by the war, considerations of territorial restructuring as well as the dissolution of cities and the resettlement of city dwellers in the countryside received a new impetus.

Fig. 8 Alster sewer system with the designation of new building land and green zones on the eastern shore. *Source* Wölfle, Karl, Hamburger Geschichtsatlas, Hamburg 1926, L. Friedrichsen and Co., p. 45

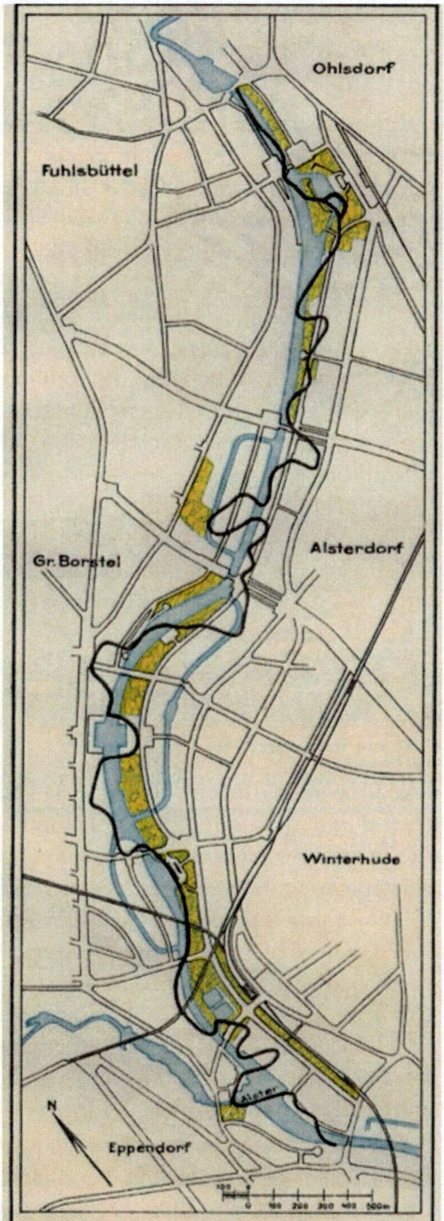

132. Die Alsterkanalisierung.
Der Anfang zur Entstehung der neuzeitlichen Gartenstadt an der Oberalster, woran sich im O. (an der Ohlsdorfer Hochbahnstrecke; Verbindung mit dem Stadtpark) das Ausstellungsgelände anschließen wird.

8 "Urban Imperialism" or "Choking Hazard"—Regional Planning

Not only issues of port development and additional land for residential development urgently required a solution with Prussia and Hamburg's neighbouring cities of Altona, Wandsbek and Harburg. In 1922, 9,000 Hamburg allotment gardeners were cultivating around 440 hectares of Prussian territory. Hamburg's director of horticulture Linne argued in favour of allotment gardens as they would aid the recovery of the city and the need to "break Hamburg's fetters and expand its area" (Linne 1922, 175). Hamburg was not only surrounded by Altona, Wandsbek and Harburg but also 300 rural municipalities that lay within the Prussian provinces of Schleswig–Holstein and Hanover. Eight different building codes applied in the wider area and conflicting issues of self-interest were jealously pursued. Workers from the neighbouring areas commuted to Hamburg and vice versa, while plans had to stop at the municipal boundaries.

Regional planning that extended beyond Hamburg's political city boundaries was considered under the primacy of port (expansion) planning. The preparatory work and surveys carried out after the First World War under Schumacher's direction—partly based on his work for Cologne—used innovative methods that should form a compendium for the establishment of regional planning (Necker/Woyke 2009, 150). Fritz Schumacher, spiritus rector of Hamburg's urban planning and master of tactical strategic argumentation, used biologistic metaphors to justify the imperative of far-sighted, coordinated regional planning after the First World War. "The problem of 'Grosshamburg', like all things in this city, has its origins in port issues, but it grows far beyond economic technical aspects and ultimately becomes a social issue of profound general importance. [...] What 'natural growth' would mean can be visualised with rare certainty when a city has such a clearly and distinctly pronounced power centre as Hamburg in its port" (Schumacher 1950, quoted in Ockert 1959, 96/97).

The central question was whether Hamburg's economic survival was as closely linked to the expansion of urban areas as was claimed, or whether the real issue was the expansion of its territorial power after all (Johe, 1998). Schumacher assumed land uses from geomorphological units and that the marshland area around the island in the Elbe River and its inlets were the "natural working area" for Hamburg. Living in the marshes he thought unhealthy while the proximity to the Elbe River, to watercourses and canals, would be ideal for cargo handling, shipyards and seaport industries. The sandy geest soils, on the other hand, would make suitable housing areas. In this respect, however, Hamburg's planned development was obstructed, since new healthy residential neighbourhoods could only be built in the north-east—but not near the workplaces.

According to Schumacher, the port as "the city's powerhouse" should give impetus for development in all directions and (healthy) residential areas should be allocated to the workplaces. Schumacher fortified the plans for the expansion of Hamburg's state territory with his powerful concept of "real" and "natural " development, which he expounded in a new memorandum in 1919.

The juxtaposition of these two schemes has been underpinned with various, often biological, terms. He mentions physical, unnatural, restricted, inorganic, crippled, inhibited, sick versus healthy, natural, organic, desirable development (Schumacher 1921, 64). The impressive scheme was quickly elevated to a "fan plan", "feather plan" or "axis plan". The "ordered" development was to keep spaces between the axes free of buildings and reserve them for green corridors. Open and green spaces formed integral parts of the concept. Schumacher viewed the growth of the open spaces at the base of the axes with great concern. He fought not only using a pen, but also—as the diagrams show—effectively employed his drawing pencils.

Schumacher's scheme was not aimed at polycentric but axial development of the region. It was to remain the key spatial model for the next decades. The message of the scheme and its minimalist presentation was easy to understand and convincing not only for experts but also laypeople. Schumacher expounded: "Because this force of natural growth does not actually work in the area of human settlement by leaving things to their own devices. [...] What arises from it is a proliferation; the result is not growth but chaos. Regional planning is a tool that serves as a surrogate for natural growth" (Schumacher 1932c, 45).

However, this did convince all opponents of the need for area reorganisation and overarching planning with Hamburg as the centre. Joseph Stübben indirectly supported the Prussian position. "The Senate is skilfully endeavouring to present its far-reaching wishes not as pursuing an advantage for Hamburg, but as a demand for the welfare of the Reich, because it is a question of maintaining Hamburg's competitiveness against the international world ports as a tool for German reconstruction. [...] The plans for Hamburg would lead directly to a trade monopoly and to the stunting of other German seaports" (Stübben 1922, 138). Schumacher, on the other hand, sought—without anti-Prussian polemics—to convince by using plans and arguments and created a memorable scheme that declared territorial reform to be a necessary prerequisite for metropolitan reform.

But after the Great Depression in 1929 and the mass unemployment that followed, an unpredicted wave of chaotic settlement development began. "In this whole urban flight movement, the 'residential shed' plays a particularly disastrous role". Thousands of families gave up their town homes and people began to migrate to the countryside. This (unplanned) allotment garden movement saw an unexpected boom (Ockert 1953, 22). A "flood-like movement" and the division of properties would lead to problematic public health conditions in the urban periphery. "All sorts of illicit settlements emerged, created with highly congenial human willpower" (Schumacher 1932b, 382) that could not be controlled by regional planning in the absence of legislation. Hamburg had become a shrinking city, its population stagnating or slightly declining from 1927 to 1936, while the Prussian suburbs recorded a growing population.

The surrounding municipalities noticed a "marked decline in the immigrants' performance". Up until 1929, the average income of the newcomers in suburbs like Sasel and Wellingsbüttel increased, but this had taken a tumble with the Great Depression. "It is the hardship that drives some of the city dwellers out of the city, the unbearable economic and also emotional hardship that those who are unemployed, or those who must fear becoming unemployed suffer. It is this need that prompts the unemployed to give up their homes in the city and make do with a shed, however inadequate it may be" (Knutzen 1933, 8). Agricultural land was sold as "Schreberland" (allotments) and the buyers pushed through the construction of makeshift homes and residential sheds. The surrounding municipalities feared that they would be faced with significant demands on public welfare. "One runs the risk, for the sake of a momentary relatively small relief of the situation, of jeopardising the whole future of the area around the cities without actually benefiting the settlements, because the harmful effects will hit them very hard" (Schumacher 1932b, 383).

Up until the 1920s, migration of tax-paying citizens to the countryside and the periphery had always been a problem for Hamburg. Had the newcomers to the surrounding municipalities been welcome taxpayers, this relation was then reversed. Poorer and unemployed people, who could no longer afford to pay the rent for their city apartments, arrived to make a living on a plot of leased land. The families and settlers who had moved there from the city were "less productive" and "to a considerable extent sooner or later became objects of public welfare" (Knutzen 1933, 105).

"Families gave up their apartments in the city by the thousands, and an unstoppable migration began from the four cities, particularly from Hamburg, to rural areas. Without any sense, cheap land was sold, no matter where it was situated, and occupied, often far from any means of transportation. Numerous scattered settlements with illicit, arbitrary parcelling emerged. […] The allotment movement experienced an unexpected boom" (Ockert 1953, 22). In 1933, Wandsbek and Stormarn had around 6,000 organised allotment gardeners. Between 1929—the year Hamburg recorded its highest number of inhabitants—and the Greater Hamburg Act in 1937, over 45,000 people left the city. " The question of regional planning arose not only in relation to port and housing issues. "The example of Hamburg best shows that cities cannot be content with open space policies on their own territory or even within their city limits. The high-level development plan and regional planning further afield must create a natural connection between city and country" (Ockert/Linne 1929, 41) The continuous growth of the dynamic city was over.

The settlers on the urban fringe and resettlers were often—according to the planner Werner Gensel (1933, 25) at the time—"demanding and spoiled by life in the city. Therefore, unfulfillable demands are made on the inefficient suburban municipalities". The fringe settlers were "carefree and hopeful". "Many of them soon experience severe disappointment. Only the best and socially most valuable sections of the population can successfully survive the struggle for existence associated with resettlement" (Gensel 1933, 25). The "locals", on the other hand, had reservations about "city matters", and aversion dominated despite love of homeland and sense of nature. The flood-like reshaping of rural areas promoted conflicts between resettlers, who had become unemployed and tired of the city, and the rural population with its traditional values. The resettlement and evacuation from Hamburg meant that Stormarn had to admit people with low incomes, often unemployed, leading to additional travel, school, homeless and welfare costs. "The most dismal settlement phenomena are the primitive and illicit estates on the outskirts, which are becoming more and more widespread with increasing economic hardship. In terms of public health, aesthetic, fire and building regulations, they are sometimes completely untenable and inhumane" (Schumacher 46). "Wild" allotment gardeners unlawfully used their sheds as permanent homes and lived in makeshift dwellings and "box houses".

In addition to urban fatigue, it was the fear of inflation and the plight of agriculture and urban unemployment that led to mass migration from Hamburg. The uncontrolled growth of scattered settlements in the form of primitive huts and makeshift buildings were untenable for reasons of public health, aesthetics, planning law and building regulations (Schubert 2012, 213). Temporary residential sheds became permanent dwellings. Municipalities in the suburbs and surrounding areas were faced with considerable additional costs for roads, schools and welfare.

9 Structural Change, Appropriation and Adaptation

Under Schumacher's direction and with Oelsner's cooperation from neighbouring Altona, the intensive work of the Hamburg Prussian State Planning Committee only lasted five years, from 1928 to 1933 (Ockert 1953). According to Schumacher (1932, 26/27), the most difficult and most important task was to secure open spaces that would form a recognisable structure. While Hamburg's open space policy is restrictive, in Altona it bears imprints of abundance. After Schumacher's retirement at the age of 64 and Oelsner's dismissal by the Nazis in 1933, Karl Köster (who had previously been chief planning councillor in Harburg) took office as chief planning and building director in Hamburg. The committee was dissolved in 1935 and replaced by the Hamburg Regional Planning Association.

It was less the systematic preparation by the Regional Planning Association that paved the way for the Greater Hamburg Act of 1937, but rather an arbitrary order from the Prussian governor Hermann Göring to the Prussian authorities. Granting Hamburg's Reich governor Karl Kaufmann powers on Prussian territory would only have caused complications. Kaufmann explained: "With a strong hand, the Führer [Hitler] has cleared up the sovereign, administrative, transport political, urban planning and economic mess in the Elbe region. The Führer has thus secured the delivery of functions of the largest port in the empire for all time. […]" (Hamburgisches Staatsamt 1937, 27).

A few months after the end of the war, Schumacher (1945, 5) spoke of the "apocalyptic events that blew over us". Almost half of the homes had been destroyed and entire districts were in ruins. After 1945, parts of the open spaces in the destroyed neighbourhoods were used to build makeshift homes and cultivate potatoes and vegetables. On Dulsberg too, areas of the green corridor—secured by wire fences—were occupied and used for vegetable production. The green corridor was later reinterpreted as an open space and redesigned in the 1950s. Meanwhile, households in the neighbourhood are being involved in discussions about new uses (street hockey, basketball, etc.).

The Stadtpark Verein has sought to maintain and adequately redesign the park since 1911. Several buildings in the park, such as the large civic hall, fell victim to bombs in the Second World War. Changing requirements for its use and careless maintenance made it difficult to maintain the park as a total work of art. In the meantime, it has been listed for preservation and the Stadtpark has become accepted and much and flexible used open space (Hesse 2014, 220; Grunert 2015, 30). In Altona, plans to convert the Volkspark into a sports and event park were averted and a maintenance concept for the ecological development of its woodlands devised.

The plans for Schumacher's system of axes and Oelsner's "green rings" were integrated into a system of open spaces at the end of the twentieth century. Schumacher's mantra of 100 years ago, "to permeate the city's mass of houses with a coherent network of green corridors", to create a kind of "ventilation system for the city" (Staats- und Universitätsbibliothek, NFS II A 3, 4) could have been uttered in 2020. New contemporary requirements and networks have been integrated. Cooperative open space development, procedural issues, civil society involvement, local cooperation and flexible uses play an increasingly important role. After 1990, environmental protection and ecology should promote a new understanding of parks and ecological issues in general. Hamburg currently has nearly 40,000 allotments, which cover an area of around 1,400 hectares and their users are organised in 311 associations. Against the backdrop of the Covid-19 pandemic, the demand for allotment gardens is increasing. They make an important contribution to biodiversity, air quality, the urban climate and a more sustainable future.

10 Fritz Schumacher's Legacy and Transfiguration

Fritz Schumacher's work is thus caught between careful preservation and heritage conservation as well as between modernising redevelopment and adaptation. In retrospect, he proves to be a cautious, deliberate while comprehensive reformer, without the radicalism that was practiced elsewhere—but often he was met with little acceptance. However, Schumacher always had the city as a whole in mind and did not limit himself to building new satellites on the outskirts, but tried to also make lasting reforms in the old city. New residential developments on the periphery with open spaces, the "green ring around Hamburg's old body", was not enough for Schumacher—he also wanted to transform the old, the existing urban fabric. Because of the ownership structure alone, this was a much more difficult undertaking than building new housing developments. Thus Schumacher's goal was to reform the city into a better city, and to achieve this goal he proved himself to be not a revolutionary, but an ambitious radical reformer.

Schumacher linked spatial scales (from apartment to urban region) and thematic references (from building materials to design) in order to reform—not abolish—the city. Now he would probably be described as a workhorse and workaholic who would be multi-tasking in the evening while his sister read to him, working on plans and amendments which were drawn up in detail by his employees the next day. He was responsible for over 100 public buildings, including more than 30 schools in Hamburg.

Schumacher had always been a visionary and pragmatist, reformer and realist all at once. After 1933, the "apolitical" planning and building director held back any comments on specialist policy, but was still undisputedly regarded as the authority in the background when he began to focus on literary activities. He studied Shakespeare, for example, and as an admirer of Goethe, borrowed his metaphor of "architecture being coagulated music" and published three of Goethe's texts in an anthology. Between 1933 and 1942, he managed to write ten books, skilfully and eloquently avoiding conflicts with Nazi censorship (Schubert 2020: 3). Like few other architects and urban planners, Schumacher documented and published his own work in a timely and continuous manner, with many commentaries by other authors. In doing so, he provided the authority to interpret his work and promoted a myth about his personality.

Perhaps a quote best describes his work. In 1932, shortly before he left the building authority, he wrote: "Don't despise me for wishing! It is a tremendous force in life. It is not the wishing of a general and indeterminate kind that aims to amass good and beautiful things like in a fairy tale, but the wishing that emanates from the peculiarities and circumstances of reality and that has the courage to consider the path of its realisation in all directions all the way to the end" (Schumacher 1932c, 44). Schumacher's œuvre and its full impact are not specific to Hamburg, but are integrated into the regional, national and international discourse on the reform of the city and the development of the discipline town planning. His ideas, his concepts for reform, his methods and his ability to ensure that the plans are implemented promptly, are still current and forward-looking. "One must shape one's intentions in such a way that they can be realised" (Schumacher 1940, 33).

References

Archives
Staatsarchiv Hamburg 311–8/211 (Fritz Schumacher)
Staats- und Universitätsbibliothek Hamburg Carl von Ossietzky Nachlass Fritz Schumacher
NFS II A 3 Grünpolitik der Großstadt
NFS IV C1 Rede als Beigeordneter in Köln (1923)
NFS VI A 17a Vortrag: Der werdende Stadtpark (1913)
Beck J, Voss R G (1999) Die Alster. Ein Fluss prägt die Stadt. Ellert & Richter, Hamburg
Bey O (2021) Der Volkspark in Altona. Ein Waldpark als städtebauliches Labor der Moderne. In: Schubert, D, Michelis, P (eds.), Hamburger und Altonaer Reformwohnungsbau der 1920er Jahre. Vergleichende Perspektiven von Modellen der Moderne. Dölling und Galitz, Hamburg, (Dölling und Galitz)
Bezirksamt Hamburg-Nord (2014) Bezirksamt. Hamburger Stadtpark 1914–2014, Hamburg
Bezirksamt Hamburg-Altona (2014) Architekten und Ingenieurverein zu Haamburg, Hamburg und seine Bauten, Hamburg 1929. Altonaer Volkspark 1914–2014, Hamburg

Burkhardt, H G (2014) Konrad Adenauer und Fritz Schumacher—Wege zur Großstadtreform in Köln. In: Schubert, D, Seifert, J, Völlmer, Th (eds.), Die Reform der Großstadt, Informationen zur modernen Stadtgeschichte, 1: 38–51
Clasen, G (1947) Die Hamburger Staatssiedlung Langenhorn und ihre Schule, Hamburg
Frank, H (ed.) (1994) Fritz Schumacher, Reformkultur und Moderne. Hatje, Stuttgart
Frank H (2014) Volkspark und Stadtlandschaft. Schubert, D, Seifert, J, Völlmer, Th, Die Reform Der Großstadt, Informationen Zur Modernen Stadtgeschichte 1:65–86
Fritz-Schumacher-Gesellschaft E.V. (ed.), Fritz Schumacher in Köln. Rückblick und Perspektiven, Hamburg 2000
Gensel, W (1933) Siedlungsprobleme am Großstadtrand dargestellt an dem Beispiel des Vorortkreises Storman im Hamburgisch-Preußischen Landesplanungsgebiet, Würzburg
Genzmer, F (1924) Köln. Entwicklungsfragen einer Großstadt, Deutsche Bauzeitung: 75:481–486
Grunert, H (ed.) (2014) Betreten erwünscht. Hundert Jahre Hamburger Stadtpark, Dölling und Galitz, Hamburg
Grunert, H (2015) 100 Jahre Hamburger Volkspark—Eine wegweisende Parkplanung für eine zukunftsfähige Stadt. In: Die Gartenkunst 1, 100 Jahre Hamburger Stadtpark: 15–36
Hamburgisches Staatsamt (ed.), Groß-Hamburg, 26. Januar 1937, Hamburg
Hesse, F (2014) Ein Gartendenkmal zum Bewohnen. Der Stadtpark heute. In: Grunert, H (ed.) (2014) Betreten erwünscht. Hundert Jahre Hamburger Stadtpark, Dölling und Galitz, Hamburg: 220–235
Hoffmann, P Th. (1929) Neues Altona 1919—1929. Zehn Jahre Aufbau einer Deutschen Großstadt. Eugen Diederichs Verlag, Jena
Jahresbericht der Verwaltungsbehörden der Freien und Hansestadt Hamburg, Hamburg 1925
Johe, W (1998) Territorialer Expansionsdrang oder wirtschaftliche Notwendigkeit? Die Gross-Hamburg-Frage. In: Landeszentrale für politische Bildung (ed.), Vom Vier-Städte-Gebiet zur Einheitsgemeinde. Altona—Harburg-Wilhelmsburg—Wandsbek gehen in Groß-Hamburg auf, Hamburg: 9–40
Knutzen, F (1933) Denkschrift. Wanderungs- und Siedlungsvorgänge im Unterelbegebiet. Umfang, Bedeutung, Folgen, Aufgaben, Notwendigkeiten, Wandsbek
Krings, I (2014) Die Erfindung vom „Haus im Garten“. Die Architektur des Stadtparks, in: Grunert, Heino (ed.), Betreten erwünscht. Hundert Jahre Hamburger Stadtpark, Dölling und Galitz, Hamburg: 46–65
Kuick von, E (2014) Hundert Jahre öffentliche Grünverwaltung, in: Grunert, H. (ed.) (2014) Betreten erwünscht. Hundert Jahre Hamburger Stadtpark, Dölling und Galitz, Hamburg: 166–187
Linne, O A (1922) Hamburgs Grünflächen, Kleingärten und Friedhöfe, in: Otto Rautenberg (ed.), Deutschlands Städtebau. Hamburg, DARI, Berlin
Lichtwark, A (1917) Eine Auswahl aus seinen Schriften Vol. II, Wolf Mannhardt (ed.), Berlin, Bruno Cassirer
Meynen, H (1979) Die Kölner Grünanlagen. Die städtebauliche und gartenarchitektonische Entwicklung des Stadtgrüns und das Grünsystem Fritz Schumachers, Mit Karten Anhangband, Schwann, Düsseldorf
Migge, L (1909) Der Hamburger Stadtpark und die Neuzeit. Die heutigen öffentlichen Gärten—dienen sie in Wirklichkeit dem Volke? Conrad H. A. Kloss, Hamburg
Migge, L (1913) Die Gartenkultur des 20. Jahrhunderts, Eugen Diederichs, Jena
Nath-Esser, M (1997) Die Entwicklung der Volksparke in Hamburg und Altona, In: Schriftenreihe des Vereins Fritz-Schumacher-Gesellschaft e.V. Sautter + Lackmann, Hamburg
Necker, S, Woyke, M (1909) Vom Achsenkonzept zur Metropolregion. Stadt- und Regionalplanung für den Großraum Hamburg seit dem Ersten Weltkrieg. In: Zeitschrift des Vereins für Hamburgische Geschichte, Bd. 95, Hamburg: 143–166
Ockert, E (1953) Der Hamburgisch-Preußische Landesplanungsausschuss. In: Architekten- und Ingenieurverein Hamburg E. V. (ed.) Hamburg und seine Bauten 1929–1953, Hamburg
Ockert, E (ed.) (1959) Fritz Schumacher. Sein Schaffen als Städtebauer und Landesplaner. Wasmuth, Tübingen

Ockert, E, Linne, O (1929) Grünfragen, Sport- und Parkwesen. In: Architekten- und Ingenieurverein zu Hamburg (ed.) Hamburg und seine Bauten mit Altona, Wandsbek und Harburg-Wilhelmsburg 1918–1929, Boysen & Maasch, Hamburg

Schubert D (2012) Die Herausbildung der Regional- und Landesplanung im Raum Hamburg. Fritz Schumachers Modell der „natürlichen" Entwicklung. Die Alte Stadt - Forum Stadt 3:211–234

Schubert D (2014) Fritz Schumacher—Pragmatischer Visionär und konservativer Modernisierer. Städtebau und Wohnungsbau in Hamburg 1909–1933. Flandziu 2:79–100

Schubert D (2020) Fritz Schumacher—Neglected German town planner and urban reformer in Hamburg and Cologne. Planning Perspectives 1:1–19

Schubert, D (2021) Großstadt und Wohnreform 1923–1933. Fritz Schumachers „Gürtel um den alten Leib" Hamburgs. In: Schubert, D, Michelis, P (eds.), Hamburger und Altonaer Reformwohnungsbau der 1920er Jahre. Vergleichende Perspektiven von Modellen der Moderne. Dölling und Galitz, Hamburg 2021: 18–51

Schumacher F (1919) Die Staatliche Kleinhaus-Siedlung in Langenhorn. Baurundschau 47(48):201–209

Schumacher, F (1917) Die Kleinwohnung, (2nd ed. 1919)

Schumacher, F (1919) Hamburgs Wohnungspolitik von 1818 bis 1919. Ein Beitrag zur Psychologie der Großstadt. (Gross-Hamburgische Streitfragen, ed. Fred S. Baumann), L. Friedrichsen & Co, Hamburg

Schumacher F (1921) Das gesunde und das kranke „Hamburg". Bau-Rundschau 5:63–65

Schumacher, F, Arnzt, W (1923) Köln. Entwicklungsfragen einer Großstadt. Georg D. W. Callway, Köln

Schumacher, F (1927) Ein Volkspark. Dargestellt am Hamburger Stadtpark, Georg D. W. Callway München

Schumacher, F (1932a) Das Werden einer Wohnstadt. Bilder vom neuen Hamburg, Hamburg

Schumacher, F (1932b) Siedlungs-Sorgen. Deutsche Bauzeitung, May: 382–384

Schumacher, F (1932c) Wesen und Organisation der Landesplanung im Hamburgisch-Preußischen Planungsgebiet, Hamburg

Schumacher F (1936) Rundblicke. Ein Buch von Reisen und Erfahrungen, Stuttgart

Schumacher, F (1945) Zum Wiederaufbau Hamburgs. Rede im Hamburger Rathaus am 10. Oktober 1945. Johann Trautmann Verlag, Hamburg,

Schumacher, F (1949) Stufen des Lebens, Stuttgart [1935]

Schumacher, F (1940) Probleme der Grosstadt. Verlag E. A. Seemann, Köln

Sitte, C (1899). Der Städtebau nach seinen künstlerischen Gesichtspunkten. Vermehrt um „Grossstadtgrün". Friedr. Vieweg & Sohn, Braunschweig (Reprint 1983)

Stübben J (1922) Groß-Hamburg. Deutsche Bauzeitung Nr. 23:137–140

Thomas D (1970) London's Green Belt. Faber and Faber, London

Timm C (1985) Der preußische Generalsiedlungsplan für Groß-Hamburg von 1923. Zeitschrift Des Vereins Für Hamburgische Geschichte 71:75–125

Wagner, M (1915) Das sanitäre Grün der Städte. Ein Beitrag zur Freiflächentheorie, Carl Heymanns, Berlin

Wulff, G (1986) Das Werden der Fritz-Schumacher-Siedlung 1919–1921, Erster staatlicher Wohnungsbau in Hamburg, Hamburg

Land-Nature and its Economic and Cultural Value: The Case of the Zuiderzee Reclamation

Herman van Bergeijk and Denise Piccinini

We have to make sure, above all, that our mind is not halved by a horizon (A. Sen, Identity and Violence, New York 2006, 186)

Abstract In a country that is situated rather low in regard to the sea level and where available land has been gained from the sea, the conception of nature and landscape, forms of living and producing—and the way that it is implemented—is of fundamental importance in the creation of new land. Especially the evolution/changes in the form and function of the landscape elements can be seen and studied through the four major polders built in the Netherlands during the past century. The role of nature and its relationship with the utilitarian land has been continuously modified. Starting with a discussion surrounding the integral design approaches of Marinus Granpré Molière and Cor van Eesteren, we propose to investigate the shift of views about city, nature, countryside relationship and spatial consequences, from blueprint plan/to process planning by looking to how the Wieringermeerpolder was conceived. Then a look will be given to the North-East polder and the way that the structure was different from the Wieringermeerpolder, a unique conception of a spreading city according to models of Walter Christaller. Finally, the polders of Flevoland will be taken in regard in which among others, Van Eesteren again had a big influence. The recent historical views are of great value for how the role of nature is perceived in contemporary urban and landscape design.

Keywords Dutch-reclaimed land · Constructed new land · Production/aesthetics Dilemma's · Landscape evolution and planning · Polder politics

H. van Bergeijk (✉) · D. Piccinini
Delft University of Technology, Rotterdam, The Netherlands
e-mail: h.vanbergeijk@tudelft.nl

D. Piccinini
e-mail: d.piccinini@tudelft.nl

C. Brisotto and F. Lemes de Oliveira (eds.), *Re-Imagining Resilient Productive Landscapes*, Cities and Nature, https://doi.org/10.1007/978-3-030-90445-6_3

1 Introduction

As is well known in the Netherlands, water is deeply and fundamentally intertwined. It is largely determined by economics, even if social issues also play a crucial role. Natural resources are scarce and the balance between both fresh and saltwater and land is a delicate one (Metz 2012). At the beginning of the past century, when the creation and reclamation of the Zuiderzee finally took place after long deliberation, elevating the productivity of the land was a primordial consideration. Intensification of agriculture and animal husbandry were the basis for many developments, considering the fact that heavy industry was not a mainstay of the Dutch economy at that time. Industrialization was never a key factor, but over the course of many years, the concept of productivity regarding the land has changed. The dimensions of productivity have augmented and profits are no longer calculated merely in terms of financial profits. Other factors such as cultural intentions and sustainability have come into play. The use and designation of polders illustrate this in a clear manner. In their struggle against the water, the Dutch population has reclaimed extensive territory from water throughout the centuries. Particularly during the twentieth century, there were several big government initiatives for reclamation projects. Not all of them came to be, but several large water bodies such as the Zuiderzee, an arm of The North Sea and part of the Rhine Delta, were turned into land after long debates and planning processes. They encompassed approximately 165.000 ha. of new land currently known as the IJsselmeerpolders due to fact that this North Sea arm was enclosed by the Afsluitdijk, the long dyke between Western and Eastern sides of Holland. The waters behind this new dyke became a freshwater lake—the IJsselmeer. The long time required for the effectuation led to a continuous fine-tuning in the objectives and the methods. From a site of mere production, the polders increasingly became a petri dish where experiments were introduced that had to do both with landscape architecture and town planning as with the creation of natural habitats. It is interesting to see how the various polders illustrate, through time, different goals and procedures on a local, regional, and national scale. Of the five planned polders, Wieringermeerpolder, Noordoostpolder, Eastern Flevoland, Southern Flevoland, and Markermeerpolder, only the first four came to be and were officially declared reclaimed from the sea waters of the Zuiderzee in 1930, 1942, 1957, and 1968, respectively (Fig. 1). The fifth never was realized although several plans were made throughout the years. Most recently, artificial islands have been built in the area that function as natural reservations for flora and fauna.

For the sake of clarity, the polders discussed in this chapter will be handled two by two, the first two polders (built in 1930 and 1942) followed by the two last built polders (built in 1957 and 1968) since they show clear conceptual differences, proceeded by a short discussion around the role of the town planners, Cornelis van Eesteren and Granpré Molière, to better contextualize the momentum caused by the Zuiderzee reclamation project. Whereas the first two polders were directly connected to the existing land causing drainage problems, the other two were conceived.

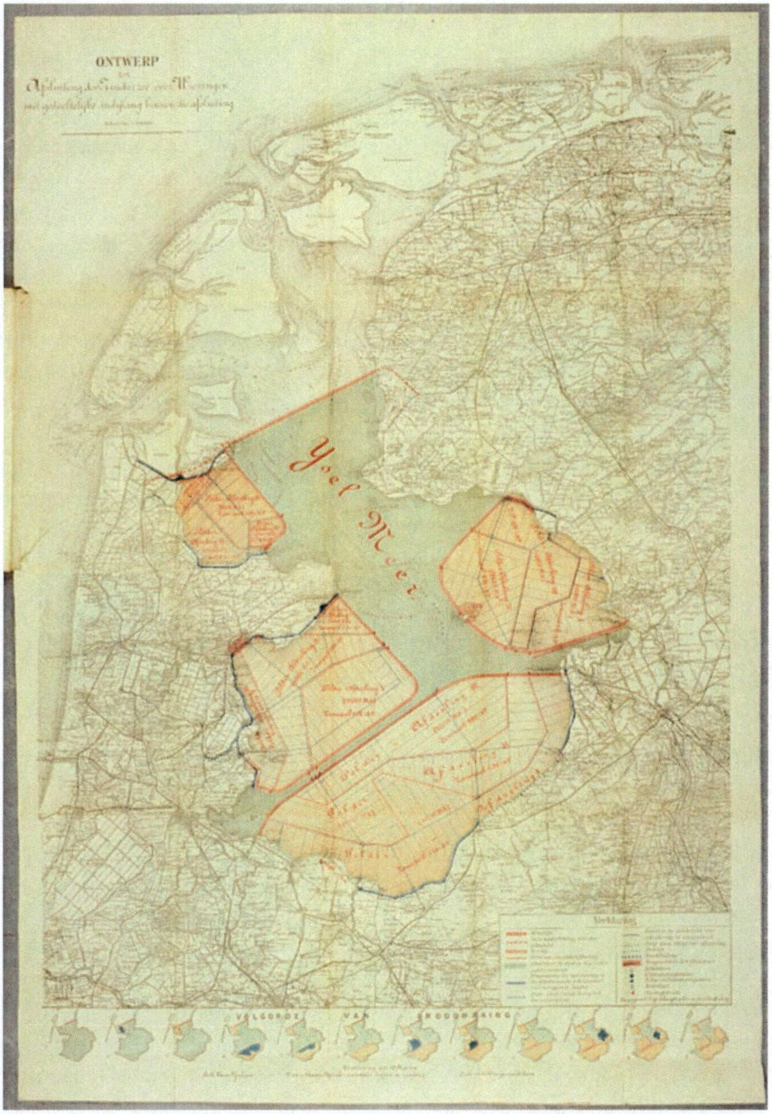

Fig. 1 Map of the Plan C. Lely, 1891 showing the original five planned polders, Wieringermeerpolder, Noordoostpolder, Eastern Flevoland, Southern Flevoland and Markermeerpolder (never realized) (Batavialand Archive, Lelystad)

They were divided as separate areas, divided from the existing provinces and shorelines but well connected with to their surroundings by bridges. Additionally, the first two polders are distributed around a central point and built as planned, while the other two polders are examples of extensive negotiation and the specific planning methods of the seventies. The different polders also show great differences in their planning related to the amount of land dedicated to the different activities or scopes. This is fettered to the changing views on productivity, the well-being of inhabitants, the role of nature and urban planning over the course of the century.

The polders, as a human creation, are an illustration of the evolving perspective on new production needs and environmental problems, as well as on issues regarding the relocation of people and an illustration of discussions about landscape compositions and aesthetics. Whereas initially, the inhabitants of the first two polders were to be farmers or people involved in agrarian productivity (and were therefore carefully selected and chosen), this slowly altered and ultimately the new settlers were meant to ease the pressure on the Randstad and in particular on Amsterdam. A different demographic became dominant and there was room for more variation in population. The reclaimed land shifts from being used as farmland (approximately 85%) and just 1% as settlements in the case of the Wieringermeerpolder (1930), to less than 50% dedicated as farmland on the final polder (1968) in Zuidelijk Flevoland. This ratio of farmland to urban settlement has continuously changed due to the dynamic character of the Dutch planning approach, and will be clarified and discussed throughout the chapter.

2 City-Countryside Debate at the Turn of the Twentieth Century and the Dutch Protagonists

The countryside and the city have always been on difficult terms. The relationship has been tense and the boundary between the city and its surroundings has often been problematic. At the end of the nineteenth century, and at the beginning of the twentieth century, green areas became a fundamental topic in town planning for the first time, leaving the space for production outside of the urban borders. In 1898, Ebenezer Howard gave space to green zones in the diagrams in his book To-morrow: A Peaceful Path to Real Reform. A year later Camillo Sitte extended his book City Planning according to Artistic Principles from 1889 with a chapter on 'big city green'. Leberecht Migge and Max Läuger talked about the radical reform of the garden (Haney 2010). The first one, Migge, gave also together with Alfred Lichtwark an impulse to the planning of a Volkspark in Hamburg. Läuger had made a design but the park would be realized under the supervision of the building director Fritz Schumacher who became one of the most influential town planners in Germany and beyond. In the Netherlands, Schumacher would make a name for himself with his remarkable plans for Cologne in which he tried to introduce green spaces to the city. The book which he wrote with Wilhelm Arntz was considered a major publication on urban planning (Meynen 1999). In September 1922, an exhibition of his plans was held in the rooms of the Art Circle in The Hague. Shortly after he was asked to give a talk at the International Town Planning Conference in Amsterdam in 1924. His topic was on 'the system of open spaces in large towns'. Although not present due to illness, Schumacher in his paper pleads for 'lay-outs of a landscape design' that 'will penetrate the body of the town as a net of open spaces and will establish a connection with outlying points' (Schumacher 1924, 105).

The opening lecture at the conference was held by Marinus Granpré Molière and was entitled 'The Modern City'. According to him, town planning 'will seek to re-unite town and country, civilisation and nature; it will introduce nature into the town and will guide in orderly ways the ebb of the crowd towards the country' (Granpré Molière 1924, 14). Granpré Molière had a natural, organic approach to town planning and the city, as was acknowledged by Fritz Schumacher who had been jury member with Granpré Molière in a competition in 1926 in Zürich and remembered that Granpré Molière had 'as much feeling for the design of nature as for the design of buildings' (Schumacher 1936, 174). This can be seen in Vreewijk, a new neighborhood in Rotterdam South where he had developed a town plan in the style of a garden city. Soon after its completion, Granpré Molière would become the chairman of the newly installed Urban Planning Council that fought for more attention to aesthetic problems related to the new embankments of the Zuiderzee. The influence of Granpré Molière would be particularly considerable. At the conference, Piet Bakker Schut discussed the 'twelfth province'. Like many others at the conference, he pleads for a regional planning approach in which the beauty of the landscape would be considered as a crucial factor. The conservationist and town planner Cleyndert already coined the term 'productive parks' (Cleyndert 1924, 267). The implications of the term would subsequently be expanded and have gained importance and have a contemporary impact.

During an excursion to Hilversum, the 200 conference participants were met by W.M. Dudok who led them around and showed him his newest work. He told them that he wanted to transform Hilversum into a garden city with a well-defined circumference. A young attendee of the town planning conference in Amsterdam was Cornelis van Eesteren. During his Prix de Rome traveling through Germany, Van Eesteren had encountered many city officials and architects, but he particularly admired Fritz Schumacher whom he visited in Cologne in October 1922. They would regularly exchange letters and when Schumacher died in 1947 Van Eesteren published a 'memoriam' in a Dutch journal (Eesteren 1948). In 1929, Van Eesteren would be appointed as head of the then established Urban Development Department of the Public Works Department in Amsterdam. Until that time, he participated in competitions, taught at several schools, had become a teacher of urban planning at the university in Weimar (Staatliche Bauhochschule) and gave lectures. In one of the lectures, he showed the 'green belt' extension plan for Kiel, made by Leberecht Migge in 1927 (Haney 2010, 151–154). Migge had combined biological urban planning and modernist open space in an interesting plan. A green belt was designed around the historical city and provided recreation areas. Yet the green fringe did not really penetrate the historical core. The planning examples of Schumacher and Migge were important parameters for Van Eesteren, who not only put his mark on the extension of Amsterdam but also on the new polders of the Zuiderzee.

The growing attention of academics for the realization of the polders: a literature review.

A considerable number of books have discussed the reclamation of the Zuiderzee and the planning of the new lands. All the projects, urban areas, and protagonists have been studied in depth. Historical and analytic studies overlap each other and hardly variate in perspective. However, few were in English. An exception is the doctoral thesis of the physical planner Coen van der Wal, In praise of common sense. Planning the ordinary (Wal 1997). He focuses mainly on the physical planning of the towns and considers them as a result of the typical Dutch sobriety. The continuity that Van der Wal indicates as being the result of the logic with which the Dutch proceeded is yet all but naturally and is due to the intrinsic discussions and conflicts that accompanied the work of reclamation and the planning of the polders. The broadest picture has recently been painted by architectural historian A. van de Woud in his study of the changing landscape during the period of 1850–1940 in which he affirms that by 1940, the Dutch landscape had been rendered more beautiful, 'Really according to our own ideal, everything neatly in drawers and boxes' (Woud 2020). The motto of his book is 'What is agriculture? A big food factory' Van der Woud explains how, in the Netherlands, the changing of the landscape was a result of making it more useful and higher in productivity. His book paints the bigger picture of the landscape in the Netherlands in general but for our topic, the book of Geurts is the most extensive as it deals with the ideas, the discussions, the decision-making and the protagonists who were involved in all issues regarding the design principles of the landscape, from the allotment plan to the distribution of barnyards, to the implementation of new forests, and the location of land art projects (Geurts 1997). Yet, in general, the production of all these studies about the recent history of the polders is an indication that there is a need for reflection on how all the interventions came to be. A proudness has been substituted by indecisiveness and criticism if the way in which the polders came about was the right way. Being built under the sea level and under the pressure of seepage and subsidence processes, the water in the polders needs to be continuously pumped out and higher dykes and other measures are necessary to safeguard the land against the rising sea level. However, despite this, the Zuiderzeepolders can be regarded as a good alternative for the green heart of the Randstad—the area between Amsterdam, the Hague, Rotterdam, and Utrecht that has been slowly sacrificed for urban development.

3 The Two First Polders, Wieringermeer and Noordoostpolder. Food Production as Condition Sine Qua Non

After years of deliberation, political complications, and preparations at the end of the second decade of the twentieth century, the reclamation works for the Wieringermeerpolder began. One of the most important events was the flood of 1916 when the coasts of the Zuiderzee were hit and a large part of the provinces along the coastline disappeared under the sea. Civil engineer C. Lely, for the second time nominated as

Minister of Water Management made the decisive argument to carry out his plan to close the sea arm, the Zuiderzee, by building the Afsluitdijk (the long dyke between Frisia and the province North-Holland) and the first IJsselmeerpolder. Another decisive argument for the reclamation was that the Netherlands needed more agricultural area to feed its population during and just after the first World War. The destination of the new polders was to be agrarian, its population would be chosen according to strict procedures, its layout and how to harmonize rural and urban issues gave way to significant controversies. By 1915, J.P. Thijsse had advocated for an approach that was different from before: 'We hope that the allotment of the new polders will not take place according to the block square system, that the land will not be exploited to the last square meter to generate monetary value. The quarter of a million people who will be living there still need something other than material prosperity and it must be protected against boredom and one-sidedness. The polders must be beautiful and, let's say, amusing' (Bosma 1993, 218–219). There was the fear for monotony and many voices asked for aesthetic consideration. The polders should not be reclaimed simply for agricultural purposes but should be thought of in an integral way, taking into consideration the people who would live there and their needs for public facilities, recreation, transport, and nature. The big question was, thus, how to organize productive land alongside urban areas in an aesthetic way, pleasant and efficient at the same time. A challenge that was part of the reclamation and design of all four polders.

3.1 Aesthetic Advisor Granpré Molière and the Future Landscape of the Zuiderzeepolders Rapport

After many discussions and interferences without direct results through the influence of D. F. Hudig, M. J. Granpré Molière was appointed aesthetic advisor of the Zuiderzee Works Department in 1927. Granpré Molière became the most influential figure in the planning of the polders. As an architect and town planner, he had become a professor at the Technical University of Delft in 1924 and from 1926 onwards, was the aesthetic advisor for the planning of these new lands in the Northwest of the country. Although the new land was a tabula rasa on which everything was possible, the power of Granpré Molière was limited. The Department of Zuiderzee Works had already established the allotment plans and initially, Granpré Molière had to deal only with the aesthetics of the bridges and villages. The advisor continuously had to fight uphill battles against the different authorities, ranging from ecological and professional institutions to bureaucrats at all levels.

Granpré Molière could, to a certain degree, realize what he already professed in his inaugural speech of 1924: 'It is a peace of mind to know that all form is a limitation and that the rapid speed of development usually concerns the upper stream of society; under the moved surface a being maintains itself, a movement takes place slowly through time; each area into which we penetrate more deeply shows us a sustainable life, a long-term growth; there is no innovation that does not include a return of values that a previous generation has carelessly ignored'. His attitude has been perceived as anti-urban but that is maybe too shortsighted. It was, after all, a project aimed at creating new agrarian lands and not new cities. The flatness and emptiness of the new land was striking and within that land, different urban 'growth centers' were planned. The fact that Granpré Molière dedicated much attention to the green side of the polder was certainly innovative because, until that moment, the polder was only considered as a place of production. This did not mean that there was much biodiversity, but this is not what received much criticism. Radical modernists disliked the fact that Granpré Molière had a view that was both directed towards the future and towards the past. Among those we can count is A.D. van Eck who was appointed in 1932 by the management of the Wieringermeer as leader of the construction office. Predictions for the polder were made on the survey that Th.K. van Lohuizen had made, but according to architectural historian K. Bosma, one of the main features in the planning by Granpré Molière was that he search for a closed townscape (Van Bergeijk 2015). He also did not wish to have long roads between villages because he would have found them aesthetically unpleasant.

He was however helped by the report Het toekomstig landschap der Zuiderzeepolders (The future landscape of the Zuiderzeepolders), which was published in 1928 by the Dutch Institute of Housing and Town Planning. The report investigates suitable layout principles, necessary functions, and their relationship and the form they could take, also based on the long tradition of Dutch land-making. The premise of the commission D. Hudig, who wrote the report, was that modern people should be able to experience the beauty of the utilitarian landscape, not necessarily because of its variation but also its sobriety, as a healthy environment and because of the care put in each of its elements. The elements proposed were mostly spatial landscape elements needed to shape this, otherwise a flat and homogeneous landscape, into a landscape with a human scale. They were also meant to connect the farmyards with the surrounding cropland and the villages among them. Ditches, waterways, planting plans, and the definition of the agrarian program all became part of the strategy to create a productive landscape that would contribute to feeding a growing population.

3.2 *Composing Urban Settlements, Agrarian Allotments, and Landscape Elements*

The idea in the planning of Wieringermeer and later in the planning of the Noordoostpolder was to create a series of villages closely related to each other and to the central village to support and complement each other in terms of facilities and social engagement. The farmyards standing along these roads formed green clusters. The farms, built with standardized concrete elements, had to look 'traditional' (Elpers 2019). One of the driving forces behind this report was P. Verhagen, a partner in the firm of Granpré Molière. Granpré Molière was capable of putting his distinct mark on

Fig. 2 Aerial View of Slootdorp in the Wieringermeerpolder. The original plan was made by M.J. Granpré Molière (Historisch Genootschap Wieringermeer)

the location and design of the various new villages and farms that had to be founded (Fig. 2). He also took care that the main roads of which one was from Amsterdam to the northeast of the country, did not go through the villages but passed them on the edge by slightly altering their course as well of the waterways and canals that served as drainage and for transportation ends (Van Woensel 1999).

He changed the monotonous and economically determined land allotment of the whole polder and made it so that the main villages were the focus. These four villages (in his former plan 12 small villages in a wreath) formed a triangle from where the roads led off into the surrounding land. Small areas, often leftover areas, were destined to be turned into woodland with high trees that could function as a windshield. However, the farmyard plantings and the ones along the road were minimal with hardly any variation. Everything, also the efforts of human work, was aimed towards the maximization of production. Productive qualities of the green zones were never regarded as of any importance. Only small patches of unsuitable ground for agricultural production gave some opportunities to create small forest pockets.

As an alternative, the landscape architect J. T. P. Bijhouwer was asked to present a planting plan. Bijhouwer, who later became a professor at the agricultural college in Wageningen, had travelled to the United States and had been deeply influenced by Lewis Mumford and the American parkways (Andela 2011). What he presented in 1937 was, in fact, a landscape structural plan based on four precepts: the habitability of the new land had to be increased by providing shelter and shade, the enormous vastness of the landscape had to be reduced to dimensions measurable to the human eye, and simple and ordered planting along the roads and the tree assortment had to be chosen in harmony with the soil type to express the landscape the best. Not much was built from his plan, instead, the usefulness of the planting was paramount and dictated the results; the only planting allowed were trees for wind protection along the roads and farms and a few small forests for short strolls on soil less suitable for farming (Aten 2007). More successfully were the designs for the different farmyards; the arable land, the meadows/grasslands, and the mixed farms as they were classified. Each was designed according to their business, creating a clear distinction in their internal layout, stables sizes, the water surrounding the yards, and planting choices. For instance, the planting of the meadows/grasslands and mixed farms had to connect with the planting of roads giving access to them and creating a visual continuity (Andela 2011). Food production in big quantities was still a major issue. Slowly that would change and especially in the Flevoland polder, biological diversity became a new target.

3.3 From Polder to Polder; Learning from Each Other. The Noordoostpolder Experiment

Wieringermeerpolder, considered a partially failed experiment was, however, fruitful. The other nascent Zuiderzeepolder, also known as the Noordoostpolder borrowed from the lessons gained in the planning and the persuasiveness of its predecessor for a better landscape composition. The planning of the Noordoostpolder initially took place more or less along the same lines as the Wieringermeerpolder, although there was a greater amount of rationalization which would facilitate the exploitation of the reclaimed land. In 1937, Granpré Molière gave a lecture in which he explained his views and stated: ‘A start has now been made with the Noordoostpolder. The surface is larger and more even, the soil is a bit better and the operation cheaper. And as here the phase of searching has largely made way for knowing, the character will probably also become tighter and more powerful. Construction payed more attention to aesthetics, it can be hoped, partly in view of better economic relations, that something will be achieved with this polder in many respects, which was not the case with the Wieringermeer’ (Westfriesch Dagblad 1937). With this Granpré Molière indicated that the views had changed and that the planning of the Noordoostpolder would be more convincing. It was definitively a process of learning by doing. Even as wartime planning did not come to a halt but continued in what had to become a mainly agricultural production project. Only gradually the space for other functions increased; the Wieringermeerpolder expended about 3% on forest and recreation, the Noordoostpolder about 6%. Over the course of time, some procedures and methods were perfected and the one-sidedness of the design of the Wieringermeerpolder, which had been criticized by architects of the modern movement, led to changes.

3.3.1 Villages

The planning of villages was based on the ideas of the German geographer Walter Christaller, who was in the Netherlands for an international congress in 1938. The different villages were planned in a concentric manner with the main urban nucleus in the middle. From the beginning of the allotment plan, a place for the 'urban care cores' (verzorgingskernen), as the villages were called, was considered, forming together with the roads and layout of the farm plots an important structuring element in the plan. For everyone, a village should be reachable at an accessible distance by bicycle or car of 7 to 8 km. In the end, the plan included ten villages around a central one, Emmeloord. The villages were meant as residences for mostly agricultural workers, servants at the farms, and other caretakers. To arrive at this decision, a discussion took place between the proponents of reducing the social distance between farm owners and farm servants. This discussion took place between Granpré Molière, his pupil C. Pouderoyen, and members of the Agricultural Department as well as those arguing for ways in which to maintain this social hierarchy. The consequence of the discussion was a plan including housing for the 'first farm worker' outside but at a close distance of the farmyard. Houses which were later, from the 60s on, becoming popular as vacation/second houses, B&Bs, retired farmers, or even permanent houses for people who wanted to move to the countryside to grow their own food as a healthy alternative.

3.3.2 Farms

Despite the fact that other functions were taking more space, the sizes of the agrarian parcels had increased considerably. The dimensions were based on the requirements to keep arable plots dry, as well as on the mechanization tools to work the land and the economically desirable size of the parcel. In the Wieringermeerpolder, the standard parcel size is 250 by 800 m, in the Noordoostpolder it has increased to 300 by 800 m. Technology had its impact. The standardized parcels formed clusters of four farmyards to assure a minimal social life for the farmers and their families showing a clear preoccupation with the well-being of the farmers who would be even more distanced from each other because of the increasing parcel sizes (Feddes 2004). The clusters themselves were a structuring element giving a certain rhythm to the perception of the landscape. Not all farm plots were, however, very big: with the introduction of fruit production, smaller plots were added near to the villages, forming a rhythm along the roads between two villages—from smaller parcels fronts, to bigger ones, to smaller ones again.

3.3.3 Planting Plan and Country Relaxation Tendencies

There were discussions around planting plans made by L. Brands Buys, C. Pouderoyen, and J. T. P. Bijhouwer during the II World War, a period when the further

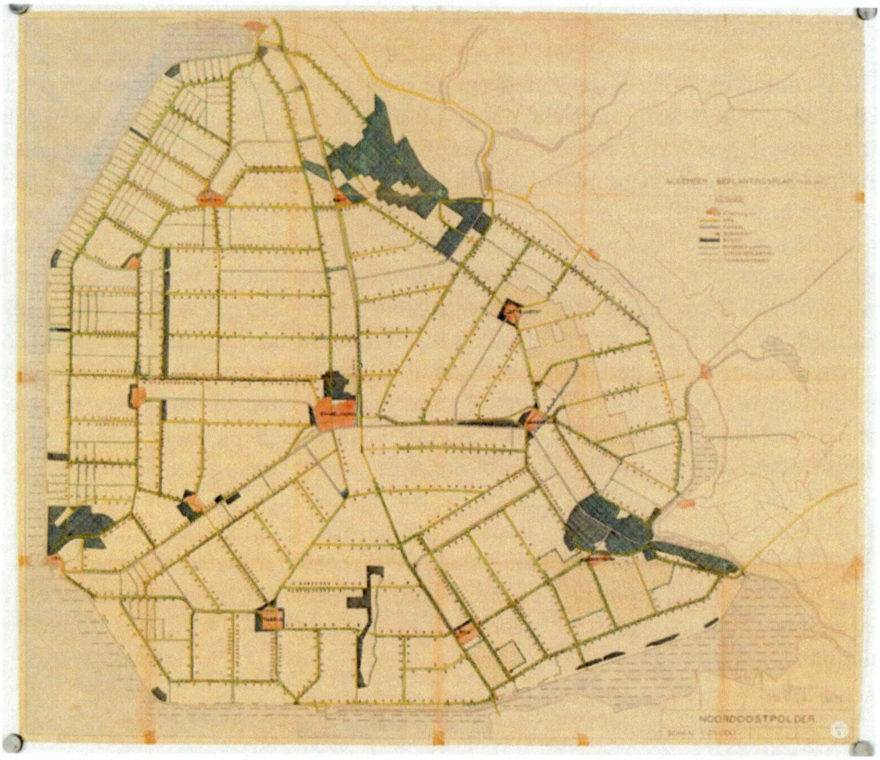

Fig. 3 Planting Plan Noordoostpolder, C. Pouderoyen and J. T. P. Bijhouwer, 1947 (Het Nieuwe Instituut, Rotterdam)

construction of the Noordoostpolder was stopped, but none of them were totally approved by the Forestry Commission because they would take up too much arable land (Fig. 3).

Pouderoyen did not subject himself to the allocation plan of the lots but dedicated much room to woodland areas and to broad wedges that would penetrate deeply into the polder and towards the central city. His plan is often overlooked or dismissed as being unimportant, but it is certainly interesting. He was the first one that clearly introduced landscape scenery elements and did not accept the polder only as an agrarian production machine. His alternative plan was rejected, leading to some minor effects. Slowly, more surface was designated to become part of the green structure but there was still much resistance, also from the farmers themselves who didn't want their agricultural plots to be shadowed by the trees along the roads and water channels. They were still totally focused on maximizing their harvest and were less interested in ecological or recreational aspects. Based on research directed by Bijhouwer, a forest of 20 ha was considered necessary for each settlement to provide the villagers space for sport, play, and other forms of relaxation. The forests, bound to the central core of Emmeloord, would be 50 ha. Half of the forest area would be used as a walking route, the other half as a recreation area with plenty of space for sports and games. An important additional function of these green pockets was to provide shelter from the wind and to create intimacy for the villages in the otherwise empty polder land. The effectively realized village forests are usually located to the west or north of the villages, and are eventually between 10 ha at Luttelgeest and 90 ha at Emmeloord. Emmeloord, the most important one in the polder, located centrally in the polder, was mainly designed by Pouderoyen (Bruggenkamp 1994).

Although during the war and shortly after a 'shaking hands' mentality was present, soon conflicts between modernists and architects with more traditional views rose to the foreground. To a large extent, the views of the traditionalists, of which Granpré Molière was the main exponent, still were the dominant factor but the modernists were able to put their mark on the village of Nagele, that forms an exception in comparison to all the other villages built in the polder. With Nagele's plan, the modernist 8 en Opbouw group wanted to arrive at openness and conceived the village as consisting of units. Wide green zones divided the small clusters of habitations that all had flat roofs, indicating their modernist pedigree. During this period C. van Eesteren slowly came into the picture. For a long period, he would oversee the design of the Flevoland polders as the most influential advisers to the responsible board. The landscape of the polders became a concern of primary importance. Different principles became the guidelines along which the new polders had to be designed. Van Eesteren continuously came into conflict with many ideas that the leaders of the agency of the IJsselmeerpolders, who were factually in control of the whole operation.

4 The Experience of the Flevopolders—Eastern Flevopolder and Southern Flevopolder, Van Eesteren and the Urbanization Ideas of the 70s. Dismissing Agricultural Use for Other Forms of Production

The reclamation of land that had to form the Flevopolders occurred in two phases. After the dykes were made, the east part was laid dry. Later the south side was reclaimed. The basis of the allotment plan was formed by cities, forests, green corridors, and land for agricultural production (mostly as compensation for the loss of agricultural areas in other parts of the country). Even though the farmland, housing, recreation, and nature development may seem to be designed as separated, individual parts, there is a high degree of interweaving among them because of the integral approach of this multifunctional landscape wherein even land-art objects along with farmland where possible. The plan organizing the different uses can be qualified as an 'open plan' or process planning, meaning it is made to give a direction to the future development different to the plans made for the first two polders, which were called 'blauwdruk' (blueprint) planning or final design planning. The experience with the first two polders had shown that it was sometimes difficult to predict the development in time, men, therefore, wanted to prevent clear-cut goals and create a more flexible and global approach along with some practical rules and a code of conduct. The other important approach was the one introduced by both Van Eesteren and Dudok, who had advocated for a more open-minded attitude towards landscape elements. They believed that the development plan should adopt a more holistic approach from the start and that the farming allotment division should be adapted to the requirements of the landscape design. Dudok visualized the interests of the standing committee and came up with a map and the recommendation to not only plan living, working, and traffic conditions but also take into account recreational areas. He also argued for parkways or recreational roads. Important was Dudok's contribution to the system of forest strips, which he saw as American parkways leading through and along the edges of the polder. He suggested that these green belts would be laid out throughout the polder more or less like veins and that the land itself would be wavy. Van Eesteren and the other advisors were charmed by this idea. This is especially important because it indicates that the differences between Van Eesteren and Dudok were not that big. Initially, Van Eesteren had been an advocate of an analytical town planning based on surveys, whereas Dudok was in favour of a more creative approach. That Van Eesteren followed the opinion of Dudok regarding the green zones is illuminating. In addition to forest belts, the Working Group proposed a range of village woodlands, very large (production) forests with irregular edges and more open spaces and ponds. In 1942, during the Monumentendag (Monument study day) on the IJsselmeerpolders, Dudok presented a memorandum on the landscape of the IJsselmeerpolders. To this memorandum, a drawing was added showing a system of massive elongated forest belts over the future Southern Flevoland. The standing committee had estimated a minimum forest at 5% of the total area, approximately 7500 ha with a width of 2 km. This would amount to a total of 35 km. That is approximately the sum of the length of all forest strips designated by Dudok in his presentation. His proposal was critiqued and Dudok responded by saying: 'The niggling on a small forest strip means that we don't have a culture like old Hellas; that specialists have too little cultural insight to venture beyond their remit. But then there must be a government that is aware of its great task' (Van Dissel 1991, 206). In general, his proposal found fertile ground. The forest belt of Dudok can be seen as a means of limiting the scale and bringing it more into line with that on the old land where recreation is the benchmark. In 1943, a memorandum on recreation, natural beauty, and tourism was published by the urban development department of Amsterdam, where Dudok's plan was further elaborated with a few differences; instead of 5% afforestation, 10% is used based on the example of the Twente landscape, an area in the East of the Netherlands. More attention was given to the landscape with a wooded character instead of just recreational value.

4.1 Modernization of Production. Up and-Downscaling

Nevertheless, all changes in approach, the farm sizes were continually under scrutiny and increased to 300 by 1000 m in the Eastern Flevoland. The layout of Southern Flevoland saw a big leap in scale with a basic module of 500 by 1700 m (Fig. 4).

In practice, this jump in size proved to be way too large, therefore a subdivision of farms was, in the end, necessary to reduce the parcel sizes. The increasing parcel sizes were possible, in part, due to the modernization of agriculture set in motion by the Dutch Ministry of Agriculture under the leadership of S. Mansholt after the second World War. The policy was also to make land equally accessible for all farmers and to render production more profitable and adaptable to constantly changing conditions (Andela 2000). The theme of urban agriculture was put forward at the moment that farmers were confronted with environmental problems among others. The necessity of the new polders for the production of goods for the national population lost its importance. At the same time, Southern Flevoland had to be considered within the orbit and the influence of the Randstad where existing cities had a decreasing possibility to expand. On the one hand, the polder became a relief of the pressure that was exerted by a growing population in Amsterdam. On the other hand, the factor of recreation and culture became even more prominent. Ecological motives were leading in a different way than they did before. The restoration of

Fig. 4 Oostelijk and Zuidelijk Flevoland, 1971—Rijksdienst voor de IJsselmeerpolder (Het Flevolands Archief, Lelystad)

old and the creation of new ecosystems became one of the objectives. A network of ecological paths for wildlife that had to cover the whole country was a goal. This mentality change led the Netherlands in another direction. Territory allocated for agrarian purposes was given to other uses. Large portions of the polder were designated as parks, landscape art sites that tried to tie the local to the cosmic or natural reservations sometimes totally closed to the public, where rewilding could take place. Tourism and entertainment were new principles that provided new sources of income and better prospects of profit. The search for more variety is not only clear from the broad spectrum of different landscapes but also from the planning of the city of Almere.

4.2 Lelystad and Almere, The Future is a Process

The main city in the east part was Lelystad for which Van Eesteren made the plans according to the principles of the CIAM. Originally, it was conceived of as an industrial town of approximately 100.000 inhabitants, a number it never reached (Geurts 1995). His plans, which foresaw a city on a magnificent bay, were eventually altered in a radical way, mainly because they were thought to be too expensive and difficult to realize (Fig. 5).

Van Eesteren wanted to raise the roads for the cars and a monumental town centre. The city, which had to be the capital of the polder, had to be constructed in a very short

Fig. 5 Photo showing the reclaimed land with the location of the future city of Lelystad, 1957

time. The various institutions were combatting each other the whole time and, in the end, only small parts of Van Eesteren's plan for Lelystad were realized. Attention shifted towards Almere that would also be closer to Amsterdam.

The first plan for the new city of Almere, which together with Lelystad had to become the main centres in the province, was conceived by Alle Hosper among many others. Because the precise destination of the plots was not established from the beginning, the plots were designed as open areas and recreation was depicted in an abstract manner. Depending on future wishes, they could be filled in as space for parks, sports fields, water, or other types of recreation. Almere, now a town with a population of 215.000 inhabitants—but set to become a city of at least 400.000 inhabitants in the next decade as governmental studies predict—has become a field of experimentation for Dutch architects, just as the polder as a whole has become a playground for new insights. Almere consists of several distinct urban districts that are divided by green wedges, a landscape structure, a zone thought to serve later spatial developments. Aside from the districts of Almere City, Almere Harbor, Almere Buiten Almere Wood, and Almere Gate, the possibility of building new cores is being discussed and planned (Fig. 6).

For instance, Almere Oosterwold, where small-scale private or community-related urban agriculture and green energy resources are located beside do-it-yourself

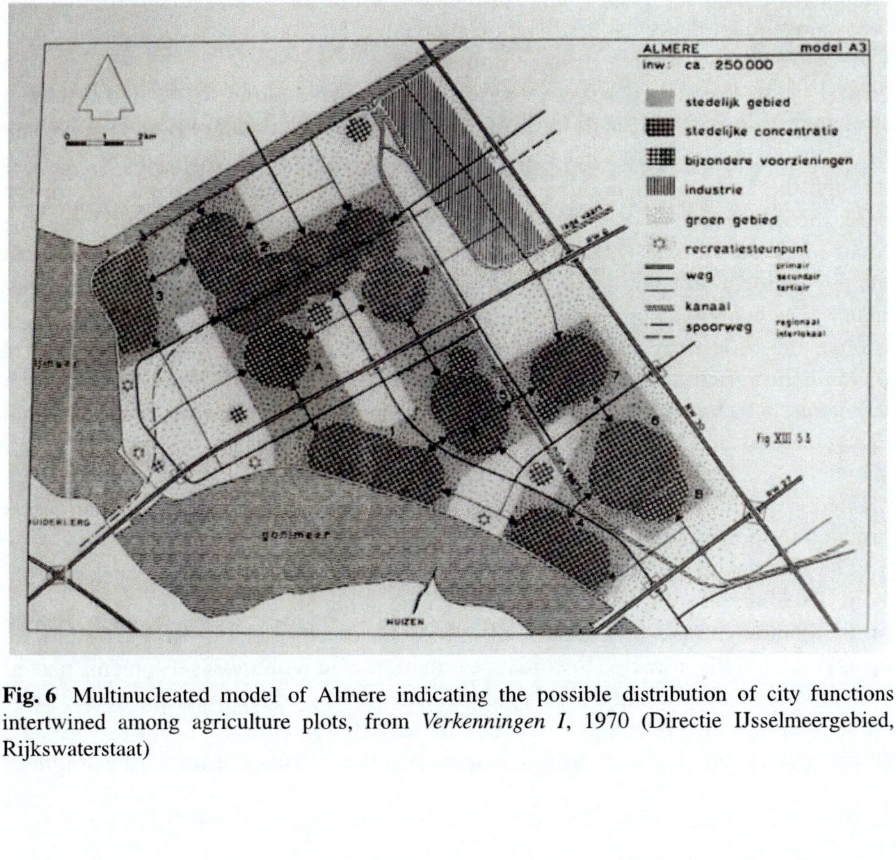

Fig. 6 Multinucleated model of Almere indicating the possible distribution of city functions intertwined among agriculture plots, from *Verkenningen I*, 1970 (Directie IJsselmeergebied, Rijkswaterstaat)

housing projects. This multi-core model, to be traced back to the layout for the villages used on the Noordoostpolder for instance, seems to be applied for Almere, however, the urban cores are bigger and closer to each other. They are also reminiscent of the garden city model of Ebenezer Howard where city and countryside should be intertwined—however, Howard's model was seen as utopian in a crucial point, that of land ownership. For the city of Almere, half of the area was programmed as a garden city, with the same amount of inhabitants (Feddes 2008).

5 Conclusion

When we look back at the four polders, an attitude towards landscape and urban planning emerges, an attitude towards integrating farmland, settlements, recreation, and new forms of nature became explicit. Reflections made by the many authors involved in inventing this new land and how to build the next polder was not necessarily a follow-up of the already built polder, but included other influences like the availability of technology, as well as the economic, social and aesthetics needs.

Most certainly the points of view of the figures involved in the design process were of great influence; while Granpré Molière, Verhagen, Bijhouwer, and Pouderoyen among others, mostly responsible for the creation of the Wieringermeerpolder and Noordoostpolder had what is often considered a more conservative and picturesque way of designing. Having designed quite introvert enclaves, distributed around a central point, their designs were very orderly, arranged along roads, and formed a strong unity, not only in terms of layout but also in complementary functions. Where the roads were clearly meant to be part of the connection between villages, going through the farm landscape, connecting enclaves. The agricultural plots form the base, the matrix where the other functions and circulation system are built on. The focus on these two polders is to create as much as possible productive land to feed the population, making a little effort to also give spatial quality and a pleasant experience. The economic and functional values were predominant here. The landscape is open and wide, only punctuated by the farmyards and the strict pattern of ditches dividing the agriculture plots. The borders of the villages are well defined having but few landscape elements intermediating between enclosed and open. The aesthetical characteristics of the flat and wide areas were hardly appreciated. Roads and water channels were made to go from A to B, for the rapid transport of goods, trees were planted as wind protection with a few exceptions for the rows of poplars and wood production forests.

The Flevoland polders, designed in the 1960s under the advice of C. van Eesteren, W.M. Dudok, T. Koolhaas, A. Hosper, and many other experts and organizations are, on the contrary, more functional and process oriented. These two polders are set up asymmetrically, directed towards the outside world with most settlements placed along the edges, bordering the old land and water with a heavily planted road. Transverse roads are also meant as recreation paths, often integrating water as well as a recreational, ecological corridors and views over the landscape. Mostly, the Southern

Flevopolder is a model of negotiation planning of the 1970s. Agriculture is no longer the all-defining land use but a set of different programs grouped together with a clear preoccupation of creating a more articulated landscape composition. The agricultural plots still covered around 50% of the surface of the polder but are no more the underlying matrix but one of the many functions displayed together. Recreational sites, forests, natural corridors, land art projects, and more recently land use manifestations, require space too. Rather than creating a fixed plan, the idea was to create spatial conditions for different environments to shape the urban cores and the surrounding landscape allowing for flexibility in use through time. The question is if this will be the case. Looking back, we see that all the plans of the more creative architects and town planners were weakened down by professionals within the administration. Concepts continued fluctuating while, in the end, it was mostly an economic factor that determined the course that was taken. Experimentation with ideas and realization with the hand on the money purse! All plans for the future, especially the most recent bold ones for Almere—need scrutiny and a deep examination of their possible effects on all levels. The last realized plans include those for the Marker Wadden, a series of islands in the Markermeer, that was initiated in order to boost the biodiversity of the ecosystem.

So, while the agricultural plots grow in size mostly because of sowing, mowing, and other technical and scientific improvements, the amount of surface designated to accommodate the farm units decreased strongly while space for settlements, displacement, recreation, and new nature has taken its place, showing a strong shift in policy and needs. The agricultural plots are compounded together, however, forming large areas of flat land between closed zones like forests, fruit production, and settlements, distributed not only according to the most suitable soil type for agricultural production but also to create the sense of flatness as being the most intrinsic characteristic of the new land and a crucial ingredient of national identity.

In a recent research initiative, a shift of perspective is noticeable. The view from the sea has become a central issue for example. The inland sea, the IJsselmeer, the wide space hidden mainly behind high dikes get more attention and are taken as a point of departure for new developments. As well, the Afsluitdijk, the dike that was constructed and finished in 1932—an event that was celebrated with a monument designed by Dudok—has the attention of planners due to the fact it has to be redesigned in order to cope with sea level rising. Considering their proximity to the metropolitan region in the Netherlands, the consequences of climate change and the need for more recreation, the polders of the Zuiderzee will remain a topic of interest in the coming decades. It is still an ongoing project for which the strategy in the future needs to be flexible and open to ad hoc developments. Policymaking has become more difficult due to the many criteria that are at play. Planners will have to deal with this. But in the end, much remains uncharted territory.

In this contribution, we wanted to reflect on the ways in which the polders came into being and to illustrate how, in the Netherlands, the issue of making new land, very often by controlling water, is a seminal topic that is subject to continuous rethinking. The Dutch verb 'polderen' is already a testimony of an attitude in which results are arrived at through negotiating and compromising. This is a natural consequence of a democratic society in which the wavering decision process has its pros and cons. This is clear also from the history of the Dutch Zuiderzee polders. Nature is menaced by the increase of leisure time, by economic reasons, and by industrial motives. Farmland is giving way to space for new housing and the development of 'new' nature. The Dutch government and the agencies responsible for the planning of the country are well aware of this. They have to reinvent the problematics connected to the reclamation projects and a way to deal with these constantly. New insights have their impact on the built and unbuilt environment. For many years, the Dutch government protected the central green zone in the Randstad, (mostly filled in with agriculture uses) the so-called Green Heart. Initially, the concept of a green heart was favored above the layout of green belts. Recently discussions have begun about the need for this. Certainly, also the 'open areas' in Flevoland have substituted this need. Through a system of belts, endangered animals could freely wander from one zone to the other, covering a great territory. Instead of dividing up the country among farmers, there is room for more different approaches. Respect for nature has become increasingly more important than financial gains or agrarian productivity. People appreciate more and more and want to enjoy the countryside, alongside the agrarian plots and far, low horizons. Also, the new neighborhood of IJburg, adjacent to and part of Amsterdam, has taken some pressure off the housing shortage in the capital. It came in the place of a woodland area that Van Eesteren had originally planned, the Uilenbos. It is a small step in the consideration of how land, or the Earth for that sake, can be saved for the future. If this does not work, the Dutch have a reputation for creating and demolishing the artificiality of their country. As far as the polders are concerned, in the twentieth-century nature slowly entered the picture. At the end of the century, it was more than ever recognized that nature could have esthetic and recreational benefits. The term productivity received a more comprehensive meaning, including nature in its broadest sense, new forms of agriculture that is very much downscaled and intertwined with housing, forming alternatives living options, has become possible. Short-term profit and maximum productivity were tempered by growing attention for sustainability standing next to extremely efficient, high-technology farming.

References

Andela G (2000) Kneedbaar landschap, kneedbaar volk. De heroïsche jaren van de ruilverkavelingen in Nederland, Bussum: Thoth

Andela G (2011) J.T.P. Bijhouwer. Grensverleggend landschapsarchitect. Rotterdam: Uitgeverij 010

Aten D. Reh W, Steenbergen C (2007) Sea of Land. The Polder as an Atlas of Dutch Landscape architecture, Amsterdam: Architectura & Natura

Bergeijk H van (2015) Van Lohuizen & Van Eesteren. Partners in Planning and Education at TH Delft, Delft: Bk Open

Bosma K (1993) Ruimte voor een nieuwe tijd: vormgeving van de Nederlandse regio, Rotterdam NAi

Bruggenkamp JW (1994) 'Emmeloord, traditie en vernieuwing', in: Jaarboek Monumentenzorg, 74–91

Cleyndert H (1924) Parks and recreation grounds in the Netherlands. In: Part I (ed) International Town Planning Conference Amsterdam 1924. Papers, Amsterdam, pp 257–271
Dissel, A.M.C. van (1991), 59 jaar eigengereide doeners in Flevoland, Noordoostpolder en Wieringermeer, Zutphen: Walburg Pers
Eesteren C, van (1948) 'In memoriam Fritz Schumacher', in: Tijdschrift voor Volkshuisvesting en Stedebouw, no. 4
Elpers S (2019) Wederopbouw boerderijen. Agrarisch erfgoed in de strijd over traditie en modernisering, 1940–1955, Rotterdam: NAi010
Feddes F (2008) De Almere Principles. Voor een ecologische, social en economische duurzame toekomst van Almere 2030. Bussum: Uitgeverij THOTH
Feddes Y (2004) 'The Designed Landscape of the Zuiderzee Polders. An Analyse of the Significance of Farmyards and Roadside Plantings', Oase, 63
Geurts AJ (1995) Lelystad. Stedebouwkundige ontwikkeling en vormgeving, Lelystad: De Twaalfde Provincie
Geurts AJ (1997) De 'Groene' IJsselmeerpolders. Inrichting van het landschap in Wieringermeer, Noordoostpolder, Oostelijk en Zuidelijk Flevoland, Lelystad: De Twaalfde Provincie
Granpré Molière MJ (1924) 'La Cité Moderne', International Town Planning Conference Amsterdam 1924, Part I, Papers, Amsterdam
Haney DH (2010) When Modern was Green. Routledge, Life and work of landscape architect Leberecht Migge, London/New York
Hemel Z (1994) Het landschap van de IJsselmeerpolders. Planning, inrichting en vormgeving, Rotterdam: NAi Uitgevers
Hudig D (1923) 'Stedebouw', Algemeen Handelsblad, 6 November
International Town Planning Conference Amsterdam 1924, Amsterdam 1924
Jarzombek M (2008) 'A Green Masterplan is Still a Masterplan', I. and A. Ruby (ed.), Urban Trans Formation, Berlin: Ruby Press
Metz T, van den Heuvel M (2012) Sweet & salt. NAi Publishers, Water and the Dutch, Rotterdam
Meynen H (1999) Die Kölner Grünanlagen, Düsseldorf: Schwann, 1979; D. Schädel (ed.), Fritz Schumacher in Köln, Schriftenreihe der Fritz Schumacher Gesellschaft, Hamburg: Christiaans
Schumacher F (1924) 'Grünpolitik der Grosstadt-Umgebung', International Town Planning Conference Amsterdam 1924, Part I, Papers, Amsterdam
Schumacher F (1936), Rundblicke. Ein Buch von Reisen und Erfahrungen, Berlin/Stuttgart: DVA
Wal C, van der (1997) In praise of common sense. Planning the ordinary, Rotterdam: 010
Westfriesch Dagblad (1937) 'De Wieringermeer—stedebouwkundige ervaringen', Westfriesch Dagblad, 21 June
Woensel JTWH, van (1999) Nieuwe dorpen op nieuw land. Inrichting van de dorpen in Wieringermeer, Noordoostpolder, Oostelijk en Zuidelijk Flevoland, Lelystad: De Twaalfde Provincie
Woud A, van der (2020) Het landschap, de mensen. Nederland 1850–1940, Amsterdam: Prometheus

Sharp's Town and Countryside

Izaskun Aseguinolaza-Braga

Abstract With the aim of searching for criteria for sustainable urban development, this chapter will try to rescue the understanding of the relationship between town and countryside by reviewing *Town and Countryside*, one of Thomas Sharp's first publications. In this book, Sharp criticises the lack of success in urban development taking place at the time. In front of an uninteresting urban landscape, British longed for a return to the countryside. The garden city and the solutions derived from it found the ideal context for their success. However, according to Sharp, the supposed return to the countryside of these solutions meant the simultaneous destruction of the natural and the urban. It is in the return to the logical antithesis of town and country that he sees the solution. He considers it particularly important to define the appropriate rural or urban character of each of the two areas as well as their boundaries. In this sense, he rejects the criterion followed by urban planning which decreases its density as it moves away from the urban centre. In any case, he acknowledges that both realities are inseparable, which means that from the city the countryside must be accessible to citizens and leads him to analyse the size of urban areas and how these should be organised in the territory.

Keywords Town · Country · Town planning · Landscape design · Civic design

1 Introduction

In the first third of the twentieth century, when the foundations of urban planning were being laid in the United Kingdom, some professionals were aware of the necessity of urban development while keeping in mind its interaction with the territory in which it is located. This article takes a look at this past, with the aim of highlighting and rescuing some ideas that are useful for truly sustainable urban development in our time.

I. Aseguinolaza-Braga (✉)
Departamento de Arquitectura, Profesora Adjunta, Universidad del País Vasco (UPV/EHU), Leioa, Spain
e-mail: aseguinolaza@ehu.eus

C. Brisotto and F. Lemes de Oliveira (eds.), *Re-Imagining Resilient Productive Landscapes*, Cities and Nature, https://doi.org/10.1007/978-3-030-90445-6_4

With this aim in mind, Thomas Sharp´s disciplinary contribution to town planning is examined. Sharp was undoubtedly one of the leading figures in British planning in the central years of the twentieth century and one of the greatest advocates of attending to the environment being planned, an attitude that was already evident in his early work *Town and Countryside. Some Aspects of Urban and Rural Development* of 1932. Through the review of this text, this article attempts to highlight Sharp's understanding of countrysides and towns, and the relationship he considers should be established between them.

The first section reviews the disciplinary context in which this book was written, followed by a brief biography and an identification of Sharp's main contributions to the discipline of urban planning. The second section shows how Sharp emphasises the need to understand the town and the countryside as two diverse and complementary realities that must be maintained in a happy antithesis. Within this framework, throughout this section, we also delve into the formal qualities that—according to Sharp—should guide actions in towns and in the countryside and—what is more decisive—the relationship between both spheres that should be reinforced, if necessary. We conclude this article by stressing that Sharp's book, beyond being a mere defence of the character of the British countryside, sets out a clear and unique way of acting in towns and in the countryside and therefore proposes the systematisation of Town and Country Planning, which he continued to defend and practice throughout his professional career.

2 British Urban Development at the Beginning of the Twentieth Century

Sharp's contribution was made in the years following the First World War, when the need to address the housing shortage produced a large number of urban development projects driven by government policy. These needs meant that the active *Garden City and Town Planning Association* (GCTPA), while trying to remain faithful to the principles of the *Garden City*, understood that these had to be adjusted to a different situation[1] and to the experience provided by the first garden city of Letchworth, in 1903.[2] At the beginning of the second decade of the twentieth century, Unwin and others had begun to develop the idea of decentralisation of the city, which was implicit in Howard's diagram. Thus, they incorporated the territorial scale into practice and moved closer to the regionalist approach of Patrick Geddes (Ward 1992, p. 11).

[1] This broadening of vision was already evident in 1909 when, following the passing of the first *Town Planning Act*, the *Garden City Association* was renamed the *Garden City and Town Planning Association* (Hardy 1991, p. 36).

[2] Already the reprinting of Howard's (1902) showed that the social reform he envisaged in, Tomorrow: A Peaceful Path to Real Reform, remained an environmental reform for the city based on the garden city model (Ward 1992, p. 4). This was confirmed in the implementation of Howard's ideas in Letchworth in 1903, where Howard's co-operative approaches had not been achieved.

However, confident that this would be a way of facilitating the construction of more housing in less time, it was a few years later, in the final years of that same decade, when these approaches to satellite towns were once again reformulated with particular force. Benjamin Purdom, together with Howard and W. G. Taylor, formed the *National Garden Cities Committee* with the aim of influencing the government to become involved in the construction of these new towns. Soon after, Patrick Abercrombie, George Cole and Frederick Osborn joined in the effort, and the group became known as the *New Towns Group.*

In *New Towns After the War*, in 1918, they proposed the application of the urban model proposed by Howard in a generalised manner to create new towns financed with public money. This model, as the *Newtownsmen* understood it, would make it possible to organise these new settlements into a rational and strategic system in the territory. The GCTPA, at its annual meeting in 1918, took on the objective of supporting *Newtownsmen*'s proposals without much debate, although they maintained their support for the implementation of Howard's ideas through the garden suburbs (Hardy 1992, p. 192; Ward 1992, p. 10).

Purdom, in the following decade—as he explains in *Town Theory and Practice* and *The Building of Satellite Towns*, texts published in 1921 and 1925, respectively—qualifies his support for the promotion of various garden cities by defending that garden cities should be understood as only one of the models that can be used in urban planning; since, in addition to garden cities, there should be other models, the important thing is that town planning should respond completely and simultaneously to all inhabitants´ requirements for a city (Purdom 1921, p. 33, 1925, p. 37).

Despite the role of these contributions as a means of tackling the housing problem and avoiding suburbanisation, the reality was far from the scenario set out in the books. Most of the urban developments built were garden suburbs included in the main municipal housing programmes, that were totally dependent on the existing urban centres and without the logic and territorial order that the GCTPA and the planners advocated as essential (Hardy 1991, p. 145; Ward 1992, p. 11).

Consequently, there was a progressive and accelerated occupation of the territory in which a lack of adequate attention to the geographical and social characteristics of the English countryside inevitably led to an accelerated loss of its rural character. Faced with this situation, throughout the second decade of the twentieth century, there was a lively public reaction in defence of the countryside.

Patrick Abercrombie, together with other professionals, founded and promoted the *Council for the Preservation of Rural England* (CPRE) in 1926; well-known planners such as Thomas Adams and Raymond Unwin were actively involved in the work of this Council. Their aim was to protect the rural character of the countryside from the effects of suburbanisation (Hardy 1991, p. 174).

The work of the CPRE, together with the support of other associations such as the GCTPA, led to initiatives that encouraged the preservation of rural environments in cities; prominent examples of this work were the approaches taken in Birmingham, Sheffield, Glasgow, Leeds and Oxford. As a measure of growth containment, these local authorities sought to create continuous green belts around their cities. To this end, they made use of the possibility provided by the *Town Planning Act of 1925*, which allowed for inclusion in town planning schemes of areas that, because of their natural values, they wished to preserve from urban development.

Concern for maintaining the character of the countryside was also expressed in numerous articles in the press and in professional journals such as *the Town Planning Review* and the Journal of the *Town Planning Institute.* Books published on this subject include Abercrombie's *The Preservation of Rural England* in 1926 and Clough William-Ellis' *England and the Octopus* in 1928. These texts attempt to respond to and redirect the way in which new developments and settlements had been taking place, in which there was usually a lack of adequate coordination of the urban and territorial phenomena that these growths entailed. It is in this context that Thomas Sharp's role in this task and the impact on urban planning practice of his *Town and Countryside. Some Aspects of Urban and Rural Development* (1932) took shape.

3 Thomas Sharp

3.1 His Beginnings

Sharp was born in Bishop Auckland, a mining and trading town in northeastern England between Darlington and Durham, on 12 April 1901. From a modest family, he seemed destined to work in the colliery like all his family members. However, a taste for writing and the landscape (Stansfield 2008), which his mother had passed on to him, prompted him to seek other paths, and he began work as an apprentice surveyor at Spennymoor (in County Durham).

After three years of this work, when he turned nineteen, he moved to Margate (County Kent), one of the first English towns to have a town plan. It was here that his direct contact with planning began. Without any training other than that provided by his own experience, in 1924, he was appointed *Planning Assistant to the City Surveyor of Canterbury.* From there, he went to London, where he collaborated with the team of Thomas Adams and F. Longstreth Thompson in the preparation of various regional plans for the British capital's surroundings. At the same time, he prepared himself for the *Town Planning Institute*'s entrance exams, which he successfully passed shortly afterwards.

In 1930, when he found that his name had been forged on a report he had prepared for the *South-West Lancashire Regional Advisory Group*, he reacted strongly, resigned from his job and took his case to court. However, this event was a result of standard practice, as such reports were signed by the *Honorary Surveyor*, who was responsible for the opinions expressed in the report, but Sharp was not prepared to acquiesce in this practice. This legal action was perhaps the first public manifestation of his peculiar, nonconformist, anti-system personality, which would be evident from then on in his work and, above all, in his texts, in which he did not avoid expressing his opinions, even though he was aware that they would provoke controversy. It was precisely in this vital and professional context that he began writing what would become his first book, *Town and Countryside.*

3.2 Approaches of the Planner

The inclusion of the study on Thomas Sharp (Standfield 1981) in Gordon Cherry's well-known 1981 text *Pioneers in British Planning* provides relevance to Sharp as a town planner (Ward 1983, p. 100).[3] This traditional historiography (Cherry 1981) emphasised Sharp's professional career as being concerned with the ways British town planning neglected the importance of urban form, and in this vein, it primarily highlighted his work for the historic cities of Oxford, Exeter and Durham in the late 1940s.

Somewhat overlooked, however, was his concern and dissatisfaction with the direction urban development was taking in the UK in the 1930s. The content of several of his books, *Town and Countryside* (1932), *English Panorama* (1938) and *Town Planning* (1940), clearly showed his disagreement with what was happening.

However, it is primarily in his first book, where in a clearer and more emphatic way, he expressed his way of understanding the British town planning discipline as characterised by an integral approach to the territory where both the town and the country played a transcendental role. The effect of his publication on the professional media is evident in the reviews published over the following years in newspapers and specialised magazines on town planning and architecture. Approximately thirty of them can be read in a notebook in which Sharp himself collected published reviews of his book. Although these reviews are generally positive in tone, they confirm that Sharp's thoughts were confrontational at the time.

In this sense, it is interesting to note one of these reviews (RIBA 1937) of a reprint of the 1937 book states that "Town and Countryside marked the end of an epoch, the first epoch, in English official planning in which one school could dominate practice without effective interference. Mr. Sharp was not the first to argue that all was not well, nor even the first to propound an ideal halfway between the garden city of Howard and the *ville radieuse* of Corbusier, but his arguments were so downright and closely reasoned and presented with such *éclat* that they the immediately attracted attention".

At this time, Sharp was still starting, but his contributions during these years were transcendental both in his professional career and in the debate over the shape of the discipline of town planning in this country, as the review mentioned above continues"…It is hardly possible to open a recent town-planning book or journal without seeing evidence of the extent to which this new critical approach to planning

[3] Although Ward (1983, p. 100) points out that Sharp is a little known urbanist, especially outside Great Britain, his texts were translated, for example, into German in 1948, *Städtebau in England*, Ernst, Berlin; into Spanish in 1947, *Urbanismo* published by Lautaro, Buenos Aires; and in 1959, *Planeamiento Urbano*, Ediciones Infinito, Buenos Aires.

is influencing development. Although it must be acknowledged that the influence is as yet more in words and consciousness than deeds".

It is not surprising therefore that the most recent bibliography on the author has rescued Sharp's figure as prominent in British planning, already in the 1930s. In this sense, both Ward (2008, p. 525) and Pendlebury (2009, p. 19) emphasise that Sharp's approaches situate his proposal as an alternative to those traditionally identified in the literature as alternatives: the garden city and the approaches of Le Corbusier.

Ward (2008, p. 526) stresses that Sharp's discourse during these years not only is focused on a strong critique of the garden city approaches, the anti-urban character of which he denounces, but also shows a defence of the characteristics of the urban ensembles built during the Georgian period. These urban fabrics were characterised by compact developments of medium densities and mixed uses, desirable objectives for Sharp and transferable to the new British urban developments of the first half of the twentieth century.

Pendlebury (2009) identifies, as a key concept in Sharp's contribution, a concern for the definition of the physical qualities of space through urban design that goes beyond the urban scale and encompasses a more holistic understanding of planning (Pendlebury 2009, p. 4) that is concerned not only with the architectural form of the place but also with its integration into the territory. In this sense, he reviews Sharp's contribution throughout his professional career, centred on his works of the 1930s and 1940s.

In this context, an analysis of *Town and Countryside* will allow us to confirm this new approach to historiography, but above all, it will allow us to identify the core of his thinking, the need to maintain the antithetical opposition between the city and the countryside, an opposition that enriches both realities and should guide any action in the territory, whether maintaining its rural character or locating urban development. This criterion, through his actions and publications, enabled him to systematise the British town planning discipline.

4 Town and Countryside: The Book

Town and Countryside is a relatively short text, 224 pages, that is structured into an introduction, two parts that are separated by an interlude and a conclusion. The introduction sets out the state of art, and the end of the text presents an overview of what is likely to happen in the country in the coming years. The first part, entitled "*Country*", deals with aspects of the countryside in four chapters, and the second, "*Town*", also in four chapters, deals with the city. Between the two central parts, "*Country*" and "*Town*", in the "*interlude*", he deals with issues common to both towns and the country, including problems caused by fumes, pollution, waste, noise and billboards. The conclusion occupies the final eight pages and is developed under the meaningful title "*Is there any hope?*".

As in Sharp's text, this study will elucidate what the countryside and the town consist of for this author, pointing out the qualities and characteristics that each of these areas should have; subsequently, we will reflect on the relationship that exists between the two areas. This shows how Sharp assesses the relationship between the town and the countryside that is being produced by urban planning practice at the time of his writing and what he believes the relationship should be.

4.1 The Countryside

For Sharp, the countryside is a geographical area that has a clear rural character. This character is the consequence of human intervention in the natural environment, an intervention driven by the need to obtain the products necessary for subsistence; the countryside is therefore the scene of economic activity that has been taking place in the country for centuries. In this sense, Sharp states that "the history of the evolution of the natural scene throughout Europe from the primeval forest and swamp to its present condition is the history of the rise, decay and supersession of the elaborate agricultural economy" (Sharp 1932, p. 16).[4]

The English rural character of the time is therefore not identified as natural and wild. The fundamental characteristic of the English countryside is its degree of humanisation. Sharp is aware that all landscapes have been humanised, but he considers that the English countryside has been shaped in such a way that places it close to a work of art: "it has been given it a pattern and a rhythm: it has been made beautiful through an artistic impulse" (Sharp 1932, p. 15). The quality of the English countryside and its friendly character are emphasised on countless occasions.

Sharp defines the characteristics of the English countryside through a historical survey of the way in which the countryside has been worked in Britain. Over time, its degree of naturalness has been reduced; the gradual shift from a "primeval forest, swamp, jungle, prairie and desolate moor" (Sharp 1932, p. 16) to the rural environment he defends has taken place. In each era, the landscape had specific characteristics that reflected the relationship of the people to their environment. These characteristics have changed and even disappeared over time. Sharp detects the beginning of the English rural scene he praises in the Tudor era, during the reign of Henry VIII and Edward VI when the *commons* and *common fields* began to disappear (Sharp 1932, p. 19) through the parcelling out of land.

Shortly afterwards, he points to another landmark, *Sylva, or a Discourse of Forest-Trees and the Propagation of Timber*, a book published by John Evelyn in 1664; for Sharp, this is another of the determining factors in the modern configuration of the English landscape. The text reports the loss of forests. However, what is relevant to our narration is that it was an invitation to arboriculture, which, according to Sharp, encouraged, in the following decades, the practice of planting and "rural adornment" (Sharp 1932, p. 24) among countrymen. In this way, the landscape hitherto shaped exclusively by agricultural activity became shaped by people who wished to create a

[4] Quotations are from the original English edition, Sharp (1932).

beautiful scene for their own enjoyment. This is the moment when people's relationship with the environment in Britain began to be productive in character and included an artistic component that would eventually develop the special landscape that Sharp praises.

The pace of land division accelerated in the eighteenth century. Each property had to be fenced, and plots were enclosed by hedges in the lowland country and by stone walls in the hill country. Moreover, each landowner carefully planted trees around the edges of the plots, individualising the landscape and giving it its own special character. They practised what Sharp called "landscape art", which was a popular tradition whereby each landowner cared for his property through "rural adornment" (Sharp 1932, p. 23). In Sharp's words, "infinite discussion and vast sums of money were expended on this art which every man could practice" (Sharp 1932, p. 24).

It was also a professional practice whose exponents would be Kent and Capability Brown and, a little later, Humpry Repton. These landscape artists were in favour of a treatment of the environment characterised by naturalness and spontaneity, as it worked through "rural adornment". Buildings and paths were treated with spontaneous artificiality by being inserted into landscapes with sufficient naturalness to make one think that they had always been there.

According to Sharp, in the second half of the eighteenth century, the modern English scene was already moulded into a single, enlightened, man-made individuality. In the second half of the nineteenth century, it reached the peak of perfection, and from then, the decline began, on the one hand, through the decline in enthusiasm for "rural adornment", which led to a lack of maintenance of the countryside in terms of replacing fallen trees, and on the other hand, as a consequence of industrialisation, which is what truly concerned Sharp.

In short, this historical review identifies the specificity of the cultural landscape of the English countryside, which has little to do with the degree of naturalness that the environment would have shown in its origins. Rather, the current environment is the result of a popular construction, and for this reason, Sharp does not understand why the countryside is not valued and is being occupied without being sensitive to its landscape value.

To remedy the loss of this distinctive rural character, he believes it is necessary not so much to preserve what exists but rather to guide new growth in a way that respects the landscape. Sharp is clear about the way forward, but he knows it is a difficult task. Perhaps this is why he sets out the criteria by which he approaches this challenge by pointing out the mistakes that are often made. Nevertheless, it is easy to identify the criteria he proposes for action in the countryside.

As previously mentioned, most of the English landscape has been simply built through hedges and trees. According to Sharp, the hedgerow is an element that provides rhythm and differentiates patterns in the landscape, reduces distances and breaks up the large scale of the countryside, making it friendly to people. With trees, something similar happens, but in this case, it is the height of the element that allows the scale of the land to be reduced, mitigating the "oppression of the sky" (Sharp 1932, p. 32).

As a general criterion, the implementation of buildings in the countryside should always ensure that the balance of rural amenity is maintained. The implementation of this general criterion, in terms of the construction of buildings in the countryside, implies that in the countryside, there are few and scattered buildings, so it would be desirable to maintain the "void". However, once it is decided to build upon and therefore to occupy that void, it must be ensured that its rural character is maintained, so the landscape must be able to absorb the buildings (Sharp 1932, p. 44). This will involve consideration of how buildings are grouped or scattered and the use of appropriate designs in scale, form and materials in harmony with the countryside.

Despite Sharp's admiration for the English rural landscape and the peculiarly natural and unconscious way in which it has been formed, he does not consider that it should be protected as it is; he sees the landscape as a constantly evolving construction. He therefore considers planning for new settlements in the countryside to be appropriate as long as their rural character is not compromised. Furthermore, he argues that action should be taken without imitating the characteristic irregularities of these fabrics, as proposed by Sitte, whom he expressly mentions (Sharp 1932, p. 66). The aim is to avoid artificiality in the new settlements in the countryside, and he points out that "artificiality in the countryside simply means incongruity" (Sharp 1932, p. 66).

Traditional "villages" provide an example of clustering for new residential settlements that is considered appropriate. In this respect, he stresses that the rural character of new settlements "does not depend on any of those things that are popularly associated with it, flowering gardens, irregular, informal, and quaint buildings, and so on. It seems to depend on much smaller and more subtle things, upon a certain modesty, a certain lack of smooth, mechanical finish of the town, and above all upon the harmony of the material of its buildings with the countryside" (Sharp 1932, p. 67). For this reason, he is receptive to the planning of new settlements in rural areas through those fabrics that are organised around a square in its formal variants, such as specifically defined quadrangular or even triangular geometries. For the shape and scale of the buildings, he shows a preference for volumes in which horizontality stands out, requesting the neutralisation of those with a tendency towards verticality, especially in those buildings that draw the skyline (Sharp 1932, p. 73).

However, his position on the materials to be used is different: if he rejects imitation in terms of form, he is against the use of materials that do not follow the rural tradition or are not in harmony with it. He considers that the variety of traditional materials is sufficient to meet this criterion. To provide criteria for harmony, he addresses the colour and texture of the materials.

Within the part dedicated to the countryside, an entire chapter is devoted to the roads that cross it. Again, as he does when discussing how buildings relate to the landscape, he compares the characteristics of the traditional road in the countryside with the road that is being laid out at the time of writing.

Regarding the traditional road, he emphasises the benefits of its irregular section, its plan layout, and the treatment of its edges, recalling and mentioning once again the hedges and walls that define the plots and outline the road, as well as the trees planted in its vicinity. These specific characteristics absorbed the impact of the road on the countryside and made driving along them an attractive experience. The chapter is therefore a suitable tool for the implementation of the new road to cross the countryside.

He concludes the chapter with a sentence that well summarises his proposal for the design of contemporary roads: "Thus can the country parkway and country road be kept in character with the traditional English lane and the traditional English landscape, and made not only means of communication but objects of beauty and ways of pleasantness" (Sharp 1932, p. 95). He recognises the need for the road to provide an attractive experience but also to be useful and to allow the flow of traffic, which has increased in recent years. He therefore reviews some of the criteria to be considered, the width of the road, for example. In this respect, he is concerned that an excessive section of the road could generate a hostile and desolate experience when passing along it and that it could produce an incoherent void in the landscape, calling into question the rural character of the countryside (Fig. 1).

Another of the problems facing road design at that time was the standardisation of the section. For engineers, the same section is possible along the entire route, whether it runs through the countryside or through the city, but for Sharp, it is inconceivable, and he proposes its variation by interspersing hedges and trees that can be planted or by respecting the existing ones that can be incorporated into the layout (Sharp 1932, p. 89). He thus seeks to maintain the non-uniformity characteristic of the traditional road and rejects the regular planting of trees, which would mean trying to give the road in the territory the character of an urban avenue.

In addition to these guidelines, criteria and tools for the proper integration of buildings and roads in the territory, Sharp is also concerned about the conservation of rural areas where no action is taken. He refers, for example, to the need to protect the existing nature reserves in the territory and shows interest in making them accessible from certain nearby towns, as he considers it necessary for people to be in contact with the natural environment.

4.2 The Town

The town is a creation with artificial forms, as opposed to the action on the countryside where humans control and mould the attributes of nature so that they are useful without ceasing to be beautiful. The city is identified with the area where people initially grouped together to "provide warmth, contact, security and cooperation" (Sharp 1932, p. 137). The real importance of this grouping is that its form is an expression of people; it symbolises them (Sharp 1932, p. 137). Sharp thus shows the importance of the formal configuration of urban space.

In the urban developments that have been carried out since the industrial revolution, Sharp detects a lack of urban spirit, which, in his opinion, means a lack of urbanity. The spirit of the period in which he writes is manifested in the developments based on the ideas of the garden city, which for Sharp are inadequate as they express neither urbanity nor proper rusticity. In this sense, he insists on the need to

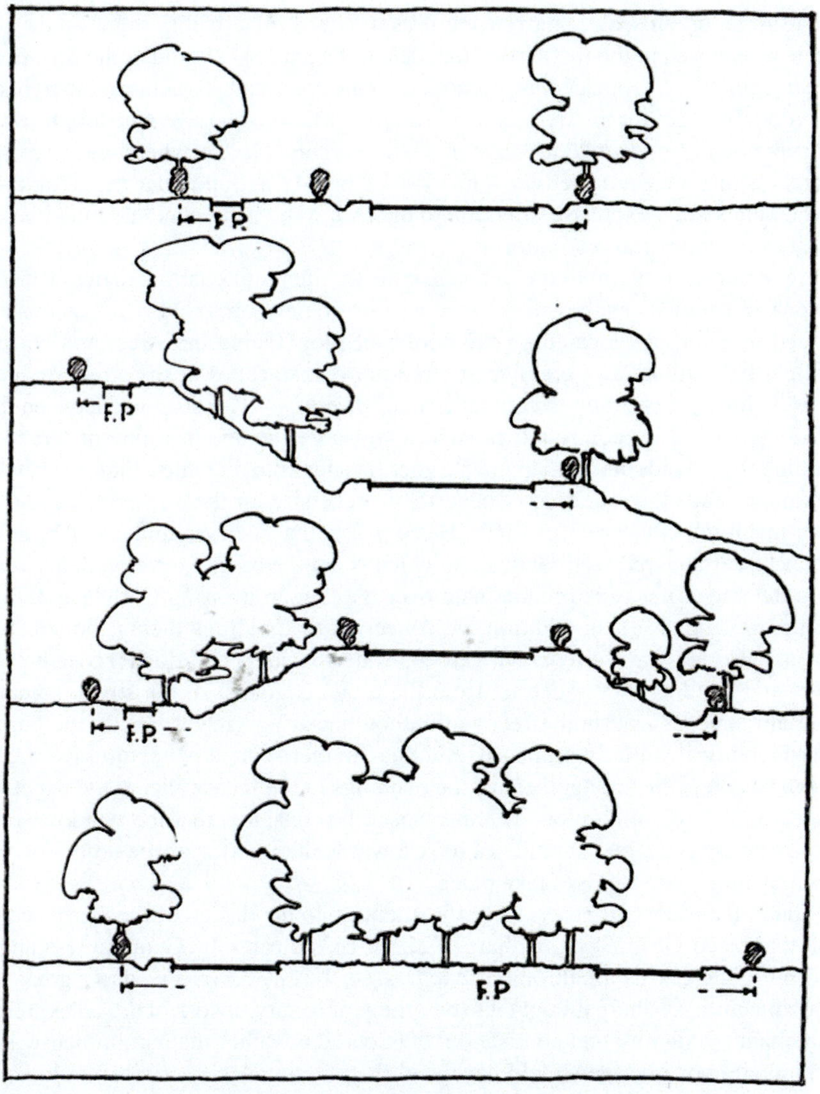

By permission of Oxford University Press www.oup.com

Fig. 1 Possible sections of the road for its proper insertion in the landscape. (Sharp 1932, p. 89)

create streets and not "Way (Meadow Way, Valley Way, Broad Way, etc.), Avenue, Gardens, Grove, Drive, Green, Ridge, Hill, Dale, Lane" (Sharp 1932, p. 150), referring to the terminology used in garden cities to name the places for relationship and communication. In this way, he shows that these roads suggest a rusticity they do not achieve in reality, thus renouncing the urbanity of the streets.

From the urban scale, Sharp recognises the need for urbanity in urban developments as opposed to the rusticity of the idea of the garden city and to the rejection of the street that Le Corbusier was promoting on the continent. To achieve the right city, he proposes a "return to Architecture" (Sharp 1932, p. 164). He recovers Edward's contribution in *Good and Bad Manners in Architecture* (1924), where the civic values of architecture are vindicated (Edwards 1924, Chap. 1), as opposed to the monotonous developments that exclusive attention to the lack of housing generated in cities as a result of the industrial revolution.

He identifies urbanity with some specific architectural characteristics, which he reviews in detail. Thus, he refers to the need to achieve urban fabric through streets formed by buildings located on the border of plots facing the street and ensuring sunlight but without an excessive separation of the façades of the street. It should be usual for each building to contain a set of dwellings. The distance between these buildings should be controlled in such a way as to maintain a preponderance of buildings over voids between them. He goes thus far as to recognise that "everywhere continuous and close building is an aesthetic necessity for the true expression of the town-medium" (Sharp 1932, p. 163). He considers that such groupings will be able to achieve a certain architectural quality, which is not the case in developments where, from the street, the gaps predominate over the "broad massing of their buildings" (Sharp 1932, p. 163). In addition, he conceives of buildings facing the street and not set back, leaving a only small garden, as he considers additional space between street and building would detract from the urban character of the street, providing a certain unnecessary rural effect and without ensuring greater health, the purpose for which it was initially proposed. Another smaller-scale architectural feature that he reflects on is the arrangement of the openings in the façade facing the street with respect to the adjoining plots. In this respect, he considers that the windows of the houses facing the street should not look towards the interior of the adjoining plots where people's private lives take place.

Sharp, therefore, proposes a specific attention to the design of the urban scene, a task that faced *Civic Design* (Sharp 1932, p. 163) during the Georgian era and for which he longs. After all, through *Civic Design*, the city is conceived as a great work of architecture, as Sharp thought it to be a thing of beauty, a work of art, an expression of humanity´s dignity and civilisation. "Or, perhaps, rather than an immense architectural composition we should have said that the town is a series of architectural compositions, streets, squares, circuses, &c.., each of which is a composed unity, a complete picture of itself, in sympathetic relationship to its neighbours whether in concord or in formal contrast" (Sharp 1932, p. 162).

In *Civic Design*, the urban scene is conceived through architecture, but roads and open spaces must also be considered. These are elements of the urban system that Sharp reviews in separate chapters, assessing what characteristics they must have to ensure the appropriate urban character that the town must reflect. In this sense, he identifies typologies of both roads and open spaces, highlighting the aspects that can endanger urban beauty and urbanity.

With regard to roads, the need to define the road network is underlined: types of roads and functional characteristics should be described in such a way that an adequate formal definition of their sections is carried out in order to care for and respect the amenity of the town without allowing the car to acquire an excessive importance that gives rise to an inadequate urbanity.

In the system of open spaces, the objective will be the same: the system will be made up of the types of open spaces necessary for the recreation of the population. These elements will be distributed throughout the city with the goal of them functioning as a system. For the design of each of the typologies identified, the criterion is repeated: although they are parks and gardens (spaces of a natural character), they must be designed to reflect an urban, civic, formalised character (Sharp 1932, p. 201). They should be elements with a civic, sophisticated and formal character (Sharp 1932, p. 198). In medium-sized towns (150,000 inhabitants), no other type of design would be justified, and open spaces should be designed in the terms mentioned above to enhance the urbanity of the town. For Sharp, "the artificial cannot reproduce the natural: it can only caricature it" (Sharp 1932, p. 203); however, he justifies the design of parks such as Hyde Park and Sefton Park in cities such as London and Liverpool because nature is not at an accessible distance for citizens.

Despite the urban character that the open spaces of the city should have, Sharp does not renounce the need of the people who live in the city to have contact with nature. In this sense, he believes that this contact should be achieved through easy access on foot from town to the countryside. This is in line with the urban model he defends and is the key to formulating his approach (Sharp 1932, p. 173) to the growth of towns; the town should be of a size that ensures easy access for the population to the countryside without the town losing its urban character to solve the lack of contact between people and nature.

This premise advocates a system of subcentralised towns overcoming the mere decentralisation produced through the suburbanisations that were encouraged at the time to control the centralisation and congestion of cities. Sharp acknowledges the rescue of an idea already somewhat briefly captured in a diagram in Howard's 1898 edition of the book; in this, a central city is seen, and six miles from its centre, six smaller settlements are arranged, all separated by cultivated areas. This early idea would later give rise to the satellite city theory explained by Purdom in his 1925 publication and which Sharp identifies with his approach to subcentralisation, which allows the differentiation between town and country he supported. The subcentres he advocates are separate from each other and from the main centre, which is the original one, and they are interrelated and in some way dependent, not only on the main centre but also on the others that are located in their surroundings (Fig. 2). Each of these urban areas will be towns and should be organised as such according to the characteristics he considers desirable for them.

4.3 *The Country and the Town: Two Antithetical Realities*

Sharp believes that the solution to avoid the loss of the rural character of the countryside and the urban character of the town lies in achieving an appropriate relationship

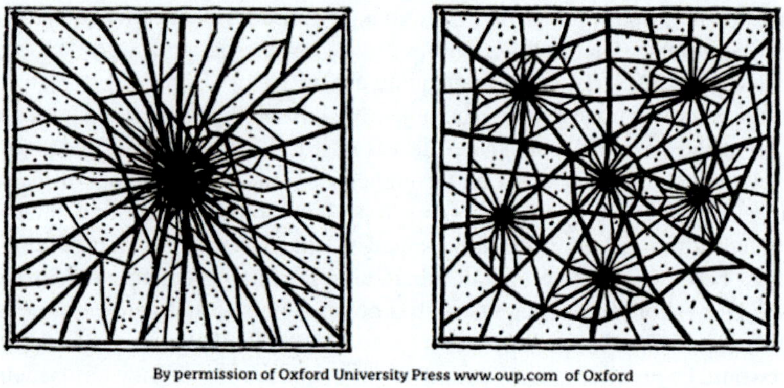

Fig. 2 Diagram of a centralised town versus the subcentralised town proposed by Sharp (1932, Figs. 26 and 27, p. 168)

between the two areas. This will be achieved by maintaining their opposing characters. In this way, the town is presented as the antithesis of the countryside, and the countryside is presented as the antithesis of the town, an idea that he expounds time and again throughout his book. The loss of the specific character of each area provokes a neutrality that, far from intensifying their characters, disfigures them, turning a rich and complex reality into something confusing and boring.

This is the basis on which Sharp addresses, throughout the book, the critique of the *town-countryside* binomial that Howard has fostered by putting forward the idea of the garden city as an intermediate solution: the third magnet. Howard "had no interest in the town as a thing of beauty, a work of art, an expression man´s dignity and civilisation" (Sharp 1932, p. 140); after all, like George Cadbury and William Hesketh Lever,[5] he was a town reformer, trying to solve a social problem, and we do not find in his proposals direct concern with improving urban form. This is not the position of Sharp, who, as an architect, criticised the social reformers from a formal point of view, an aspect that they had neglected in their theories, which, in Sharp's opinion, meant that it was impossible to provide a scenario that would ensure the success of the social conditions sought.

The unprecedented need for housing in cities at the turn of the century (nineteenth to twentieth) led to an uncontrolled construction of urban developments in streets where dwellings were crowded side by side. At the same time, the quality of architecture was also devalued, and new streets in the cities showed minimum standards (Sharp 1932, p. 139). In the face of the situation, some solutions were adopted; concessions, such as the front garden of houses were limited to 5 feet deep

[5] Lord Leverhume and George Cadbury are two well-known English businessmen who from the mid-nineteenth century were concerned to improve the conditions of their workers, both in terms of the work itself and their accommodation, and promoted model villages for them and their families (Meakin 1905, pp. 416–443).

(approximately 1.5 m) which, of course, did not solve the situation. Rather, the situation provided a breeding ground for dubious urban alternatives and totally opposite urban forms to evolve.

Sharp places in this context the publication and the reception of Howard's book *Tomorrow a Peaceful Path to Real Reform* (1898). This is considered "in the full sense of the word an epoch-making work" (Sharp 1932, p. 139), since the "Town-Country Magnet" (Title of the first chapter of Howard's book) concept included in the book that would, according to Sharp, starting a few years after its publication, forever change the character of the development of the English towns (Sharp 1932, p. 140): "So from Letchworth to the beginning of the War the story of town development is the story of universal adoption of the garden-city ideals applied to the garden-suburbs, garden-villages, garden-this-and-garden-the-other" (Sharp 1932, p. 144).

Planning legislation also supported developments under the concepts established by Howard so that, as Sharp explains, they were already part of the way the country was being developed (Sharp 1932, p. 144). After the First World War, open development became the urban model used: it was characterised by the use of large plots, densities not exceeding 12[6] dwellings per acre, single-family or semidetached typologies, and the setback of buildings from the public space with a garden at the front of the plot (Sharp 1932, p. 145).

The development characteristics sought to move away from the degradation of the town, at the turn of the century and the beginning of the twentieth century, towards a romantic appreciation of nature to provide a renewed character to the town (Sharp 1932, p. 11). However, from Sharp's point of view, this renewed character implied a neutral kind of beauty that could not be identified with either the countryside or the town. The city has become a collection of cottages which, the looser and more dispersed they are ("more countrified" as Sharp would think), the more successful they are to the public, but "in them there is no real civic expression and, therefore, they do not encourage urban life or contact between" the citizenship (Aseguinolaza-Braga 2009).

This character of the binomial that Howard proposed led to the loss of the proper character of the town and countryside: "Two diametrically opposed, dramatically contrasting, inevitable types of beauty are being displaced by a drab, revolting neutrality" (Sharp 1932, p. 11).

In this sense, Sharp proposes the total abandonment of the *town-countryside* binomial defended by Howard as a solution. The aim is to recover the logical antithesis between the countryside and the town, an antithesis to which he continually alludes, recalling in the book the interest of the English countryside as well as the various urban settlements in the country. He particularly deals with this question in the part of the book dedicated to the countryside, when analysing the insertion of buildings in the landscape, and when explaining what urbanity means in the part dedicated to the town. In the introduction, he is already categorical in this respect: "The town is town: the country is country; black and white (...). Only through the preservation of these distinctions is there any salvation", but he continues, "Only through the preservation

[6] A density of 30 dwellings per hectare.

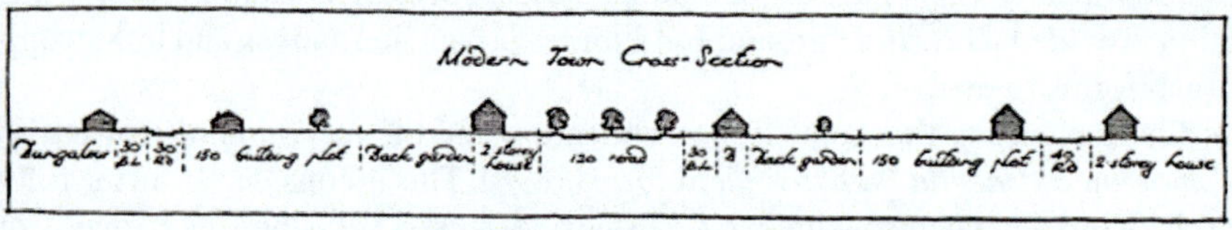

Fig. 3 Gradation of the town in its approach to the countryside, a solution used at the time Sharp writes and which he criticises. (Sharp 1932, Fig. 23, p. 157)

of the town as town can the countryside be saved; and only through the limitation of rurality to the country can the town be preserved" (Sharp 1932, p. 11); that is to say, the beauty of each area implies interrelationships between the two that must not endanger their own identities.

In the previous sections, we analysed the physical characteristics that each area must have to ensure its antithesis. It is now time to reflect on what it means to translate the antithesis presented by both areas into reality. In this sense, the *town-countryside* binomial supported by Howard has led to the use of a gradual section of the city in its encounter with the countryside that seeks to cushion it as far as possible.

Sharp is categorically against the possibility of using a gradation of urban character as one approaches the countryside. In his view, the town should be a town in the centre and at the edges, just as should be the case in the countryside. For him, the possibility of gradation is inconceivable (Fig. 3), and in this respect, he states that "the country should end and the town begin emphatically, unmistakably and finally" (Sharp 1932, p. 164).

Contrast is the criterion Sharp advocates for defining the relationship between these two spheres. However, Sharp advocates for the use of urban design tools in the countryside as long as they do not call into question the character of the countryside. Contrast is not a problem as long as the specificity of the area is respected. However, when the contrast is repeated too much, this specificity is endangered because the contrast itself imposes a new character. In Chapter IV, about roads in the countryside, he develops an example by analysing the way in which trees can be planted on the edges of a road that runs through the middle of the countryside: if the trees are arranged as in an avenue at a certain point along the road, this may be adequate; however, if the trees are arranged in this way along the whole route, the contrast is lost, and becomes an inadequate strategy.

4.4 *The Practice of These Ideas, the Missed Opportunity*

The book also contains a strong critique of British town planning legislation, which had a negative impact on the practice of *Landscape Design* and *Civic Design*. In this sense, he considers that the first *Housing and Town Planning Act* (1909) could have saved the practice of *Civic Design*; however, on the contrary, this possibility had not been exploited, and in practice, an exclusive focus on hygienic factors had found its way into the idea of the garden city, which was thus consolidated precisely because it had been drafted on the basis of the sociological foundations of the garden city reformers, which had nothing to do with the creation of urban beauty, the primary objective of *Civic Design*.

In this sense, he uses the words of Trystan Edwards to affirm that "the *Art of Civic Design* has been killed by the *Science* of Town Planning" so that "modern town planning is hardly more of a science than it is an art" (Sharp 1932, p. 220). However, science has its foundation in reality, and modern town planning has virtually none; he adds "not only it is not town planning, it is the negation of any kind of planning" (Sharp 1932, p. 220). In fact, he considers that "if their [town planning] schemes can be classified as 'planning', words no longer have any meaning" (Sharp 1932, p. 221).

The abandonment of *Civic Design* meant the end of the town and the beginning of the tragedy of the countryside: living habits had been altered so that the desire to live in the town gave way to the desire to leave it and live in the countryside; but logically, these people do not know what it means to live in the countryside and adopt a hybrid way of life that is neither urban nor rural.

The *Town Planning Act of 1925* could have made it possible to plan large areas of rural land and thus achieve "the general objective of securing proper sanitary conditions, amenity and convenience" (Sharp 1932, p. 220). In practice, however, its implementation meant the construction of new settlements with the faded ideals of the garden city: towns that are neither urban nor rural. In this sense, it calls for a breath of fresh air in the legislation so that it definitively addresses the problems that have been detected.

The *Town and Country Planning Act*, which was still going through parliament in 1932 when Sharp was writing this text, is one possibility, but he fears that it is just another missed opportunity precisely because once again the new law would consolidate the model of occupation of the garden city, which would hinder the recovery of *Civic Design*. In this sense, he states about the text being debated in Parliament that "the Town and Country Planning Bill is sterile from the beginning" (Sharp 1932, p. 223).

As far as *Landscape Design* is concerned, however, he recognises the progress that the new draft law represents for the proper management of the countryside, as it provides some useful instruments for this purpose, such as financial compensation as a response to the prohibition to build on land. He is optimistic in this respect but adds some other proposals for more effective control.

In this regard he points out various possibilities; on the one hand, he suggests creating a new independent central body, which he proposes to call the *Rural Development Board* or *Board of Scenery*, so that rural land planning would be carried out by another ministry unrelated to the development of garden cities and would ensure the suitability of the various activities that the ministries carry out to direct rural development; on the other hand, he suggests establishing sanctions for local authorities who act inappropriately and that the members who direct the activities in the various administrations should be qualified professionals for this purpose.

5 From Town Planning to Town and Country Planning

Throughout this analysis of *Town and Countryside,* it has become clear that Sharp's objective goes beyond the problem of the destruction of the values of the English countryside and the lack of quality of the new settlements and reaches the formulation of the criteria to be applied when intervening in the landscape, understood as the sum of the countryside and the town.

Sharp's analysis of the English situation helps identify the key formal aspects that make the countryside, countryside and town, town. In the English case, the traditional practices of *Landscape Design* and *Civic Design* provided the key decades ago. Both practices succeeded in physically characterising the countryside and the town, in their different locations, with their different geographical conditions over centuries. Historically, the people who have inhabited each area have adapted it according to their needs. Sharp thus recognises that each area has a specificity according to its geographical location and according to the circumstances that, at any given time, have led to its evolution.

It is thus clear that landscape is a cultural construction that shows the changing relationship that the people who inhabit it have with their natural environment, as identified for example some decades ago by the *European Landscape Convention* (2000).

For the relationship between the countryside and the town, as has been made clear in the course of this article, Sharp argues that the two spheres should be antithetical in character but considers that the two realities are inseparable, as people need to keep in touch with nature. This relationship with the natural environment is diametrically opposed to Howard's viewpoint and consists of nature located at a reasonable distance from the town, allowing citizens to walk to it at any time. This appreciation, which at first sight may seem minor, has a great impact on Sharp's approach, as it doubly conditions the urban model to be developed. First, a crucial issue is as follows: for Sharp, the city cannot develop without the countryside, so both concepts are inseparable; second, this condition imposes a size limit on the town that ensures the pedestrian accessibility of the countryside from the town, which leads Sharp to support the subcentralised system of cities.

In this sense, for Sharp, it is inescapable to move from town planning to town and country planning, as the legislation already seemed to foresee. The practice of planning must include attention to the whole of the territory in which one wishes to act, that is, to both the country and the town. To this end, he is committed to recovering the practice of planning in the tradition of the arts of *Landscape Design* and *Civic Design*, so that attention to the countryside and the town is duly attended to, in the terms in which each reality requires it, adding that their possible and unique relationship is that of contrast.

References

Abercrombie P (1926) The preservation of rural England, etc. Hodder & Stoughton, London
Aseguinolaza-Braga I (2009) Planning in Sharp's town and countryside. RA Revista De Arquitectura 11:81–90
Cherry GE (1981) Pioneers in British planning. Architectural Press, London
Edwards T (1924) Good and bad manners in architecture. Philip Allan & Co, London
Evelyn J (1664) Sylva, or a discourse of forest-trees and the propagation of timber. J. Martyn and J. Allestry, London
European Council (2000). European Landscape Convention. Florence
Hardy D (1991) From new towns to green politics: campaigning for town and country planning, 1946–90. E and FN Spon, London
Hardy D (1992) The garden city campaign: an overview. In: Ward S (ed) The Garden City part, present and future. E & FN Spon, London, pp 88–106
Howard E (1898) Tomorrow a peaceful path to real reform. Swan, Sonnenschein & Co, London
Howard E (1902) Garden cities of tomorrow. Swan, Sonnenschein & Co, London
Meakin B (1905) Model factories and villages: ideal conditions of labour and housing. Fisher Unwin, London
New Townsmen (1918) New Towns after the war: an argument for the Garden Cities. J. M. Dent and Sons, London
Pendlebury J (2009) The urbanism of Thomas Sharp. Planning Perspectives 24:3–27
Purdom B (1921) Town theory and practice. Benn, London
Purdom B (1925) The building of satellite towns: a contribution to the study of town development and regional planning. J. M. Dent & Sons, London
RIBA (1937) English planning critically examined. RIBA Journal. November.
Sharp T (1932–1934) Town and countryside. Thomas Sharp papers. Newcastle University special collections and archives. https://archiveshub.jisc.ac.uk/data/gb186-ths/ths21. Accessed 3 Jun 2021
Sharp T (1932) Town and countryside: some aspects of urban and rural development. Oxford University Press, London
Sharp T (1938) English panorama. J. M. Dent & Sons, London
Sharp T (1940) Town planning. Penguin Books, Harmondsworth
Sharp T (1947) Urbanismo. Lautaro, Buenos Aires
Sharp T (1948) Städtebau in England. Ernst, Berlin
Sharp T (1959) Planeamiento urbano. Ediciones Infinito, Buenos Aires
Standfield K (1981) Thomas Sharp 1901–1978. In: Cherry G (ed) Pioneers in British planning. The Architectural Press, London, pp 150–177
Stansfield K (2008) Thomas Sharp as I knew him. Planning Perspectives 23:535–543
Ward SV (1983) Planning History in Britain. A Review Article. J Urban History 10:93–102
Ward SV (1992) The Garden City. part, present and future. E & FN Spon, London
Ward SV (2008) Thomas Sharp as a figure in the British planning movement. Planning Perspectives 23:523–533
Williams-Ellis C (1928) England and the octopus. Pub Geoffrey Bles, London

Tinctured Polarities: Town and Country Planning for Post-war London

Fabiano Lemes de Oliveira

Abstract This chapter analyses Patrick Abercrombie's views on town and country as the two domains of human planning activity and the question of urban productive spaces through his most significant writings and two of his wartime plans: the County of London Plan 1943 and the Greater London Plan 1944. In contrast to a dichotomist reading of his positions, the central argument reflects upon Abercrombie's understanding of such polarities as necessary for achieving a dynamic balance between the urban and the rural in regional planning. Since such polarities were to be tinctured by one another and considering the wartime as a moment of significant crisis, the chapter also interrogates how urban productive spaces have been included in these plans. A reflective reading of Abercrombie's conceptualisations of town and country, employment of the green belt and green wedges as mediatory elements, defence of the protection of agricultural land, localisation of food production, and resilient approach regarding the capacity for adaptation of green spaces for food production can contribute to contemporary discussions on territorial dynamics and the sustainability of food systems.

Keywords Abercrombie · City · Country · Urban planning · Rural · Regional planning

1 Introduction

Patrick Abercrombie is arguably the most renowned British planner of the twentieth century. Although the bulk of his activities took place in the first half of the century, his work set the ground for much of the regional planning theory and practice developed in Britain in the post-war period. Much of this is due to his wartime *County of London Plan 1943,* with J. H. Forshaw, and the *Greater London Plan 1944,* which also served as significant references for reconstruction plans internationally. Abercrombie's planning activities started at the University of Liverpool. In

F. Lemes de Oliveira (✉)
Department of Architecture and Urban Studies (DASTU), Politecnico di Milano, Milan, Italy
e-mail: fabiano.lemes@polimi.it

C. Brisotto and F. Lemes de Oliveira (eds.), *Re-Imagining Resilient Productive Landscapes*, Cities and Nature, https://doi.org/10.1007/978-3-030-90445-6_5

1909 he became a research fellow at the newly created Department of Civic Design and the founding editor of the *Town Planning Review*. A few years later, in 1915, he succeeded Stanley Adshead as professor of civic design and in 1935 took up the position of professor of town planning at University College London. By the time he retired, in 1946, Abercrombie's work had spanned nearly four decades in both planning teaching and practice.

Planning history literature has extensively addressed Abercrombie's work, at times identifying his pioneering and forward-thinking approaches (Dix 1981), at others setting his role and legacy in a broad context (Hall 1995), be it related to activism in the protection and positive planning of the countryside (Dehaene 2004), the structuring of regional planning (Dehaene 2005; Hewitt 2011), in reconstruction and replanning studies (Bullock 2002; Hasegawa 1999; Larkham 2011) and in the development of green space planning thought (Lemes de Oliveira 2014, 2015). However, despite such broad coverage, a more nuanced reflection upon Abercrombie's views on the relationships between town and the country and the roles of productive urban landscapes have not been sufficiently examined. This is unsurprising for two main reasons. First, Abercrombie's explicit differentiation between the urban and rural domains of human activity can suggest a dichotomist reading of his positions. Second, because Abercrombie's conceptualisation of the essence of town planning as grounded in civic design, with evident inclinations for a compositional approach (albeit anchored in a Geddesian tradition of the survey), may indicate a position of departure that denies rurality outside the countryside. This chapter instead explores how, at his major works done at arguably the most severe attack to the country's independence and very existence in modern times, Abercrombie's visions for a balanced relationship between the town and the countryside are at the core of an envisaged harmonious and prosperous future.

Furthermore, it investigates what he called "tinctures" of one domain into the other, in particular by analysing the presence of "urban rurality" in these plans. The central argument draws upon Abercrombie's understanding of the city and the countryside as distinct polarities of human artifice and the need for a continuous and dynamic balance between the urban and the rural, not conceived as fixed entities but as domains in which flows are mutual and exchanges variable. It also points to the early resilience approach presented in his visions regarding the capacity for adaptation that particular urban forms and typologies of open spaces, such as the green belt and green wedges, would provide in regard to sudden shifting needs, such as food production.

The chapter draws from archival material from the London Metropolitan Archives and the National Archives in the UK, as well as from Abercrombie's publications, mainly his 1933 book *Town and Country Planning*, and the two London plans. Initially, the chapter addresses how the urban and rural challenges and competing agendas of the interwar period consolidated Abercrombie's views about taking the region as the most appropriate planning unit. The following section reviews his conceptualisations of such "polarities" (as well as his seemingly particular understanding of "nature"), the interplay between them and how such understanding should be translated into planning. Section four investigates the concept of "tinctures", especially regarding the presence of productive landscapes within urban areas. The *County of London Plan 1943* and the *Greater London Plan 1944* are analysed in both sections three and four in their roles as potential manifestations of Abercrombie's theories and aspirations for a balanced relationship between urban and rural areas, as well as on the presence of productive landscapes within cities, after the Second World War. The final section concludes by suggesting how Abercrombie's regional considerations, sought after equilibrium across the domains of planning, understanding of the benefits of grounding urban areas and society near the country and "tinctures" as opportunities for flows and transitions can offer insights to contemporary planning.

2 The Urban–Rural Question in England in the Interwar Period

Exacerbated since the industrialisation in the nineteenth century, the contentious relationship between town and country is a marker of British planning. This is, for instance, at the core of the very conceptualisation of Howard's garden city idea (1902). The Town Planning Association, founded in 1899 to give practical effect to the garden city principles, came into being precisely to promote "the proper planning of all land, whether urban or rural" (Abercrombie 1910). Besides, before the First World War, at the professional foundation of town planning, the regional scale incipiently appeared as the integral unit able to forge a harmonious relationship between societies and the environment. Patrick Geddes's (1911) call for the civic survey as a method for understanding the relationship between people and the land would set the foundations for the renewed attention to regionalism in the interwar period. Abercrombie (1912) could see that little could be accomplished without such perspective regarding the contention and positive planning of urban development and accompanying green spaces. Already in 1912, he was concerned that the lack of a strategic and coordinated planning approach across boundaries and scales would break up any possibility of a cohesive park system for London, hampering both the city's environmental quality and the countryside beyond (Lemes de Oliveira, 2014).

The interwar period in Britain saw the growth of conflicting interests and agendas. The regional question framed the discussions on both the nature of cities and rural regions, and what they should become. From the urban side, this was a period of intense unplanned sprawl and densification of inner urban cores (Hall 1996; Ward 2004), which did not only pose problems to urban areas but also challenged the physical structure, characterisation and social construction of the English countryside (Williams 1975). The more cities grew, the more inner urban dwellers felt further away from the countryside. If in small towns a short walk would be enough to get one out in the country, this was no longer the case for many living in cities. The deterioration of urban living conditions, mostly noticed in industrial cities, including congestion, precarious housing and sanitation and lack of access to green spaces fuelled the public's drive towards the countryside. For long, the attachment to the countryside had been associated with Englishness and a sense of national identity (Hoskins 1955; Matless 1998). A forced severance from the land thus meant a significant psychological displacement that required compensating. For those who could afford it, the expansion of public transport and growing car ownership (O'Connell 1995) facilitated access to the country to visitors and settlers. As Abercrombie would put it, the rural areas came under invasion from the outside, through "unconscious" urban development—such as suburbanisation, ribbon development and proliferation of "weekend cottages"—and pressure to live and recreate in the country from urban dwellers, while also disintegrating from within due to the fragmentation of land ownership. With the end of subsidies for agriculture in the early 1920s, and the subsequent intense process of mechanisation of the 1930s, rural poverty was a reality for many, which in turn drove migration to urban areas. As such, for him, not only was the interwar period concerned with the question of urban planning, but also with the "place of agriculture in planning" (Abercrombie 1933).

Although already concerned with the rural question at the beginning of his career, Abercrombie ramped up his leadership on these matters after the First World War. In 1926, he published *The Preservation of Rural England: The control of Development by Means of Rural Planning*. In the same year, he became one of the founders of the Council for the Preservation of Rural England. Two years later, he would write an epilogue to William-Ellis' *England and the Octopus*, a severe criticism of uncontrolled urbanisation of the countryside. Finally, in 1933, a year after the passing of the Town and Country Planning Act, Abercrombie published his very own book *Town and Country Planning*. There he further developed his visions on the essence of the urban and that of the rural and the issue of comprehensive planning at the regional scale.

3 Polarities: Opposition and Complementarity in Town and Country Planning

If it is certain that many contemporary planners to Abercrombie outside Britain would too be concerned with the lack of balance between the town and the country, it is no less arguable that the word "nature" would find a much more relevant position in their discourse. The concerns in planning circles in western Europe about the advances of urbanisation upon rural land tended to evidence more precisely the loss of the natural world, connoting forests and wilderness (Harrison 1993). Abercrombie often remarked positively on the views professed by Raymond Unwin that we should look at the region as a background of green upon which well-distributed developed areas would be placed (GLRPC 1929; Miller 1989). This green background was the countryside, not "nature" if understood as an imagined wilderness. Considering the woodland cover in England at the beginning of the twentieth century as of only 5% (Forestry Commission 2018), the countryside became a proxy for "nature".

Abercrombie presented no reticence in often using "nature" and "country" interchangeably: "the town should indeed be frankly artificial, urban; the country *natural, rural*" (author's italics). Although campaigns for the creation of National Parks took momentum in the 1930s, and eventually led to the National Parks and Countryside Act of 1949, for Abercrombie "nothing, not even the extremest mechanisation of farming, can prevent the country from being the town-dwellers' chief contrasted recreation and relaxation". Abercrombie here comes close to what Williams termed as the myth of the idyl country being operationalised as memory. Real or not, this shared memory was not to be lost. Still, he was not dismissive of wilderness (Abercrombie 1933). He defined two types of countryside: the "Wild Country" (or "wild nature"), including "unenclosed mountain, moor, valley and the coastline" and the "Tame", referring to the agricultural areas (Fig. 1). The countryside is not a subset of "nature", but instead the container within which wild and agricultural areas are conceptualised. In his discourse, little on ecological considerations was expressed beyond the need for preservation for "amenity" purposes of the "Wild Country". Although Abercrombie (1945) did bring up the need to educate the public "to rejoice that some few places are left free for wild life", he was sceptical of large-scale afforestation since "land is too valuable agriculturally, or if of low fertility, it is needed for open space". Notwithstanding this, he recognised that "it is, of course, dangerous to attempt to define where nature ends and where human artifice begins" (Abercrombie 1933).

The view that agricultural areas, amenity and forestry have their own separate predominant zoning is in line with Abercrombie's broader perspective of the all-encompassing role of planning as an organising force across the region. The claim being for the planning of both the urban and the rural. This was a fundamental act of reason since "planning occurs when mankind in the group makes a definite and conscious attempt to model or mould his environment", the opposite being "natural human growth", which in turn would take place when "mankind is unconscious of or unconcerned with its general form". In his view, planning would be to organisation as "natural growth" is to "complete muddle". Drawing from his civic design

Fig. 1 Examples of Abercrombie's concept of the "wild" and the "tame" country. *Source* Abercrombie (1945)

background, Abercrombie saw the act of regional planning as composing a "social organism and a work of art" (1933). Indeed, he is conscious that the countryside had been as much subject to human planning and action across time as cities have (Matless 1993). As such, the countryside is too an object of human intention and action, and therefore artifice.

In *Town and Country Planning,* Abercrombie highlighted the contrasting essence of town and country "as representing opposite but complementary poles of influence". While their identities and distinction were to be sharply marked, the country remaining rural and the town being the place *per excellence* of urbanism, on a similar vein to Thomas Sharp's views; there should, however, be some controlled "tincture" of one into the other. If on the one hand, Abercrombie is categorical in saying that "in the country the open space is farmland", the background upon which villages and small towns would sit; on the other hand, open space in urban areas would mostly be seen as recreational and parkland. Yet, for the green "tinctures", Abercrombie suggested degrees of formality and artificiality, which would decrease from the urban centres towards the outskirts. He is here aligned with planning and landscape theory of the time, sharing this transitional approach with important names such as T. H. Mawson, landscape architect and colleague at the University of Liverpool, J. Stübben, the foremost German planner of the *Städtebau* tradition and W. Hegemann whose work on civic design in the USA soon became a critical reference internationally.

Considerations on agricultural policy and post-war replanning and reconstruction started to run in parallel. With the Green Belt Act 1938, which allowed municipalities to buy land to be set aside for green belts, the possibility of enforcing the differentiations of the domains of planning presented previously came a step closer. In 1940 the Royal Commission on the Distribution of Industrial Population, the Barlow Report, pointed to the need to decentralise industry and reduce the population from dense urban areas. This implied regional planning and would mean reorganising the "tinctures" both in the country, which would need to accommodate the displaced urban dwellers, and in urban areas where more open space would become available. Such rebalancing was met with concern since preserving agricultural land was seen as a priority. This view was not only evident in the Scott Report from 1942, the Report of the Committee of Land Utilisation in Rural Areas, but also in subsequent governmental communications. In the Memorandum by Agricultural Departments from 10th March 1943, for instance, the need to stop uncontrolled sprawl and ensure that agricultural land was preserved was reiterated. The argument was that if food policy aimed to centre on the idea "that all reasonably good agricultural land should be maintained in a state of fertility and productivity", hence it should follow that "the great bulk of the land which was being used for agriculture before the war should be so maintained." For that to happen, it would be necessary to stop "indiscriminate speculation and unplanned urban and industrial development" (War Cabinet 1943).

Discussions at the London County Council (LCC) started contemporarily. In April 1941, Abercrombie was appointed to draw up a plan for the County with J. H. Forshaw, the LCC chief Architect and Abercrombie's former student at Liverpool. In September the same year, it was agreed that he would also develop a plan for the whole of Greater London. Much of his academic reflections published throughout the preceding years regarding regional planning and open spaces, especially in the *Town Planning Review* and in his book *Town and Country Planning*, would form the basis for much of what Abercrombie would propose in these plans. Both plans considered the prospect of decentralisation set in the Barlow Report, which, coupled with war destruction, set the conditions for developing "ideal" plans (Lemes de Oliveira 2015). In the County of London Plan 1943, despite its restricted limitation of influence to the County's boundary, the authors still reiterated that "the Region is the only satisfactory basis for co-ordination", especially regarding the definition of an open space system and the urban area's relationship to the countryside. Thus, there should be a two-way restructuring, that of urban areas and that of the countryside. For the former, presenting an ideal plan that could resolve traffic congestion, depressed housing, mix of incompatible uses such as housing and industry, contain sprawl, and address the insufficiency, inadequacy, and maldistribution of open spaces was paramount. Concerning the latter, the plan affirmed that "its general planning has been kept constantly in mind" since the loss of agricultural land and suburbanisation were "no less deplorable" a phenomenon (Abercrombie and Forshaw 1943).

The Scott Report assumed that green belt land would accommodate two primary uses: agriculture and recreation. This dual-function characteristic would be carried forward in both the County of London Plan 1943 (Fig. 2) and the Greater London Plan 1944. The belt was intended to prevent further sprawl and, given that "what happens to the countryside around London is of great importance to the County for it forms the main place of weekend recreation", would make up for the deficiency in green space provision in the park system devised for the inner areas. The standard of seven acres of open space per 1,000 of the population was defined as appropriate, and although bomb damage and demolitions would free up large areas, it was not enough. Hence, four acres would be established within the County and an extra three acres in the green belt or green wedges.

The Greater London Plan 1944 built upon the themes initiated on the LCC plan, now with a free hand to fully encompass the regional scale. One of its initial premises was that approximately one million people would be decentralised, hence reinforcing the need for conjoined town and country planning. It defined four rings: (1) the inner ring, the highly urbanised overspill of the County; (2) the suburban ring, the area that epitomised London's sprawl and defined by a radius of 12 miles from Charing Cross; (3) the Green Belt ring, predominantly of recreational use; and finally, (4) the Open Country, where farming would be prevalent and new towns be developed accommodating the decentralised population (Fig. 3). The latter two would be permanently reserved as agricultural land and eventual woodland still left over,

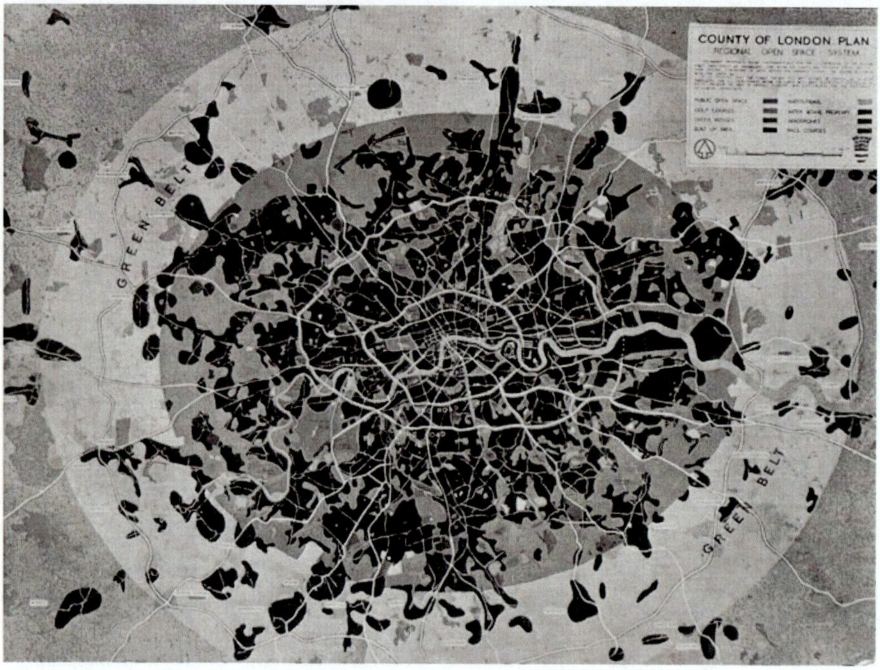

Fig. 2 The Regional Park System of the County of London Plan, with green wedges and the green belt. *Source* Abercrombie and Forshaw (1943)

and conceived from Unwin's perspective of the green background with dotted spots of building. The green belt remained a transitional zone where the urban influence exerted through the urban edge and pressure outwards for recreation were equalised by the agricultural land buffering the outer zone. The green wedges continued being the backbone of the regional park system and were brought in from the edges of the Green Belt ring to the heart of London.

As such, the green belt was both an enforcer to restrict urban growth—and, in so doing, a protection of the "open country" beyond—and a space for "rural" recreation. Within its limits, its recreational and productive functions would co-exist, albeit the former predominating. The green belt was the mediator between the city and the region, the last transitional zone before the open country was to be fully manifested. Although much present in British planning discourse since the garden city idea, compared to green wedges, Abercrombie said little about it prior to his London Plans (e.g. there is only a passing note in the second edition of *Town and Country Planning*). Conversely, Lemes de Oliveira (2015) showed how Abercrombie was a proponent of the idea of green wedges since his very first publications and plans from the 1910s, culminating with his proposals for London during the Second World

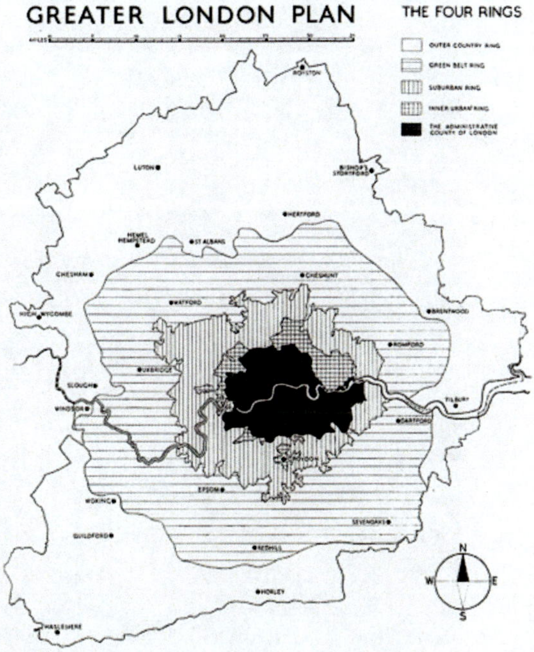

Fig. 3 The four rings of the Greater London Plan 1945. Source: Abercrombie (1945)

War. Since Abercrombie identified the radial plan (or spiderweb) as the primary form of planning activity for existing cities, as opposed to the gridiron plan as typical of the planning of new cities, such urban axes pushing out into the country would also correspond to country wedges penetrating the urban areas. The green wedge idea for him would become a quintessential model for a balanced relationship between urbanisation and green spaces within cities as much as a connecting and transitioning device mediating between the urban and the rural (Fig. 4).

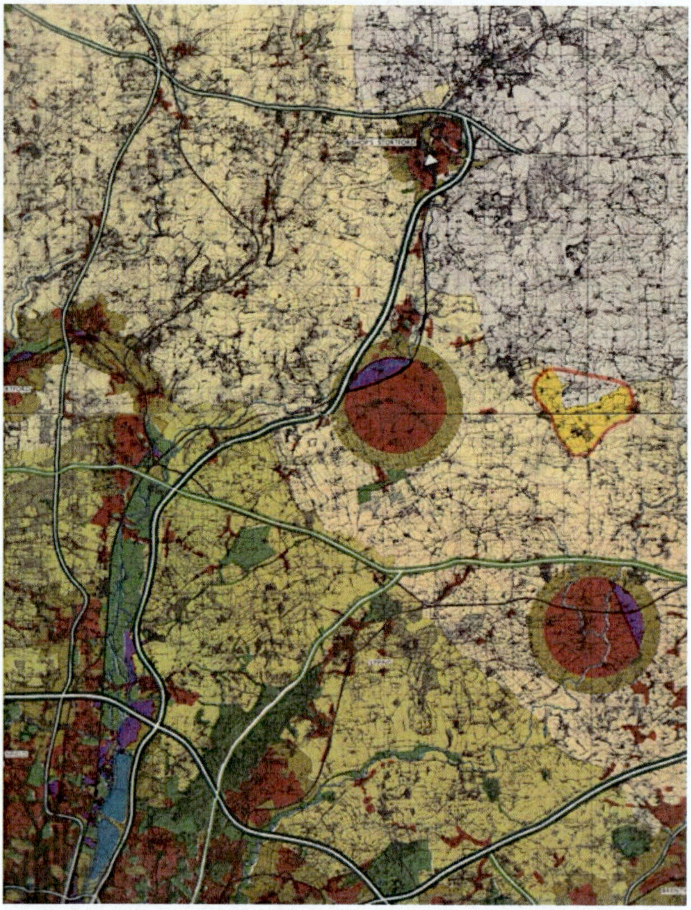

Fig. 4 Map showing the various denominations of open land in the Greater London Plan 1945. The Lea Valley Green Wedge is on bottom left, crossing a five-mile-deep zone of the metropolitan green belt of high recreational value and arriving at the agricultural land depicted in yellow. The map also shows local green belts around residential areas. *Source* Abercrombie (1945)

4 Tinctures: Rus in Urbe

Having addressed Abercrombie's views on town and country, the question of the tinctures of green within the urban areas and, in particular, the place of productive urban landscapes remains. Across his career, he defended that open space plans in cities should be done in the form of a park system. As previously discussed, connections out into the open country were essential. Green spaces "should interpenetrate the urban mass", be permanent and predominantly of public use. The ideal form of such a park system would combine the concentric and radial approaches, with parks, parkways, green wedges and the green belt (Abercrombie 1933). It is worth remembering that campaigns such as those by the National Playing Fields Association for enlarging the availability and distribution of green active recreation spaces much influenced the increased granularity of the functions of open spaces in urban plans of the period.

Abercrombie's writings before the London plans primarily refer to open space in cities from a social perspective, with recreation and "amenity" outweighing other uses. Yet, however strong this was, social views on private and public spaces in relation to food production were—much influenced and determined by political decisions—shifting in the second half of the 1930s. As the political atmosphere in Europe worsened, preparations were put in place in the event of another war. The situation was particularly concerning since Britain imported 70% of the food it consumed. In 1936 the government set up the Food (Defence Plans) Department, which became in charge of developing plans for the maintenance of supplies and controlling and distributing food in possible forthcoming conditions of war. With the outbreak of the Second World War, a significant focus was placed on food production, distribution, and the roles the countryside would play in the country's future. A few months before food rationing was introduced at the beginning of January 1940, the campaign "Dig for Victory" was initiated, encouraging people to grow vegetables in their gardens and allotments. The campaign showed significant results, helping raise the number of allotments from 930,000 before the war to 1.7 million by 1943, and the number of cultivated private gardens from three to five million (BBC 2020; Ginn 2012). Many parks were also converted into temporary allotments during the war. In the minutes of a War Cabinet meeting from 2nd October 1939 titled *Food Situation in the UK*, the directive was not only for rural areas to increase production but also that vegetable growing should too take place within towns. It reported on the launch of a "Food Production Campaign" aiming to "plough up something like 1 ½ million acres in the United Kingdom" with the help of the Women's Land Army. In addition, urban local authorities were told to increase production from private gardens and allotments and to appoint "Horticultural Committees" to support people in this task (War Cabinet 1939).

Abercrombie's London plans come into this context of a grave challenge to the country's identity and very autonomy. The "return to the land" is seen in the policies related to the maximisation of large-scale food production and private vegetable growing (Fig. 5), as well as the propaganda generated to boost morale and a vision of Englishness grounded on the land (Matless 1998).

For Abercrombie (1943), London was a "living and organic structure" that had lost its way. Its replanning would "stimulate and correct its natural evolutionary trends". It would reposition it in the course of a harmonious relationship with the land, both in terms of the countryside and with its inner open spaces. The question of the countryside has already been previously discussed. The replanning of London was to be done "spaciously", with open spaces at the core of its restructuring. By way of comparison, the implementation of the seven-acre standard of green space provision proposed in the LCC plan would represent more than doubling the existing green spaces at the time (London County Council 1944). In this plan, the park system proposed would

Fig. 5 Aerial photograph of Welwyn Garden City used in the Greater London Plan 1945 to illustrate an example in which "architecture is pleasantly combined with its natural background". On the bottom left allotment gardens can be seen. Source: Abercrombie (1945)

link up a range of green space typologies, based on a model initially proposed by G. Pepler (1923) and reproduced in *Town and Country Planning*, combining radial and concentric elements: "the Green Belt and surrounding countryside need bringing more into the centre though green wedges formed by the existing undeveloped and public land, the parkways along the ring roads giving access from one wedge to the other". The larger open spaces would determine the formal identity of the various communities in the plan and reach the local open spaces within the residential areas. It should become possible "to get from doorstep to the open country through an easy flow of open space from garden to park, from park to parkway, from parkway to green wedge and from the green wedge to the Green Belt" (Abercrombie and Forshaw 1943).

Although a substantial increase in open spaces was being laid out and a seemingly effective spatial strategy for their distribution being achieved, observations made by various interest bodies after the publication of the plan were mixed. While the Minister of Town and Country Planning was generally happy, the question of allotments raised concern. For instance, the Ministry of Agriculture and Fisheries was critical that the plan had not dealt "adequately with the provision of allotments" and that they should be considered in addition to the four-acre standard and the Royal

Institute of British Architects (RIBA) thought that four acres were reasonable, but recommended that more areas be added in the outer zone, including ¾ acre for allotments (London County Council 1943). The Greater London Plan 1944 envisaged an even more generous provision of 10 acres per 1,000 people. Its open spaces, linked up to form a park system, besides the green belt and green wedges referred to above, presented various traditionally seen typologies such as squares, parks, parkways and various types of playing fields.

In terms of productive urban landscapes, the LCC plan was explicit in the inclusion of allotments. The plan proposed "an equal amount of open space for playing and amenity purposes". Allotments were included alongside large amenity parks, parkways and riverside "pleasaunces". These types of open space would amount to 1 and 2/3 acres per 1,000 population. The maps produced, however, do not detail how these different types of open space would be distributed. For instance, the communities and open space survey plan only mapped "private" or "public" open space. This general approach is followed in the open space plan, which categorises existing or proposed open space; and "private open space suitable for public", and more tentatively "institutions with substantial open space". The open space system would have two main scales: the open space plan for the County and the regional park system. In the former, they are bundled into the category of "proposed public open space", while the latter groups existing and proposed open spaces in three main categories: public open space, golf courses and green wedges. Allotments are prescribed to be dispersed in the residential areas. When detailing the reconstruction areas, the text reiterates the inclusion of allotments as a part of their new open space provision but do not evidence the various specific typologies on the maps (Fig. 6).

Although recognising the benefits to growers of food production, Abercrombie was not convinced about its aesthetic qualities. Groupings of allotments were considered to have "negligible, if any, amenity value". Hence, their distribution, besides being convenient, would also cause "a less disturbing effect" (Abercrombie and Forshaw 1943). Food was also to be produced in private gardens. Communal gardens, however, were to be the preferred solution, as they would give a "greater sense of openness" as compared with private gardens "with their encumbering fences" (Abercrombie and Forshaw 1943). Yet, the need for private gardens and allotments would also be met, especially for medium-sized and large families and located at the rear of the dwellings, and with access from the communal gardens. While most of the housing would be distributed in buildings kept to four storeys, some large tower blocks would also be available. These would require large ground spaces between them to allow for adequate air circulation and penetration of sunlight indoors, which could also include allotments.

In the Greater London Plan 1944, allotments were considered in similar lines, but the intensity of the need for food growing spaces in peacetime was more strongly put: "as public spaces and off vacant building sites are returned to their normal uses there will be an urgent demand for additional plots near to the gardener's home". However, contrary to the LCC plan, no specific allowance for allotment acreage was considered in the open space deficiency calculations, and that this would need to be added as an

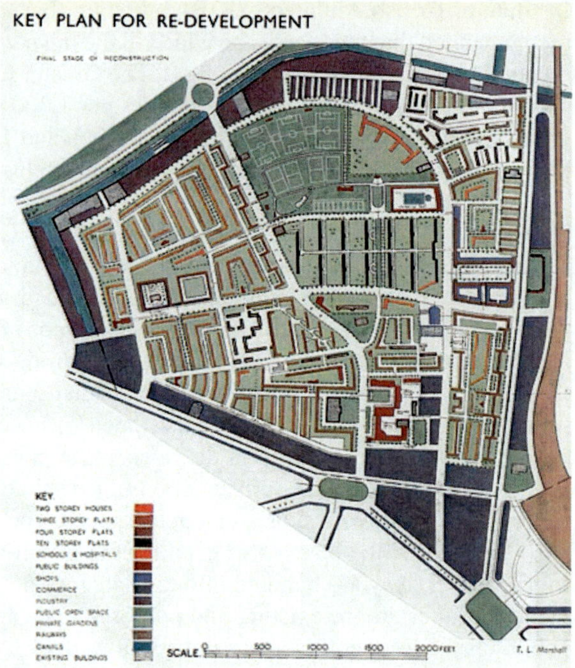

Fig. 6 A model of redevelopment in the County of London Plan 1943. Green areas are categorised as public open space or private gardens. Allotments would be near residential buildings. *Source* Abercrombie and Forshaw (1943)

additional land use "for which provision must be made" (Abercrombie 1945). Post-war commentators highlighted further the need for including urban food production in town planning. Purdom (1949), for instance, reflected upon the premise that in the new towns proposed in the Greater London Plan, it would be possible for every house to be near an allotment and that the "experience during the two wars showed what an enormous advantage it would be if every family had a piece of ground from which to keep itself supplied with vegetables". This alone would ensure that a town up to 50.000 inhabitants could be almost self-sufficient in regards to vegetables and fruits. He also suggested that this mode of production would generate a certain amount of surplus, which could be redistributed to those who did not possess an allotment or were unable to produce certain products they could not grow themselves.

Therefore, if in the County of London Plan 1943 the green "tinctures" related to food production did not acquire particular significance beyond being a war necessity, the regional plan offered a relatively broadened perspective on the topic. In the densest areas of London, covered by the LCC plan, given the identified need to fight poor living conditions and its associated problems, enhancing the provision of green space and rebalancing its distribution meant a focus on recreation, improved aesthetics and air quality, access to sunlight and noise abatement. By the time the LCC plan was to be published, the difficulty in the acquisition of open spaces after the war considering all priorities was becoming increasingly concerning, especially in London's denser areas, taking a toll on the initial idealism of the plan and forcing a compromise in provision with an interim standard of 2 ½ acres. At community level, finding areas for school playing fields took priority alongside provisions for public green spaces for amenity and recreation. For the Greater London Plan 1944, in turn, the question of the countryside was more strongly posed and could be addressed more directly. The need for more inner-city parkland diminished in proportion to residential areas' proximity to the green belt and the green wedges. In addition, by the time of the Greater London Plan 1944, the political scenario pointed to the importance of keeping urban food production on the planning agenda.

In both cases, planning spaciously and having green wedges crisscrossing the urban area would allow for flexibility in the use of green spaces and evidence a resilient approach to planning. Should the need to use these spaces to grow food ever returned, they would be available in large quantities and distributed adequately. Although it could be said that Abercrombie underestimated the significance of the urban food growing trend after the war, the plans envisaged, still in peace times, food growing across scales: in private and communal gardens, allotments, in the green wedges, the green belt and out in the open country.

5 Conclusions

As two dimensions of planning humankind's settlement needs, the urban and the rural are posited as interlinked, complementary and necessary conditions. The imbalance of such domains in the interwar period, with deteriorating conditions in large cities on the one hand, and the disintegration of the ruralness of the countryside on the other, have in turn too triggered the need for complementary positioning in planning discourse: to preserve and to proactively plan. For Abercrombie, the two polarities, the town and the country, needed to be maintained. This meant a strong call for the qualities of urbanity in cities and those of ruralness in the countryside, albeit with mutual "contaminations", as both extremes were to be avoided. Excessive concentration in cities would lead to congestion and all its woes, degenerating urbanity as much as lack of culture, use and consequent care for the countryside would do to its landscape. Thus, there is a firm attempt in Abercrombie's work to find a position of balance, albeit a dynamic one. Proactive planning of urban areas involved restabilising such a balance by reorganising the socio-cultural organism of the city into an environment that was at the same time healthier and more connected to the land. Within the urban fabric, much of the land destined for green spaces was to be reconfigured in the form of a system of interconnected areas. Beyond in the country, proactive planning meant resettling the overspill population from the densest urban areas and planning agricultural production, the use of rural land and its new identity. In-between spaces or transitional areas were conceptualised, such as the green belt and green wedges. In the case of the London plans here discussed, the first was a transitional zone captured between the need to provide urban and rural recreational space, while buffering the outer countryside. It is a mediation between the scales of the city and the region, and the polarities of the town and the country. Green wedges on the other hand, given their radial dispositions, were elements of hybridism, providing opportunities for gradation in the transition between urban and rural, formal and informal, artificial and "natural".

The wartime plans were not only considered as opportunities to reimagine London, but also its relationships with the territory. On a symbolic level, as the war loomed, the call for a return to the land could be seen as an attempt to grip to the country's identity and independence and instil resilience, as mobilised by the war propaganda at the time. On a more pragmatic level, in a moment of limited resources and food, green spaces of all scales were potential growing spaces. Planning spaciously embedded such resilient trait since open space uses could be rapidly changed in order to adapt to shifting conditions.

The influence from societal changes in gardening towards food production in those years dented orthodox planning and landscape architecture thought regarding green spaces' roles and functions. Urban green space planning had emerged in the nineteenth century entangled with public health concerns and the need to provide intra-urban recreative spaces. Although Leberecht Migge had already championed in Germany the potential of parks as productive landscapes in times of conflict, the Second World War marked another moment of a shift from ornamental gardening towards a long-lasting increase in food growing in private gardens and an enhanced interest in productive urban landscapes overall. If Abercrombie (1933) had already noted that "the house in a garden is the symbol of our individualism and the sign of the penetration of the country into the town", this was even more so as food growing took a further hold in urban areas. As seen, urban allotments were to be part of the park system of both the LCC and the Greater London plans and be dispersed in residential areas.

Abercrombie's claims for the differentiation of town and country resonate with today's concerns over suburbanisation processes and infrastructural planning leading to ultra-fragmented landscapes, the indeterminacy of the urban–rural border, and putting to check the urbanity of peri-urban development and the rurality of the countryside. In addition, the global spread of food production, distribution chains and consequent increased disconnection between people and food is leading to reactions for more sustainable and circular processes, for more localised sources of food production, and closer proximity between consumers and production source. Abercrombie's attempt to safeguard agricultural land near cities, localising thus food production and consumption, remains significant. Recovering his work from these perspectives invites us to reflect upon the domains of human transformation of the planet and our relationship with the land not from a dichotomist opposition but from an understanding of the need to intensify our research on the conceptualisation of urban, peri-urban and rural landscapes, their specificities, dependencies, and interactions, as well as the consequences of such for both people and the environment.

References

Abercrombie P (1910) A Comparative Review of Examples of Modern Town Planning and Garden City Schemes in England Town Planning Review 1:18–38

Abercrombie P (1912) Town Planning in Greater London: The Need for Co-operation. Town Planning Review 2:261–280

Abercrombie P (1933) Town and country planning, vol 163. Oxford University Press, London

Abercrombie P (1945) Greater London plan, 1944 : a report prepared on behalf of the Standing Conference on London Regional Planning. HMSO, London

Abercrombie P, Forshaw J (1943) County of London Plan prepared for the LCC, 1943. Macmillan, London

BBC (2020) Victory gardens: A war-time hobby that's back in fashion. BBC.

Bullock N (2002) Building the Post-War World Modern architecture and reconstruction in Britain. Routledge, London

Dehaene M (2004) Urban Lessons for the Modern Planner: Patrick Abercrombie and the Study of Urban Development The. Town Planning Review 75:1–30

Dehaene M (2005) A Conservative Framework for Regional Development: Patrick Abercrombie's Interwar Experiments in Regional Planning Journal of Planning Education and Research 25:131–148. https://doi.org/10.1177/0739456x04270369

Dix G (1981) Patrick Abercrombie 1879–1957. In: Cherry GE (ed) Pioneers in British Planning. The Architectural Press, London, pp 103–130

Forestry Commission (2018) Area of woodland: changes over time. 2021

Geddes PS (1911) The Civic Survey of Edinburgh. In: RIBA (ed) Town Planning Conference, London, 10–15 October 1910. Transactions. [With maps and illustrations.]. RIBA, London, pp 537–574

Ginn F (2012) Dig for Victory! New Histories of Wartime Gardening in Britain Journal of Historical Geography 38:294–305. https://doi.org/10.1016/j.jhg.2012.02.001

GLRPC (1929) First Report of the Greater London Regional Planning Committee. Greater London Regional Planning Committee, London

Hall P (1995) Bringing Abercrombie Back from the Shades: A Look Forward and Back The. Town Planning Review 66:227–241

Hall P (1996) Cities of tomorrow : an intellectual history of urban planning and design in the twentieth century, Updated. Blackwell, Oxford

Harrison RP (1993) Forests: The Shadow of Civilization The University of Chicago Press, Chicago

Hasegawa J (1999) The Rise and Fall of British Reconstruction in 1940s Britain Twentieth Century British History 10:137–161

Hewitt LE (2011) Towards a greater urban geography: regional planning and associational networks in London during the early twentieth century. Plan Perspect 26:551–568. https://doi.org/10.1080/02665433.2011.601608

Hoskins WG (1955) The making of the English landscape. Hodder & Stoughton, London

Howard E (1902) Garden cities of to-morrow. [New edn. Faber, London

Larkham PJA, D. (2011) The post-war reconstruction planning of London: a wider perspective' (Working Paper Series No. 8). Birmingham City University, Birmingham

Lemes de Oliveira F (2014) Green wedges: origins and development in Britain Planning Perspectives 29:357–379 doi:https://doi.org/10.1080/02665433.2013.824369

Lemes de Oliveira F (2015) Abercrombie's green-wedge vision for London: the County of London Plan 1943 and the Greater London Plan 1944 Town Planning Review 86:495–518 doi:https://doi.org/10.3828/tpr.2015.30

London County Council (1943) County of London Plan. Summary of Observations of Metropolitan Borough Councils and other bodies on green space proposals. London Metropolitan Archives, London

London County Council (1944) County of London Plan - housing density. Housing and Public Health Committee. Joint report by architect, chief assistant, valuer and comptroller of the council vol CL/TP/1/34. London Metropolitan Archives, London
Matless D (1993) Appropriate Geography: Patrick Abercrombie and the Energy of the World. J Des Hist 6:167–178
Matless D (1998) Landscape and Englishness. Reaktion Books, London
Miller M (1989) The elusive green background: Raymond Unwin and the greater London regional plan. Plan Perspect 4:15–44. https://doi.org/10.1080/02665438908725671
Pepler G (1923) Open Spaces Town Planning Review:11–24
Purdom CB (1949) The building of satellite towns. Dent, London
War Cabinet (1939) Food situation in the United Kingdom vol cab/68/1/12. TNA, London
War Cabinet (1943) Post-war agricultural policy vol cab/66/35/3. TNA, London
Ward S (2004) Planning and Urban Change. Sage, London
Williams R (1975) The country and the city. Oxford University Press, Oxford

The Roots of the City–Countryside Relationship: A Lesson from Post-war Town Planning in Italy

Luca Lazzarini

Abstract The chapter investigates the origins of the urban/rural relationship concept in post-war town planning in Italy. It focuses on the profile of Italian planner Giovanni Astengo (1915–1990) and one of his major works, the Plan of Assisi (1955–58), where he applies an innovative methodology to analyse the local agrarian economy and the food system. The objective of the Plan is to achieve a balance between resources and income in the locality and ensure local community's wellbeing. The relevant contribution of the Plan to post-war planning discourse in Italy lies in its complex understanding of the productive and landscape features of agricultural areas and the implications on the issues of food self-sufficiency and city/countryside functional interdependence.

Keywords Urban/rural relationships · Food self-sufficiency · Giovanni Astengo · Municipal plan · Assisi

1 Introduction

The concept of city/countryside relationship has emerged as a reaction to the problems associated with the urban/rural dichotomy that were dominant in European cities at the end of the nineteenth century. In describing its genealogy, Davoudi and Stead (2002) trace the very first theoretical contributions on urban/rural relationships in some British planning practices of the mid-twentieth century. Until that time, spatial policies were treating town and country as separate entities, each with its own investments and development programmes, and any attempt to develop an integrated approach to urban and rural areas was seen as leading to a "degenerate mixture" (Ibid.: 2; see also: Sharp 1932). Yet, the new vision did not emerge from scratch but there were some elements of continuity in the British debate between some theoretical contributions on town and country planning of the early twentieth century and the post-war reflections on urban/rural linkages (Cloke 1989). A notable example

L. Lazzarini (✉)
Department of Architecture and Urban Studies (DAStU), Politecnico di Milano, Via Bonardi, 20133 Milan, Italy
e-mail: luca.lazzarini@polimi.it

C. Brisotto and F. Lemes de Oliveira (eds.), *Re-Imagining Resilient Productive Landscapes*, Cities and Nature, https://doi.org/10.1007/978-3-030-90445-6_6

is the 1910 RIBA Town Planning Conference held in London during which a lively discussion on city models and green spaces took place and important contributions on radial development of the city, including its integration with green open spaces, were presented (Lemes de Oliveira 2014). Most of these contributions were framed within a discussion on regionalism that involved some important protagonists of the planning culture like Ebenezer Howard, Patrick Geddes and Patrick Abercrombie, including the influence that this discussion generated on the activity of the Town and Country Planning Association in UK between the 1940s and the 50s (Gaeta et al. 2013).

During that period, two main city models were confronting each other in a vibrant discussion characterized by some conceptual overlaps as well as manifold mismatches and divergences. The first is the *linear city model*, first conceived by Spanish engineer Soria Y Mata (1844–1920). Alongside this model, the representation of the city diverges from the traditional centre/periphery dualism for proposing a networked settlement organization which urbanizes a large territory through a linear pattern comprising a central backbone made of technological and mobility infrastructures. One of the main advantages of this model is its open form, not affecting subsequent configurations, and allowing an incremental and modular implementation through the aggregation of single lots (Gabellini 2001; Viganò 2010). The success of the linear city model is given by its future applications (though some of these significantly diverging from the solution of Soria Y Mata), such as Le Corbusier's model of Linear Industrial City developed in 1942–43 (Le Corbusier 1945), or the famous Amsterdam Plan of 1935 (Aa 1934) that has inspired a number of future city expansion plans based on the idea of organizing the urban growth along some major axes branching out from the city center to the countryside, such as the Copenhagen Five Fingers' Plan of 1947 (Hall and Ward 1998). While the Copenhagen Plan combined the radial city growth with the retention of the green wedges between the "fingers" of the growth, the planning of linear extensions from the urban core to the countryside with green spaces strongly integrated in the settlement structure was surely another fertile ground of debate for modern town planning. A relevant contribution on this comes the work of Ludwig Hilberseimer who during his career in the United States elaborated some diagrams of linear cities, following the awareness that the process of city decentralization was an inevitable trend. Hilberseimer adopted a ribbon structure experimenting a hybrid settlement condition between agriculture, housing and industry, and enabling the city to physically "enter the landscape" (Hilberseimer, 1944, 1949). Not just the work of Soria Y Mata, at the basis of Hilberseimer's reflection there was also a deep investigation of the theories of Patrick Geddes and his model of valley section (Geddes 1915; Ferraro 1998; Lazzarini 2020), as a device capable of developing a new understanding of the city-countryside link based on the representation of the reciprocal dependencies between the different parts of a region.

The second model is the *radio-centric city*, whose roots can be traced back in the theories on city growth of the last decades of the nineteenth century based on the need to control densities, create new satellite towns for containing the overcrowding of central areas and provide urban areas with adequate green spaces. A milestone in this debate is the publication in 1898 by Ebenezer Howard (1850–1928) of the diagrams of the "Garden Cities of Tomorrow". Howard's contribution led to a lively discussion on the problem of city scale and of the relationships between built and open spaces (Howard 1902; Gaeta et al. 2013), and on the management and protection of green areas around the city (Sturzaker and Mell 2017; Lazzarini 2018). This debate will find a further development in the following decades, especially in the process leading to the elaboration in 1944 of the Greater London Plan (Abercrombie 1945) where the opposition between green belt and radial parks is overcome in favour of a scheme made of a combination between the two models (Lemes de Oliveira 2015). Few years after Howard's book, the radio-centric model found an original declination in the work of Erich Gloeden, a Jewish architect living in Berlin and executed in 1944 by the Nazi regime. In Gloeden's diagrams (1923), the urban growth takes place by gemmation around and at a certain distance by the initial nucleus. Compared to the Garden Cities model, in this case the organization has a higher density and each cell, differently specialized and not exceeding 100,000 inhabitants, has its own physiognomy: "the gemmation process is potentially infinite" (Viganò 2010: 84). The goal of Gloeden was to develop a horizontal and decentralized model of city expansion, devoid of any hierarchies, where the relationship between the cells was articulated through the public transport network crossing the open spaces between the settlements where parks, sport facilities and farmland were located.

Tracing the extent to which these theoretical reflections have influenced the planning debate in Italy is not an easy task. Though it is likely to imagine that in the post-war period some of the already mentioned contributions on the form and nature of city expansion were being discussed in the Italian scientific and academic environments, the debate was somehow still dominated by the old rationalist ideas and models coming from the early twentieth century. By describing the debate of those years, Italian planner Giovanni Astengo (1915–1990) stated that the discussion was permeated by "cultural anxieties and hindrances implicit in rationalist schematism, the utopian longing and the engineering technicality of the 1930s" (Astengo 1966: 600). Hence, the attention was mainly oriented to the problem of the organization of the urban expansion, rather than on the shape of open spaces and on the spatial and functional interdependencies between urban and rural areas (Di Biagi 2010; Orioli 2012).

Alongside this framework, the main objective of this chapter is to describe the roots of the urban/rural relationship concept in post-war town planning in Italy and to investigate one example of municipal land-use plan—the Plan of Assisi made by Giovanni Astengo in 1955–58—that includes an original understanding of the relationship between city and countryside. In particular, in the Plan of Assisi the urban/rural dichotomy was overcome and a different approach based on shaping stronger spatial and functional interdependencies between urban and rural spaces and on working on the food self-sufficiency of the municipal area was experimented.

The chapter is organized in four sections. Section 2 presents a brief overview of the post-war town planning regulatory framework in Italy, focusing on the main problems and challenges related to the process of city expansion. Section 3 introduces the profile of Astengo, one of the protagonists of twentieth century town planning in Italy, by highlighting some biographical information (Sect. 3.1), the main aspects of his theoretical and design approach (Sect. 3.2) and an overview of his contribution on regional planning and, specifically, on the problem of self-sufficiency of cities (Sect. 3.3). In Sect. 4, the plan of Assisi elaborated by Astengo in 1955 is presented and analysed, with a specific emphasis on the analysis of the local food system and how this analysis frames an innovative idea of urban/rural relationship. The last section includes some concluding remarks and offers some key lessons for contemporary planning debate.

2 Italy's Post-War Planning Framework: Problems and Challenges

The already mentioned late development in Italy's planning debate of the issues and problems related to city/countryside interdependencies is due to a number of reasons. In addition to the persistence of the rationalist cultural inheritance already mentioned, one reason lies in the planning framework introduced in Italy in early 1940s for guiding the urbanization and the infrastructure development of the Country. The hypothesis here considered is that the national planning law, rather than supporting new models of urban development, had contributed to reproduce the dominant vision of the city-countryside relationship, exacerbating the existing dichotomy (Orioli 2012).

The reference goes to Law n. 1150, the first and main national town planning law—still in force in many parts, that was approved in 1942. The law introduced a hierarchical system of plans, from the municipal to the territorial level, to be elaborated on the basis of specific multi-level administrative procedures involving different institutions and technical expertise (Salzano 1998). The law also required municipalities to produce a municipal land-use plan (Piano Regolatore Generale), a local planning policy instrument for regulating a priori the functional and physical transformation of municipal areas and for equipping all inhabitants with adequate public services and facilities. Although the law required municipalities to take into consideration in their land-use plans the whole municipal area, until the 1950s the attention was merely oriented to plan the expansion or restoration of the built environment, due to the need to reduce overcrowding, improve the hygienic conditions of existing settlements and fight the housing shortage which in this period was particularly problematic (Salzano 1998; Di Biagi 2010). According to Patrizi (1983), a sort of ambiguity characterizes the prescriptions on open spaces because they required municipalities to generically divide the whole non-urbanized territory in zones, though not providing any planning regulations or guidance for agricultural areas. Moreover, citizens and developers were required to present a building permit just for the constructions taking place within the city, and no building permit was needed for building new residential or industrial structures in agricultural areas (Orioli 2012).

This condition of overlooking of farmland by planning policies ended in mid-1960s when the Law n.765/1967 was promulgated, the so-called Bridge Law ("Legge

Ponte") from its attempt to bring the old planning regulatory system to a new phase. The push came from a catastrophic event, the landslide in Agrigento, happened in July 1966, which was caused by a building overload of more than 8.500 constructions built in the past years without following the planning regulation. The law required all municipalities to adopt a local plan (until then just those included in a list made by central government were required to adopt it), and it introduced some limitations to those municipalities that did not comply with this prescription. Also, the law predisposed as mandatory the building permit for all constructions happening within and outside existing built-up areas. Another important innovation was the introduction of a specific planning zone (called zone E) for agricultural areas with the objective of preserving farmland from the chaotic expansion of the city.

A result of this process is that until the 1960s agricultural areas in Italy have seen no effective land-use containment policies able to comprehensively regulate and control the city expansion in the countryside. This normative lack has indeed contributed to accelerate what has been interpreted by a number of researchers as a real "explosion of the city in the countryside" given by the extensive utilization of territory and resulting in the loss of productive and valuable agricultural and green areas (Indovina et al. 2005; Agnoletto e Guerzoni 2012).

Just in very few cases a different approach to urban/rural relationships has emerged. For example, in the Municipal Plan of Siena (1956) made by Luigi Piccinato, Piero Bottoni and Aldo Lucchi, agricultural areas were safeguarded by a set of prescriptions based on rural landscape constituent features (e.g., vegetation, soil, historic settlements). Another significant example—which will be analysed in Sect. 4—is the Plan of Assisi made in 1955–58 by a group of architects and professionals coordinated by Astengo. In these two examples, the different interpretation by planners and designers on agricultural space was framed by an attention towards the landscape and environmental articulation of the open space and the spatial and functional relationships linking the countryside to the city. The intention was to avoid the perpetuation of the existing city/countryside separation and to achieve new interdependencies between different functions and spaces. The two plans show a sensitivity to the values and vocations of the countryside which is uncommon in the Italian post-war planning debate. Only the researches of the agrarian economists like Arrigo Serpieri or Manlio Rossi Doria had oriented the same attention to the countryside constituent features, though their focus was mainly oriented to investigate the economic, social, agronomic structure of the rural territory, and seldom they explored both the relationships with the urban domain and the implications on planning policies (Rossi Doria 1953; Lanzani 1996).

3 Giovanni Astengo, Between Rationality and Reformism

3.1 A Biography of Civic Commitment Crossing Academia, Administration and Profession

Giovanni Astengo (1915–1990) is known as one of the protagonists of twentieth-century town planning in Italy. Born in Turin in a family of manufacturers, he studied architecture at the Turin Polytechnic University, where he graduated in 1938 with a design thesis on a city museum supervised by Giovanni Muzio. After the war, he actively contributed to city's cultural and political environment by participating to the foundation of the Giuseppe Pagano Modern Architects Group in Turin and the establishment of the regional branch of the National Institute of Urban Planning (INU) of Piedmont and Aosta Valley. In fall 1944, he began to work with Mario Bianco, Nello Renacco and Aldo Rizzotti on the Piedmont Regional Plan, the ongoing results of which were publicly presented at the first National Convention for Reconstruction held in Milan in 1945.

In 1946, at the invitation of the Director of the Research Council Gustavo Colonnetti, the plan was exhibited in Rome and it obtained a wide interest and appreciation from the audience. This led the Ministry of Public Works to give Astengo and his research group a formal commission to elaborate the Territorial Coordination Plan of the Piedmont Region, the first planning commission obtained by Astengo. In 1948, the participation to the National INU Congress represents a key stage in Astengo's career as he met Giuseppe Samonà, architect and director of the University of Architecture in Venice (IUAV). Samonà invited him to teach at the IUAV University, a position that he will hold for 36 years. On that occasion, his relations with the INU were further consolidated as he became a member of the Board of Directors as well as editor-in-chief and then—after the resignation of Adriano Olivetti who became director of the Institute—director of "Urbanistica", the official journal of the Institute. Astengo directed the journal for over 30 years, from 1952 to 1977, turning it into the main professional and academic channel through which he brought to the attention of the public opinion and scientific community issues and problems deriving from his professional and research activity (Fantin and Fregolent 2010). In 1955, Astengo received the commission for a new municipal land-use plan of Assisi, in Umbria, central Italy, a commission he defined as "sudden, not sought". He also stated that "the impact with the historic Umbrian city could only be shocking. [...] I worked with enthusiasm, immersing myself in the environment, historical and human, and I conceived the idea that to demonstrate what could have been done [...] a plan complete and consistent in all its parts was needed" (Astengo 1991).

The experience in Assisi inaugurated a series of commissions for land-use plans in a period during which Astengo combined the academic teaching with a range of professional and research activities. In the second half of the 1950s he elaborated the municipal land-use plan for Gubbio (1956) and then in the 1960s he was appointed to those of Saluzzo (1961), Genoa (1963) and Bergamo (1965).

The chance to systematize the knowledge and the history of urban planning came from the proposal by Bruno Zevi during the IX INU Congress in Cagliari to write the item "Urbanistica" for the Universal Art Encyclopedia, which will be published in 1966 (Astengo 1966). 1966 was a crucial year for Astengo (Di Biagi 2002) as he was appointed Councilor for urban planning in the center-left council of the Municipality of Turin and he also joined the Commission created by the Minister of Public Works Giacomo Mancini following the Agrigento landslide in 1966, a tragedy that shocked the public opinion, shedding light on the problem of illegal constructions and the lack of planning regulations in many areas of the Country (Salzano 1998).

However, the professional experiences and the scientific reputation gathered by Astengo also went through some difficulties. During the work in Gubbio, in 1967 he underwent an indictment for presumed irregularities in the procedure for elaborating the plan. This indictment led him to resign from the councilor office in Turin, though later on he was finally acquitted from the accusation (Astengo 1968). In 1970 Astengo achieved an important result after a long academic debate on the need to create a specific academic course on town planning autonomous from architecture: the IUAV launched the first degree in Urban Planning of which Astengo was president. In 1975 he was appointed Councilor for Urban Planning and Management of Piedmont region. It is during this experience that he promoted and coordinated the elaboration of the first regional planning law (no. 56/1977). He also reorganized the technical structures and competencies of the department and guided the elaboration of many land-use plans in the regional territory. During the 1980s and in the last part of his career, after retiring from university, Astengo continued his professional activity uninterruptedly with major plans and planning projects including the Pistoia plan (with Luigi Airaldi) and the environmental impact assessment of the Livorno-Civitavecchia motorway (with Giuseppe Campos Venuti).

3.2 *Organic Rationality in Astengo's Career*

The long and dense biography of Astengo just mentioned outlines a multifaceted and complex profile of planner in which multiple theoretical and cultural references and design approaches coexist and combine together. The research program developed by Astengo throughout his career is not monolithic but undergoes an evolution which incorporates new planning theories and approaches applied to issues and problems related to the post-war reconstruction and expansion of Italian cities. Also, his reflection establishes some relevant links with international theories and models, which constitute a critical background to which he contributes with innovative reflections. One of the *leitmotifs* we find throughout his career is certainly the attempt to define and consolidate the theoretical and institutional field of intervention of the "Urbanistica" discipline, endowing it with a rigorous scientific statute, expanding its space of influence in society and improving its capacity to guide the urbanization processes throughout the country (Di Biagi 2002).

This happened not only through the already mentioned 30 year-long direction of the journal "Urbanistica", the main channel for disseminating planning ideas, tools, and approaches in Italy, but also through its extensive publishing activity. An important example is the item "Urbanistica" in the Universal Art Encyclopedia. Here, Astengo not only defined and built a large synoptic framework systematizing the town planning knowledge and claiming its scientific status as the other social disciplines (Palermo, 2004), but he also denounced a range of unresolved urban problems and challenges, and indicated to planners some frontiers on which they should have worked in future. Astengo also identified the main areas of planning intervention, meaning them not "as fields for sectoral and independent actions, but as interdependent focal points participating to a necessarily unitary process" (Astengo 1966: 614). As Di Biagi (2002: 274) pointed out, when looking at those areas (new residential districts, historic centers and urban renewal, business and commercial centers, industrial settlements, urban traffic, organization of leisure time), a clear reference to the four functions of the Athens Charter emerges (Le Corbusier 1957). Although this "seems to confirm the late diffusion in Italy of one of the main manifestos of modernism", it is evident that rather than endorsing this orientation, he opted to proceed towards an organicist planning approach aimed at achieving "efficient and reciprocal interrelationships between the structural elements of the system and [...] their spatial distributions through infinite possible combinations" (Astengo 1966: 637). As also noted by Lanzani (1996: 64), the organicist conception "allows the complex socio-territorial reality to be decomposed into simple elements" so that they can be analysed and then recomposed into a single image. This conception is also visible in the projects for new residential neighbourhoods made by Astengo, which follow a careful combination and compositional distribution of building blocks, a unitary characterization of the whole system, and an importance assigned to the open space inside the neighbourhood, as the case of the INA-Casa Falchera neighbourhood in Turin clearly shows (Astengo 1954; Zeier Pilat 2014; Di Robilant 2019).

Astengo's conception of town planning sees the experimentation of a rational and scientific path which leads through a comprehensive and accountable way to plans and projects. These have the purpose to generate "a common platform of understanding, on which decisions have democratic foundations on which referring to for gradually evaluating the results achieved so far" (Astengo 1966: 639), thus pursuing the expectations and interests of the local community. Astengo's ideas of the future are therefore contaminated by the empirical approach and by the "culture of municipal engineers, very much alive and vibrant in the Turin context in which Astengo was trained, and by the influence of the historical and geographical studies of the French school" (Palermo 2004: 112). From the French school, Astengo took the scientific approach opting to investigate the organic units through statistical and social analysis procedures (Lanzani 1996) for detecting and classifying them, studying their spatial and temporal distribution in the territory and describing their interrelationships.

However, as also previously mentioned, Astengo's approach is not monolithic but it evolves over time. During the first planning commissions, such as the Assisi Plan (see Sect. 4), the land-use plan was conceived as a tool to work towards a given future, as if the objective of the urban planner was to "unveil an image of the future still harnessed in present" (Di Biagi 2002: 287–288). In the following experiences, the plan becomes the creation of a future image which is able to orient hopes, expectations, interests and policies, becoming a sort of "guiding image for the society". As also pointed out by Moroni (1997), at the base of this evolution there is probably a more mature adherence of Astengo to the concepts and methods of the utilitarian planning approach. A clear example of this transition is the Bergamo Plan elaborated in 1965 which represents perhaps the most mature attempt of bringing the planning decisions to adhere to a scientific method which, avoiding arbitrary and inductive procedures, pursues the choices capable of maximizing the overall social benefits and economic returns of the urban community.

3.3 The Rise of Regional Planning and the Problem of the Food Self-sufficiency of Cities

Astengo's reflections on the problem of food self-sufficiency of cities are framed within the contribution on regional planning that he developed at the turn of the 1940s. His interest in regional planning derived from the post-war reconstruction process and from the widespread awareness that good regional policies and plans would have positively contributed to the problem of managing the urban growth in a territorially balanced way, through adequate collective infrastructures and rationally located new settlements for hosting the growing amount of population migrating from rural to urban areas.

The opportunity to apply these reflections to a concrete planning process came in 1946 when the Ministry of Public Works gave Astengo and his team the commission to elaborate the Territorial coordination plan of the Piedmont Region (Astengo et al. 1947). The Plan represents the first attempt to produce a regional planning policy in Italy (Giudice 2019). The analysis of the agricultural and food problems was included in what Astengo defined as "the representation of the status quo of a region for planning purposes" (Astengo and Bianco 1946). This study was carried out by Astengo together with Mario Bianco (a member of his team) and the results were published in 1946 in a book titled "Agricoltura e Urbanistica" (Viglongo, Turin) and constituted the analytical part of the regional plan. The book cover reports some of the central questions the authors were trying to address:

> How much land must be cultivated to feed a person? Is it possible to evaluate the agricultural domain of a city? How many farmers does a city need to feed itself? (Astengo and Bianco 1946: cover).

In the introduction of the book, the authors underlined that the agricultural problem is closely connected to the economic problem of improving the yield of agricultural crops. The policies included in the plan concern a range of issues dealing with agricultural methods such as agricultural rotation, increase in forage production, transformation of stable dry meadows, crop replacement. By working on these aspects, Astengo and Bianco identified two main objectives to pursue. The first one is the adjustment of the workforce to the average yield in order to increase the number of people fed by their agricultural work. The second is an agrarian reform for reviewing agricultural contracts and dealing comprehensively with the problem of the farms' size, which according to the authors was an indispensable condition for implementing the reclamation works and the construction of rural buildings and facilities needed to increase the agricultural production (Astengo and Bianco 1946; Astengo et al. 1947).

One of the issues through which Astengo and Bianco framed the relationship between agriculture and planning is the problem of the location of a new town and/or the creation of new industrial settlements. They pointed out that the surplus of agricultural production of a given area, capable of feeding the new population without altering the pre-existing agricultural situation and of guaranteeing a balanced relationship between production and consumption, represents the main criterion for determining the location of a new center (Astengo & Bianco, 1946). Two arc the assumptions that the two authors specify for describing the agricultural situation of a given region: on the one hand, all conditions being equal, it is always economically convenient to produce a food product of daily consumption as close as possible to the market place, as long as a condition of ubiquity characterizes the product marketed and consumed. On the other hand, any intervention in the agricultural field must be oriented to achieve the food self-sufficiency, i.e., the economic balance between imports and exports of agricultural products. This aspect, despite being framed by Astengo through issues of economic convenience linked to transport costs and the relationship between food demand and demographic trend, appears to anticipate a sensitivity to the sustainability of food systems that will appear in the scientific and policy debate only in the following decades (Egziabher et al. 1994; Mougeot 1999; Pothukuchi and Kaufman 1999).

The objective of the analysis carried out by Astengo and Bianco is the quantitative analysis of the nutritional balance of a given territory, that is to say the relationship between the agricultural production of a certain context and the average consumption of the population residing in that context. The method adopted is based on the calculation of the so-called *Nutritional Unit*, defined as the average elementary nutritional area needed and sufficient to feed an average person, which Astengo and Bianco defined FED as an homage to Feder who was the first to establish the problem of nutritional area of a city (Feder 1939; Astengo and Bianco, 1946). This unit can be calculated if the average consumption of resident inhabitants and the average productivity of the localities are known. The nutritional balance obtained by multiplying the FED by the number of inhabitants serves to determine the food relationship between cities and their agricultural hinterlands and it allows to trace the boundaries of a balanced nutritional area, i.e., an area with a balance between food production and consumption. Therefore, the analysis should guide regional planning to establish interventions for achieving a food balance between the main center, where there is a concentration of population and industrial activities, thus a negative nutritional balance, and the small centres and villages in the rural hinterland having a prevalent agricultural vocation and characterized by a positive nutritional balance.

According to Astengo and Bianco (1946), the representation of the agricultural situation of a territory through the use of FED and the nutritional balance not only serves to represent the agricultural efficiency of a region, but it should also guide the redefinition of the administrative units. Accordingly, these would be transformed to match with the precise relationships of economic interdependence that link together large urban centres to contiguous small and medium-sized towns or villages. Astengo and Bianco (1946: 25) stated that

> the administrative boundaries of a province must include, if possible, at least the territory sufficient to feed the inhabitants residing in the capital centre and in the smaller inhabited centres.

This would lead to specific advantages, such as the elimination of unnecessary transport and time costs, the use of infrastructures proportional to the influx of products and, above all, the creation of functional interdependences between cities and their rural hinterlands.

4 The Municipal Plan of Assisi

We have already mentioned that the methods developed by Astengo and Bianco for describing the agricultural situation of a region have been applied to the analytical framework of the regional plan of Piedmont. However, it is during the planning commission received from the municipality of Assisi in 1955 that Astengo had the chance to apply them to a municipal land-use plan. Assisi is a small town in Umbria, in Central Italy, a famous destination for religious tourism due to the large basilica which includes the remains of St. Francis, patron saint of Italy.

The experience of Assisi is significant in the post-war town planning in Italy for several reasons. First of all, because here we find one of the most consistent and successful attempts to carry on that process of institutionalization of the planning discipline (Urbanistica) to which many planners and designers active in those years were contributing (Di Biagi and Gabellini 1992). Astengo's work in Assisi is precise and rigorous: the local problems of a small town with a predominantly agricultural vocation are addressed by tackling one of the main challenges for town planning in the post-war reconstruction phase: the link between regulation and development (Palermo 2012). The fundamental innovation underlying the plan is "the creation of a strong idea of public project"—and therefore of development—"without which the rationale of the new plan would have seemed weaker" (Ibid., 28). Thus, by investigating this link Astengo carried on his attempt to consolidate and institutionalize the discipline.

A second reason lies in the pedagogical significance that the plan had for the future generations of Italian urban planners and urbanists. The extensive publication of the plan documents in the issues 24–25 of the "Urbanistica" journal in 1958 is based on the awareness that the plan could have served as a guide for the other land-use plans being produced in those years after the approval of the new national planning legislation in 1942. Astengo was well aware of the methodological opportunity related to the plan of Assisi, defined in the article presentation as a "remarkable comprehensive experience for consolidating a suitable methodology of planning investigation" (Astengo 1958: 11). Following this interpretation, the Assisi plan provided a valid "educational path" for integrating some local problems into an operational method able to test the planning expertise, identify the key-issues, the lexicon and concepts for addressing these issues, and articulate objectives, devices and procedures for guiding the planning process.

A third reason lies in the planning culture that forms the background of the plan, a culture that expresses links with the already mentioned municipalist tradition very much present in the geographical context Astengo was coming from, and the pragmatic-instrumental approach based on the already mentioned organic rationality (see Sect. 3.2). According to Lanzani (1996: 72), the plan of Assisi has at its basis "a radical political project to be implemented through a dirigiste reformism that from the distrust to the self-organizational abilities of society leads to a strong and pyramidal concept of planning". A basic mistrust and pessimism that Astengo explicitly manifests when he describes the social and economic situation in Assisi:

> In Assisi there are few cultural and economic animators, whose presence in a city or region has recently been considered as a condition for wellbeing [...] We should logically deduce serious doubts about the possibility that the city has to save or renew itself (Astengo 1958: 12).

The plan is therefore interpreted as an "intervention of a necessary superior control", as an exogenous and top-down measure aimed at conveying the social and economic development of the local context, otherwise destined to a progressive involution.

However, there is a fourth reason that justifies the exemplarity of the Plan of Assisi with respect to the Italian planning context of the post-war period. The plan can be considered anticipator of a sensitivity towards the issues of food self-sufficiency of cities and of urban/rural relations planning that will find only in the following decades a concrete and transversal development in the policy and academic debate. This aspect in Astengo's reflection is connected to the awareness that every planning intervention in a given territory must be based on the investigation of the relationship between population and economic resources. In Assisi, Astengo based the plan on the rigorous analysis of the local demographic and economic profile which led to highlight a situation of substantial stagnation: ageing and demographic decline, a local economy mainly oriented to agriculture, processes of economic impoverishment and environmental degradation of some agricultural contexts, deep territorial disparities between the plains and the mountains, were just some of the problems that the analysis revealed.

Within this framework, Astengo proposed a specific agricultural transformation program, applying the methodologies already developed alongside the regional plan of Piedmont (see Sect. 3.3). The analysis is based on the calculation of the elementary nutritional unit (FED), already defined as the area cultivated sufficient to feed an average person. The objective is to compare the total agricultural production with the local needs in order to calculate surpluses and shortages (Astengo 1958: 27). By calculating a FED value for each of the three geo-morphological contexts of the municipal territory (plains, hills and mountains), Astengo demonstrated that the local agricultural production not only amply satisfied the needs of the resident population—to which he also added an estimate of tourists and religious residents—but it also generated an active surplus of about 1000 units, exported in the form of wheat, oil, wine and beef and pork. Although probably the outcome of wide approximations and simplifications also due to the limited availability of reliable data on the local agricultural situation, Astengo's merit was that of having demonstrated the food self-sufficiency of the city on the basis of a method combining nutritional requirements, farm characteristics and soil productivity.

In each of the three territorial contexts, Astengo described the agricultural organization, the crop types, the size and structure of farms, and the pedological characteristics of land. He highlighted how in the mountain areas the agricultural economy was mainly intended for family subsistence, while in the plain there were more favorable soil conditions allowing families to form income and reinvest it through development projects. In the normative section of the Plan, these aspects result in a number of guidelines that follow the objective of achieving a greater balance between resources and income in all agricultural contexts of the municipal area. The most relevant guidelines are three.

The first is a "strong propulsive action of public or semi-public consortia or cooperatives for waking up the torpor of the less diligent and less skilled farmers and helping those laborious who, deprived of means and guidance, with only single forces could not undertake far-reaching transformation projects" (Astengo 1958: 82). Here the aforementioned top-down and dirigiste nature of Astengo's approach is evident as the main goal was to offer support to those who were lacking means, resources and will, that needed some help for improving their social and economic conditions. Astengo highlighted that this transformative action could have been promoted by a so-called "Municipal Agronomic Observatory" which, under the guidance of experts, could have produced a range of cognitive surveys aimed at detecting precisely the state and potential of the local agricultural economy (Astengo 1958).

A second action concerns the crop replacement, to be implemented throughout the municipal territory and especially in those areas occupied by less productive crops. This action was accompanied by the planning and construction of new works and facilities for regulating and capturing underground and flowing waters.

Finally, a third action had to do with the problem of disadvantage and marginality of mountainous farmland and concerned the increase in forestry and ancillary forage production in the most disadvantaged mountain contexts, "consolidating the steepest soils, increasing the safety of sub-lying soils, to stabilize the microclimate and favor the absorption of rainwater" (Ibid.: 82).

Ultimately, it is necessary to underline a final aspect which shows the innovative contribution of the Plan of Assisi. As Dolcetta also stated (2012: 71), "for the first time in the Assisi plan the rural space gets a specific, significant and complementary dignity compared to the urban context". The maps of the plan represent the rural space with a specific articulation, a complex spatial organization made up of thicknesses and stratifications, proximity and relationships. The choice to represent accurately the different crop types (Fig. 1) corresponds to the willingness to provide the urban analysis with an adequate cartographic repertoire, which is obtained by redesigning the information included in zenithal photos on cadastral maps. This repertoire has very few equals in contemporary planning experiences, where the rural contexts were often represented as white areas according to the dominant interpretation framing them as mere locations of new urbanization and not as spatial contexts endowed with a specific physiognomy (Della Rocca and Lapadula 1983; Erba et al. 2010). The accurate representation of the agricultural territory in continuity with the built fabric and the new expansions (Fig. 2) has the merit of overturning the vision that sees a separation between urban and rural spaces: in the municipal area farmland is perfectly integrated in a polycentric urban system where the existing historic centre, the small rural nuclei, the new residential and industrial developments, the new infrastructures and the agricultural, forestry and public open spaces form a continuous and organic system made of dense urban/rural interconnections.

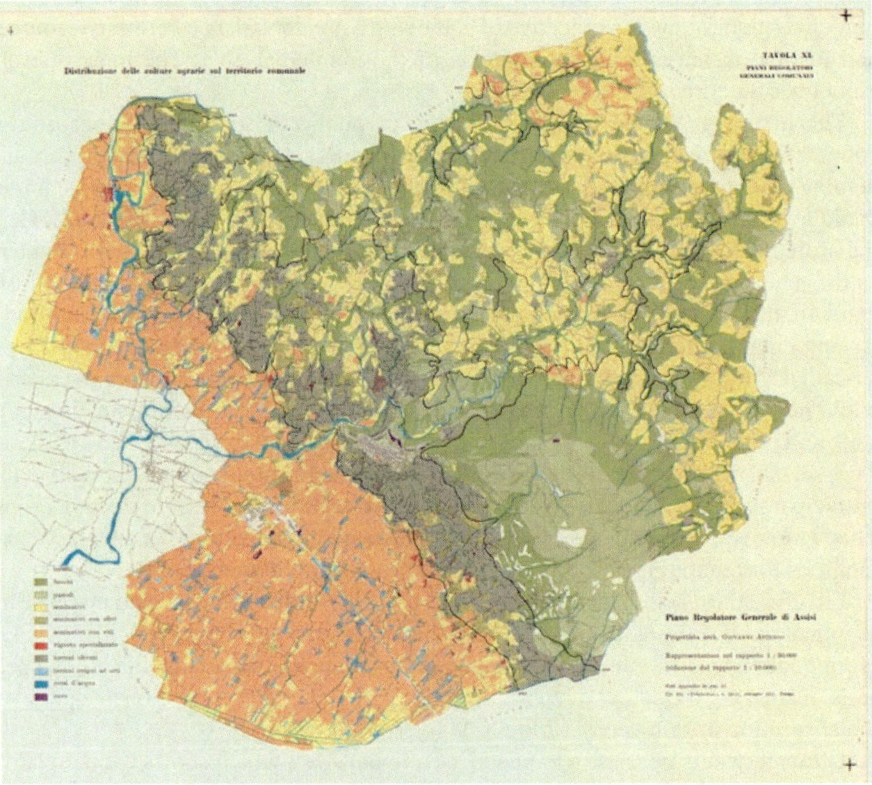

Fig. 1 Distribution of agricultural productions across the territory of Assisi, Municipal Land-use Plan of Assisi (1955). *Source* Università IUAV di Venezia, Archivio Progetti

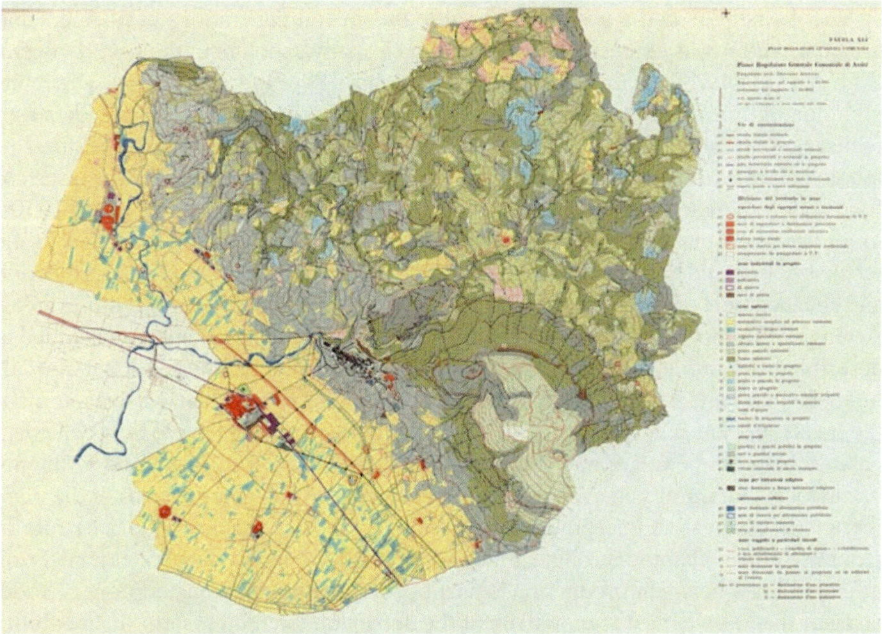

Fig. 2 Planning forecasts and division of municipal territory in zones, Municipal Land-use Plan of Assisi (1955). *Source* Università IUAV di Venezia, Archivio Progetti

5 Conclusion

In the mid-1950s Astengo developed a planning methodology which followed an original and autonomous perspective in the post-war planning debate in Italy. He took a series of models and concepts from the international discourse and translated them in the national disciplinary discussion on the institutionalization of the "Urbanistica" discipline and the consolidation of its corpus of methods, tools and approaches. Astengo believed that the conditions of institutional uncertainty and the speed of the urban growth processes (Astengo, 1956a) required a re-foundation of the discipline to endow it with scientific credibility and authority (Avarello 2010). In this re-foundation he interpreted as central the role of planners as they had to carry out the process of plan-making by taking into account "from the very beginning the most minute details of the various manifestations of the life in the municipality, unravelling the tangled skein of problems to find the common thread of a more rational development of the community, enlightening administrators and officials" (Astengo 1956b: II). A vision on the role of planning that we have already defined top-down and dirigiste but endowed with a transformative impulse, aimed at guiding communities to improve their living conditions, a pragmatic approach which must be practiced day after day, with commitment and responsibility.

The Assisi plan had a pedagogical value for several generations of planners and urbanists and it was extensively investigated by many scholars in the past (Dolcetta et al. 2015; Dolcetta 2012; Palermo 2012; Lanzani 1996; Indovina 1991). The interpretation outlined here is different from those adopted in many of these studies as it focuses on investigating the parts of the Plan related with the productive features of municipal territory and looking at the implications that the Plan has for the issues of food self-sufficiency and city/countryside functional interdependence. In the 1950s, in a context of planning practices, models and approaches mainly oriented to the reconstruction and expansion of cities with a scarce attention given to the protection and enhancement of the productive and landscape characteristics of agricultural areas, Astengo's attempt in Assisi is outstanding: he built an analytical apparatus aimed at describing the productive, occupational and social characteristics of local agricultural economy and, starting from these, he elaborated a comprehensive evaluation of the levels of food self-sufficiency of the town. The central objective is to achieve a balance between population, resources and territory in order to ensure the wellbeing of the local community in every territorial context of Assisi. At the basis of the plan, there is an innovative interpretation of urban/rural relationships where the role of planning lies in shaping stronger functional ties in the local food system, ensuring effective links between production, transport, marketing and consumption of food, an issue that, as we have seen, will enter the debate on the food system sustainability at city-region level only few decades later (Rodríguez-Pose 2008; Jonas 2012).

Finally, the contribution of Astengo is relevant for the contemporary planning debate for several reasons, two of which are briefly recalled here. The first reason lies in the fact that the work by Astengo shows a significant confidence in the ability of planning to develop a pragmatic impulse and work so that the impulse can steer tangible processes of urban and territorial transformation towards a reality alternative to the present one. Despite the manifold differences in terms of social, institutional and economic dynamics between the post-war period and the current age, now it emerges the same need to endow planning with the pragmatic and transformative impulse that mainstream policies and practices have progressively lost in recent decades. The reference goes to the capacity of planners to practice a form of knowledge that uses the intellectual and spiritual energies to produce scenarios and images of the future (Polak 1973).

A second reason relates to the Astengo's understanding of the functional economic spaces and relationships and their experimentation in regional and local planning processes. In the case of Assisi, Astengo successfully developed the notion of functional geography describing spaces and flows in the local food system of the municipal area and he applied it to a local planning policy. Still today the notion of functional space is a contested one and few are the attempts to develop its potentials in regional and urban planning (Bengs and Schmidt-Tomé 2006; Rodríguez-Pose 2008; Allmendinger and Haughton 2014). One reason lies in the fact that there are "no easy and uncontested methods to identify the relevant spatial boundaries of these functional spaces and the associated populations" (Atkinson and Pacchi 2020: 273). In this context, a rediscovery of Astengo's contribution on functional geographies in today's scientific and policy debate would be meaningful for identifying those reflections, concepts, and methodologies that could be employed for improving the understanding of current reality and, more importantly, for describing the social, economic and conceptual overlaps and convergences as well as the differences between the post-war period and our contemporary age.

References

Aa Vv (1934) Algemeen uitbreidingsplan van Amsterdam. Stadsdrukkerij, Amsterdam

Abercrombie P (1945) Greater London Plan 1944. A Report prepared on behalf of the Standing Conference on London Regional Planning by Professor Abercrombie at the request of the Minister of Town and Country Planning. HMSO, London.

Allmendinger P, Haughton G (2014) The evolution and trajectories of English spatial governance: Neoliberal episodes in planning. Plan Pract Res 28(1):6–26

Astengo G, Bianco M (1946) Agricoltura e urbanistica. Viglongo, Torino.

Astengo G, Bianco M, Renacco N, Rizzotti A (1947) Il Piano Regionale Piemontese. Metron 14:3–76

Astengo G (1954) Falchera. Metron 53–54.

Astengo G (1956a) Presentazione. Urbanistica 18–19:3

Astengo G (1956b) Piani comunali in formazione. Urbanistica 20: II-III.

Astengo G (1958) Il Piano Regolatore Generale di Assisi. Urbanistica 24–25:10–132

Astengo G (1991) Assisi: un'esperienza. In: Indovina F (ed) Le ragioni del piano. Giovanni Astengo e l'urbanistica italiana. Franco Angeli, Milan.

Astengo G (1966) Urbanistica. Enciclopedia Universale dell'Arte, XIX. Istituto per la collaborazione culturale, Venice-Rome.

Astengo G (1968) Urbanista sotto accusa a Gubbio. Arti Grafiche Rosada, Torino

Atkinson R, Pacchi C (2020) In search of Territorial Cohesion: An elusive and imagines notion. Social Inclusion 8(4):265–276

Avarello P (2010) "Ogni piano è un tormento". In: Fantin M, Fregolent L (eds) Astengo 1. Editoriali di urbanistica dal 1949 al 1976. INU Edizioni, Rome.

Bengs C, Schmidt-Thomé K (2006) Urban-rural relations in Europe. ESPON 1.1.2. Final Report. Centre for Urban and Regional Studies, Helsinki University of Technology.

Cloke P (ed) (1989) Rural Land-Use Planning in Developed Nations. Routledge, New York

Della Rocca A, Lapadula F (1983) Rapporti tra agricoltura e urbanistica nello spazio periurbano. Cedam, Padova.

Di Biagi P (2002) Una voce enciclopedica tra scienza e utopia. In: Di Biagi P (ed) I classici dell'urbanistica moderna. Donzelli, Roma, p 269–292.

Di Biagi P (ed) (2010), La grande ricostruzione. Il piano INA-Casa e l'Italia degli anni cinquanta. Donzelli, Rome.

Di Biagi P, Gabellini P (eds) (1992), Urbanisti italiani. Piccinato, Marconi, Samonà, Quaroni, De Carlo, Astengo, Campos Venuti, Laterza, Rome-Bari.

Di Robilant M (2019) Falchera public housing neighborhood on 'Urbanistica' 7, 1951 and 'Metron' 53–54, 1954: narration and iconographic presentation. Atti e Rassegna Tecnica della Società degli Ingegneri e degli Architetti di Torino, year 152, LXXIII n. 1, april 2019.

Dolcetta B (2012) Una definizione di paesaggio. In: De Luca G, Sbetti F (eds) Le eredità di Astengo. INU Edizioni, Rome, pp 65–71

Dolcetta B, Marin S, Maguolo M (eds) (2015) Giovanni Astengo urbanista. Piani, progetti, opere. Il Poligrafo, Padova.

Egziabher AG, Lee-Smith D, Maxwell DG, Memon PA, Mougeot L, Sawio C (1994) Cities feeding people: an examination of urban agriculture in East Africa. IDRC, Ottawa

Erba V, Agostini S, Di Marino M (2010) Guida alla pianificazione territoriale sostenibile. Maggioli, Rimini.
Fantin M, Fregolent L (eds) (2010) Astengo 1. Editoriali di urbanistica dal 1949 al 1976. INU Edizioni, Rome.
Feder G (1939) Die neue Stad. Springer Verlag, Berlin
Ferraro G (1998) Rieducare alla speranza. Patrick Geddes planner in India. 1914–1924. Jaca Book, Milan.
Hall P, Ward C (1998) Sociable Cities. The legacy of Ebenezer Howard. Wiley, Chichester.
Hilberseimer L (1944) The New City – Principles of Planning. Paul Theobald, Chicago
Hilberseimer L (1949) The new regional pattern: industries and gardens, workshops and farms. Paul Theobald, Chicago
Howard E (1902) Garden Cities of Tomorrow. Sonnenschein & Co., London
Jonas A (2012) City-regionalism: Questions of distribution and politics. Prog Hum Geogr 36(6):822–829
Gabellini P (2001) Tecniche urbanistiche. Carocci, Rome
Gaeta L, Janin Rivolin U, Mazza L (2013) Governo del territorio e pianificazione spaziale. Città Studi Edizioni, Milano.
Geddes P (1915) Cities in evolution. Williams & Norgate, London
Giudice M (2019) Luci e ombre della pianificazione regionale. Narrazioni e riflessioni di alcune esperienze, INU Edizioni, Roma.
Gloeden E (1923) Inflation der Grosstädte. Der Zirkel Architecktur Verlag, Berlin
Lanzani A (1996) Immagini del territorio e idee di piano 1943–1963. Franco Angeli, Milan
Lazzarini L (2018) The role of planning in shaping better urban/rural relationships in Bristol City Region. Land Use Policy 71:311–319
Lazzarini L (2020) Alcune considerazioni sull'utilità del camminare nell'insegnamento dell'urbanistica. In: Lazzarini L, Marchionni S (eds) Spazi e corpi in movimento. Fare urbanistica in cammino. SdT Edizioni, Firenze, p 29–40.
Lemes de Oliveira F (2014) Green wedges: origins and development in Britain. Planning Perspectives 29(3).
Lemes de Oliveira F (2015) Abercrombie's green-wedge vision for London: the County of London Plan 1943 and the Greater London Plan 1944. The Town Planning Review 86(5):495–518
Le Corbusier (1957) La Charte d'Athènes. Éditions de Minuit, Paris
Le Corbusier (1945) Les Trois établissements humains. Éditions Denoël, Collection ASCORAL, Paris.
Moroni S (1997) Etica e territorio. Prospettive di filosofia politica per la pianificazione territoriale. FrancoAngeli, Milan.
Mougeot L (1999) Urban agriculture: definition, presence, potentials and risks. Paper presented at the Growing Cities Growing Food Congress, Havana, 1999.
Mumford L (1961) The City in History. Its origins, its transformations and its prospects. Harcourt, London.
Orioli V (2012) "Verso la campagna. Alcune considerazioni sul ruolo del territorio rurale negli strumenti urbanistici". In: Agnoletto G, Guerzoni M (eds) La campagna necessaria. Un'agenda di intervento dopo l'esplosione urbana, Quodlibet, Macerata, p.51–61.
Palermo PC (2012) Attualità e inattualità di Giovanni Astengo per l'urbanistica italiana. In: De Luca G, Sbetti F (eds) Le eredità di Astengo. INU Edizioni, Rome, pp 27–33
Patrizi L (1983) Normative in zona agricola. In: Della Rocca A, Lapadula F (1983) Rapporti tra agricoltura e urbanistica nello spazio periurbano. Cedam, Padova, p 153–190.
Polak F (1973) The image of the future. Elsevier, London
Pothukuchi K, Kaufman J (1999) Placing the Food System on the Urban Agenda: the role of municipal institutions in the Food Systems Planning. Agriculture and Human Values 16.
Rodríguez-Pose A (2008) The rise of the 'city-region' concept and its development policy implications. Eur Plan Stud 16:1025–1046

Rossi Doria M (1953) "I rilevamenti agronomici ed economico-agrari nella preparazione dei piani di bonifica e dei piani regionali". In: INU, La pianificazione regionale, Rome.
Salzano E (1998) Fondamenti di urbanistica. Laterza, Bari.
Sharp T (1932) Town and Countryside: Some Aspects of Urban and Rural Development. Oxford University Press, Oxford
Sturzaker J, Mell I (2017) Green Belts: Past; present; future? Taylor and Francis, London
Viganò P (2010) I territori dell'urbanistica. Il progetto come produttore di conoscenza. Officina Edizioni Rome.
Zeier Pilat S (2014) Reconstructing Italy: the Ina-Casa neighborhoods of the postwar era. Ashgate, Farnham

Étienne de Groër: Planning the Lisbon Garden City Region

Teresa Marat-Mendes

Abstract This chapter examines the contribution of Étienne de Groër in the planning of a garden city region for Lisbon. It reveals an outstanding example of Ebenezer Howard's concept of a Garden City applied in Lisbon and its surroundings, set out between 1936 and 1950, while envisioning a productive and socio-economic inclusive territory. At a time when urban planning urges for solutions to tackle food security, social cohesion, and sustainability threats, revisiting Groër planning approach emerges as an important contribution to support future planning practices to reimagine more productive landscapes.

Keywords Garden city · Productive territory · Rural areas · Sustainability · Urban planning

1 Introduction

The history of urban planning in Portugal throughout the twentieth century remains understudied. Works by Silva (1994), Lobo (1995) and Pereira (1994, 1999) do remain influential references, notwithstanding new perspectives and research gaps promoted by recent investigation. Access to previously inaccessible planning materials, so far undetected by the scientific community, and the usage of urban planning history as an operative tool to discuss contemporary issues such as sustainability, food security and urban resilience, constituted two main contributions for the above identified research advances.

Access to planning documentation produced by Etienne de Groër (1882–1974) (Galvão 2020; Camarinhas 2009), until recently inaccessible, allowed for a more comprehensive perspective on twentieth-century urban planning history in Lisbon. Étienne de Groër was a Polish-Russian architect-urbanist who worked in Portugal for nearly two decades, between 1933 and 1951. Marat-Mendes' (2009) initial research on Urban Planning in Lisbon focused on the work produced by Groër for Lisbon and

T. Marat-Mendes (✉)
Iscte – Instituto Universitário de Lisboa, DINÂMIA′CET, Lisbon, Portugal
e-mail: teresa.marat-mendes@iscte-iul.pt

C. Brisotto and F. Lemes de Oliveira (eds.), *Re-Imagining Resilient Productive Landscapes*, Cities and Nature, https://doi.org/10.1007/978-3-030-90445-6_7

its nearby surroundings, on Costa do Sol. Research material was found in the archives of the Direcção Geral do Território (DGOTDU) and the Municipality of Lisbon. Yet, these proved to be insufficient as they did not cover the overall planning activity developed by Groër for the Lisbon region. Therefore, considerably more documentation from Groër may not have yet been identified or catalogued, thus, inhibiting a proper analysis of this architect-urbanist, who has yet to receive full recognition and acknowledgment within twentieth-century Portuguese architecture and urban history. Nevertheless, the work produced by Groër was the subject of analysis through research dedicated to: (i) the architects with whom he collaborated such as Faria da Costa (d'Almeida 2015) or Marques da Silva (Marques 2015, 2017, 2018; Silva 2018); (ii) the specific plans developed by Groër (Marat-Mendes 2009; Galvão 2020, Marat-Mendes and Sampayo 2015), (iii) the study of twentieth-century Portuguese urban planning (DGOTDU 2005; Lobo 1995; Pereira 2009; Almeida 2009; Camarinhas 2009; Camarinhas and Brito 2007; Marat-Mendes and Oliveira 2013); (iv) the analysis of Portuguese governmental archives dedicated to land-use and urban planning issues (Dinis 2015); or (v) the study of the IUUP (Institut d'Urbanisme de l'Université de Paris) and their former personnel (Carrasco 2018), adding valuable sources of information about Groër and his work. Namely, Marat-Mendes (2009) analysis of the PUCS, an example of Ebenezer Howard's (1850–1928) Garden City vision, placed particular attention on the role of rural areas, forests, parks and open areas, for the establishment of a continuous greenbelt for Costa do Sol, in its 'Hinterland', guaranteeing the ecological, social and economic productivity of the west region of Lisbon (Marat-Mendes 2009).

Recent research conducted on the Lisbon region examined the existing urban–rural dichotomy within Portuguese urban planning. The optimization of current mixed land-use soils was identified as a potential strategy to counteract such urban–rural division and promote a sustainable transition of the current food system with the Lisbon region (Marat-Mendes 2021a). Examining such a dichotomy from a historic perspective confirmed its implications on the planning of urban and rural areas over time (Marat-Mendes 2020). While rural spaces have consistently been excluded from design, urban areas have received greater attention on that front, mostly by public authorities. In addition, at present, in what regards regulations, rural areas are mostly managed by national and regional authorities, while urban areas by municipal authorities. Nevertheless, rural areas are also part of the municipal domains, which leaves them absent from a communal socio-economic strategy (Marat-Mendes et al. 2020). Furthermore, as observed by Marat-Mendes (2021b), current food-related activities and initiatives operating in the Lisbon Metropolitan area, for which urban, rural and agricultural areas are all important, are generating the creation of new jobs, healthy food markets and land productivity, on a microscale, therefore developing more sustainable socio-ecological systems.

The work provided in this chapter is based on previous research developed by its author (Marat-Mendes and Oliveira 2013; Marat-Mendes 2009; Marat-Mendes and Sampayo 2010, 2015; d'Almeida and Marat-Mendes 2019; Marat-Mendes et al. 2020, 2021a, b) but also on materials recently made available at the municipal archives in the Lisbon Metropolitan Area. The analysis discloses how the Garden

City vision was conceptually professed and materialized in a regional plan for Lisbon and its surroundings by Étienne de Groër. To do so, after this introduction, the chapter moves to the work developed by Groër during his professional life in France and Portugal, where he collaborated and/or coordinated major urban development projects and city extensions. But most importantly, the chapter analyzes the urban plans produced by Groër for the Lisbon Garden City Region, which were approved by the respective municipalities, including the *Plano de Urbanização da Costa do Sol* (PUCS), *Plano de Urbanização do Concelho de Almada* (PUCA), *Plano Director de Urbanização de Lisboa*[1] (PDUL), the *Plano de Urbanização de Sintra* (PUS) and the *Plano Regional Moscavide-Vila Franca de Xira,*[2] while depicting Groër's vision of the Garden City, as well as his planning, design and technical solutions to increase landscape productivity and promote firmer urban–rural links. Groër's adopted structures in his urban plans and his idea of rural and urban agriculture within the overall region are also here examined.

2 Planning the Region Through the Garden City Theory

The establishment of urban planning as a scientific disciple in Portugal, occurred during the first half of the twentieth century, notably through knowledge learned from English and French influences (Marat-Mendes et al. 2014). Adopting the Garden City as an international or universal paradigm occurred before World War I. It is rooted in the foundations of the International Federation for Housing and Planning (IFHP), established in England, and led by Ebenezer Howard between 1913 and 1928. Promoting the work initiated by Howard at the British Garden Cities and Town Planning Association (GCTPA), the IFHP was created and supported by international Garden City followers, representing their own national groups, including for example America, Belgium, England, France, Germany, Netherlands, Poland, Spain and Russia (Geertse 2012).

Although British members were encouraged to administrate this association from its very beginning, through Ebenezer Howard's leadership, the idea of the importance of such an institution diverged among the international members. In particular, the French members, followers of the *Musée Social* and the IUUP, valued the didactic responsibility of such an international association to support European municipalities to create their own garden cities. Yet, the British would favor the contribution of such an association to increase their influence internationally and protect associations devoted to the garden city idea (Geertse 2012). However, the unstable political atmosphere between the two World Wars, prompted vicissitudes in this association, exposing opposite organizational issues but also contrasting points of view regarding

[1] The PDUL was approved by the Municipality of Lisbon but was never officially approved by the Portuguese government.

[2] This specific urban plan has not yet been totally identified, including all its written and graphical materials. Therefore, our analysis is mostly focused on the PUCS, PUCA PDUL and PUS plans.

the idea of the garden city to be adopted. Nevertheless, both British and French members were keen supporters of the garden city and of its regional approach.

The French conception of town planning acknowledged the potential of garden suburbs as part of a modern town planning, as a reformative socio-economic and planning solution to solve contemporary problems (Geertse 2012). Grounded on cooperativity fundamentals, housing would receive greater attention and encourage the promotion of several suburban garden city habitats (Guerrand 2002). Led by IUUP members who supported the Garden City internationalization such as Georges Benoit-Lèvy (1880–1971) and Marcel Poëte (1866–1950), French town planning aimed for an operative scientific urban approach, grounded on the development of a new discipline, *Urbanisme*, pursuing the notion of Civics as postulated by Scottish biologist Patrick Geddes (1854–1932), as well as his pedagogical and methodological approach: survey, diagnosis and plan, based on direct observation and data gathering (Bardet 1947: 50, Poëte 1929). Regional urban analyses were then promoted as important operative instruments and published in French journals, including *La vie Urbaine* and *Urbanisme* (see for example Groër 1921, 1933c, 1934). Marat-Mendes and Oliveira (2013), Marat-Mendes and Sampayo (2015), André et al. (2012), Lebre (2017) and d'Almeida et al. (2020) presented how Portuguese architects came to think about the twentieth-century city, through key urban international influences, as was the case in the Musée Social and the IUUP, which were also very influential on Groër's work.

The British concept of the Garden City was also strongly related to a reformist idea, as expressed in the book 'To-morrow. A peaceful path to real reform' by Ebenezer Howard (1898). Its second edition in 1902, under a new title 'Garden Cities of Tomorrow' added extra attention to the important relationship between the urban and the rural. Promptly, England housed the construction of the first garden city, Letchworth Garden City (1903–1909), implying a specific aesthetic. However, Howard's Garden City vision was not a project solely commanded by spatial motives, but also by a strong political cause. Provisions from local associations were necessary to engage residents, planners and local authorities. However, when in 1932 the Town and Country Planning Act gave local authorities the power to 'zone', there were inadequate local administrative structures to manage such process (Meller 1997). Consequently, Howard's political vision lost its potential position. Meanwhile, between 1920 and 1930, modern architecture clearly defined its aims, giving form to modernity. Consequently, from the very beginning the Unwin and Parker style, associated with the British Garden City vision would be refused by the modern architects (Meller 1997).

In 1937, the IFTP 15th congress was hosted by the International Exhibition held in Paris, together with the 5th Congress of International Modern Architects (CIAM), and both dedicated their attention to regional planning. The Portuguese Pavilion at the Paris Exhibition was designed by the young Portuguese architect Francisco Kei do Amaral (1910–1975), after winning the competition for its design. In 1938 Amaral was contracted by the Municipality of Lisbon to work on several urban projects for Lisbon, mainly for the design of new green areas and urban parks (*Parque de*

Monsanto, Parque Eduardo VII and *Alameda do Campo Grande*).[3] Remarkably, Amaral's time working for the Municipality of Lisbon would coincide with the period when Étienne de Groër, a former Professor at IUUP, would assist the Lisbon municipality, first in the study of its urban development and expansion and later in the elaboration of the *Plano de Urbanização da Cidade de Lisboa* and its respective *Plano Director Municipal*.

The Paris International Exhibition was an opportunity to publicly expose the *Plan d'aménagement de la région parisienne* (PARP) concluded in 1934 by Henri Prost (1874 –1959), a prominent French planner teaching at the IUUP. Together with Léon Jaussely (1875–1932) and Alfred Agache (1875–1969), Prost promoted the internationalization of the French *Urbanisme*, alongside Georges-Benoît Lévy (1880–1971), Henri Sellier (1883–1943) and Auguste Bruggeman (1868–1943). In 1937, Bruggeman invited Groër to substitute him at the IUUP and teach the English Garden Cities course.

The Greater Paris plan followed the Garden City paradigm, promoting, however, the densification of the suburbs and the promotion of new motorways in the outskirt landscape of Paris. In France, projects of satellite towns would be abandoned due to land expropriation difficulties and financial uncertainty associated with the war context. British garden city supporters, then directed by Raymond Unwin (1863–1940) and George L. Pepler (1882–1959) at the IFHTP, however, favored the development of satellite cities, green belts, and an industrial decentralization, as already tested in Letchworth and Welwyn Garden City.

Meanwhile, Portuguese urban planning activity was being developed through the specialization of Portuguese architects in international Schools of Urbanism, particularly at the IUUP, or through the collaboration of foreign architect-urbanists professionals contracted by the Portuguese government to work in the main Portuguese municipalities of Lisbon, and Porto (Marat-Mendes and Oliveira 2013); namely, by contracting Jean-Claude Nicolas Forestier (1861–1930), Alfred-Donat Agache and Étienne de Groër, who worked for the Municipality of Lisbon, who followed the garden city paradigm and the Musée Social and Patrick Geddes approach; or by the British, Richard Barry Parker (1867–1947) who participated in the Letchworth Garden City plan and assisted the Municipality of Porto in the creation of garden suburbs for the city.

Recent access to information retrieved from the IFHTP has revealed how the Garden Cities theory was adapted and fashioned around the world throughout different cultural and geographic contexts (Geerste 2012, 2016; Graham 2011; Wagner 2016). Several countries were identified as the home for specific national Garden Cities associations, including England, France, Belgium, Germany and Russia. Thus far, in the case of Portugal, there is no available information on the existence of such institutions. Nevertheless, Portugal witnessed the presence of planners advocating for the Garden City vision, as was the case of Groër. The impact of such situation is not yet fully understood, including the conceptual and methodological

[3] Amaral worked for the Municipality of Lisbon until 1949.

approach that was adopted, but also the scale of its adaptation to the Portuguese geographic and political context.

Studying the development of the Garden City in Portugal is therefore an urgent task. The best way to do it is by investigating the work of those who have implemented it, as is the case of Étienne de Groër here under analysis.

3 Étienne de Groër

Étienne de Groër was born in 1882, in Warsaw, Poland. He was the son of a Polish catholic doctor, Franciszek von Groër (1848–1910), and a Russian orthodox mother, Maria Grigorievna von Groër (1851–1941).

In 1897, he enrolled in high school in Saint-Peterburg, Russia. In 1917, he obtained his diploma in Architecture at the Imperial Academy of Fine Arts. In 1918, he was contracted as an architect for the Municipal Office of Saint-Peterburg (Galvão 2020: 165). In February of 1920, Groër emigrated to France, where he obtained his French nationality in 1936. Initially he settled in Péronne, together with his wife Gabrielle de Groër (1891–1983) and their three children, Leon de Groër (1914–2014), Nikita de Groër (1915–2001) and Ariadna de Groër (1920–1928). The family later moved to Paris, where they lived between 1920 and 1938. It was in Nice, on May 30th, 1974 that Groër would finally rest after his death.

In France, Groër developed an intense activity both in academic and planning practice. Academically, Groër contributed with several scientific manuscripts, publishing at least in five French journals, *La Vie Urbaine, Le Maître d'Oeuvre, Urbanisme, Travaux* and *Le Monde Souterrain* (Groër 1921, 1921a, 1931a, b, c, d, 1933a, b, 1934, 1936a, b, 1937a, b, 1938) and participated in international conferences discussing the importance of urbanism, as at the 'Les Cités du monde', conference which took place in 1930 in Montereall (Rioux 2013: 292, 326). In 1925 Groër joined the Société Française des Urbanistes and in 1930 he become a member of this association (Anon 1930). During his first years in France, Groër met the French Planner Donat-Alfred Agache (1875–1959), working at his office on several urban plans (Galvão 2020).

In 1937, he was integrated as a Professor in the IUUP, teaching the first and second-year courses '*Principes de la Cité-jardin et leur application en Angleterre'*. Alongside William Oualid (1880–1942), an Israelite lawyer, who taught the 'Economic organization of cities' course, Groër substituted Professor Bruggeman in the Garden City course. At the IUUP, Groër supervised a final thesis in architecture, titled 'Le Problemme de La Ville Moderns: la Cite-Jardin' elaborated by the architect Jeanne Andrée Auguste Boulfroy in 1939 (Boulfroy 1939). Both Groër and Oualid were contracted by the IUUP between 29 June of 1937 and 1940, under the French Jewish Statute of October 3, 1940 (Carrasco 2018). Groër's academic network included professors from the IUUP and contributors to preeminent French journals and planning institutions, such as Gaston Bardet at *Le Monde Souterrain* or Bruggeman who was the director of the *l'Union des Villes et Communes de France* and the IUUP. At the IUUP Groër had the chance to meet notable architectures and planning masters, including Hendrik Petrus (1856–1934, Daniel Burnham (1846–1912), Tony Garnier (1869–1948), Ebenezer Howard (1850–1928), Hermann Josef Stübben (1845–1936) and Raymond Unwin (186–1940) (Marat-Mendes and Sampayo, 2015).

In France, Groër practiced as an urbanist from the very beginning of his arrival. His first job, coordinated by Henri Moreau, involved the planning for the reconstruction of Perrone, a city located in the north of France, which had been devastated during World War I (Galvão 2020: 35). In 1922 he submitted the *Plan general pour la reorganization et l'angrandsissement de la Ville de Belgrade* (Camarinhas 2009: 679). In 1924 he collaborated with Agache in the *Plan d'ámenagement for Bezons, Val-d'Oise,* and between 1928 and 1930 in the *Plano de Extensão- Remodelação-Embellezamento do Rio de Janeiro,* in Brazil.

Meanwhile, between 1933 and 1936, Agache was invited by Engineer Duarte Pacheco (1900–1943), the Portuguese Ministry of Public Works and Communications, to coordinate and develop a study for the west region of Lisbon, to promote the touristic coastline area and develop a new marginal road. However, Agache's study integrated a much vaster territory, including Lisbon and its surroundings, on a regional scale. Additionally, his study integrated urban design proposals for several settlements and for the first Portuguese motorway along Lisbon's west coastline (Agache 1936; Marat-Mendes 2009; André et al. 2012; Lobo 1995). Between 1933 and 1936 Groër was called by Agache to collaborate in this plan. In Agache's office, Groër would meet João Guilherme Faria da Costa (1906–1971), a young Portuguese architect, from Lisbon, who was also working for the Agache Lisbon plan.[4] The plan was submitted in 1936, yet it did not receive the approval from the Portuguese central government.[5]

In 1938, Agache was again invited by the re-elected Portuguese Ministry of Public Works to continue his work for the west region of Lisbon. However, due to World War II, he refused the invitation and suggested Groër, his former collaborator, to substitute him in the task. This was in the same year that Groër had the opportunity to visit Welwyn Garden City and Letchworth in England (Galvão 2020), denoting his interest to support the work that he was lecturing and supervising at the IUUP about

[4] In October 1933, Faria da Costa applied to the Portuguese *Junta de Educação Nacional*, for a grant to study at the IUUP, to support his specialized studies in Gardens. In January 1934, the Portuguese Ministry for Education awarded Faria da Costa with a three-month grant. In Paris, Faria da Costa also collaborated in Agache's studio between Abril and June 1934, as reported in his report delivered to the *Junta de Educação Nacional* in 1934, available in the archives of the Instututo da Alta Cultura under the reference 1315/23 João Guilhereme Faria da Costa IAC 1933 10 13 1934 06 14. Additional grants by the IUUP were conceded to Faria da Costa to complement his studies in Paris. At the IUUP, Faria da Costa completed a specialization in *'Jardins et l'art d'aménanger less villes'* for two years, concluding his studies in 1937. His thesis *'Le Plan d'aménagement, d'embellissement et d'extension de la ville de Figuiera da Foz et sa région, Portugal'* was supervised by Georges Sébille.

[5] In 1936, Duarte Pacheco was fired by the Portuguese Government and the work produced by Agache was consequently placed apart. Yet, on January 1st, 1938, Pacheco would be elected as the president of the Lisbon Municipality and on the 25th of May, of that same year, he was invited to work as the Ministry for Public works of the Portuguese Government.

Garden Cities, exposing his deep interest into this exclusive topic, both conceptually and also in practice.

During the twenty years that Groër worked in France, he participated in at least 16 plans for cities (see Table 1). Nevertheless, Groër's planning practice was rather irregular, subjecting him and his family to regular unstable financial situations. With the evolution of World War II, in 1940 when France was invaded by Nazi Germany, Groër and his family applied for a Visa to the Portuguese Consul in Bordeaux. It was signed by Aristides de Sousa Mendes, on the 14th of June, just a few days before his destitution by the Portuguese dictator, António de Oliveira Salazar, as he had defied

Table 1 Written work published by Étienne de Groër

Date	Document title	References
1921	l'Urbanisme en Russie	Groër (1921)
1929	La Circulation et les transports à Rio de Janeiro	Groër (1929)
1932	Le Gratte- Ciel est-il Nécessaire?	Groër (1932a)
1932	Urbanisme au Canada	Groër (1932b)
1932	Urbanisme au Canada	Groër (1932c)
1932	L'Habitation haute en Amérique	Groër (1932d)
1933	Urbanisme au Canada	Groër (1933a)
1933	Le Zoning et la Législation Urbaine de la Région Parisienne	Groër (1933b)
1933	L'Habitation haute en Amérique	Groër (1933c)
1934	Le Urbanisme au Canada	Groër (1934)
1936	Porquoi et comment un plan d'amenagement doit se faire	Groër (1936a)
1936	Porquoi et comment un plan d'amenagement doit se faire	Groër (1936b)
1937	Porquoi et comment un plan d'amenagement doit se faire	Groër (1937a)
1937	L'Urbanisme et la Défense Anti-aérienne	Groër (1937b)
1937	Les Leçons du Congrès International de l'Habitation et de l'Amenagement des Villes de Vienne	Galvão (2020: 207)
1938	Quelquees Mots sur les Ilots Insalubres de Paris	Identified in Galvão (2020: 207)
1938	Nécessité d'un zoning dans Paris	Groër (1938)
1945	Le tracé d'un Plan d'Urbanisation	Groër (1945)
s.d	L'Urbanisme au Brésil	Identified in Galvão (2020: 206)
1948	Anteprojecto de Urbanização e Embelezamento e de Extensão da Ci-dade de Coimbra	Groër (1948b)

express orders to issue such visas.[6] Consequently, Groër and his family become part of the list of Sousa Mendes Jews, who were able to enter Portugal and were saved from the Nazi concentration camps and persecution.

In 1940 Groër and his family moved to Portugal, where they lived for 11 years. First, at Santo Amaro, in the municipality of Oeiras,[7] and in 1947 in Lisbon.[8] In Portugal, he worked by invitation of the Portuguese minister of Public Works, who also accumulated the presidency position in the Lisbon Municipality, where Groër served as a technical adviser to support the development of the *Plano de Urbanização da Cidade de Lisboa*, while accumulating the study for the Lisbon west region—Costa do Sol.[9]

In Portugal, Groër maintained his contact with Faria da Costa,[10] and would collaborate in several urban plans, including the *Plano de Urbanização da Costa do Sol*, but also in the *Plano de Urbanização da Cidade de Lisboa*, as well as for other Portuguese cities (see Table 1).[11] After concluding his studies in Paris, Faria da Costa returned to Lisbon as the first Portuguese Urbanist and was integrated in the Municipality of Lisbon in 1938, in the *Direcção Geral dos Serviços de Obras Públicas* (d'Almeida 2015).

The following tables compile the written work produced and published by Groër as well as his urban plans, indicating those produced in France and Portugal, as well as who collaborated with.

From Table 2 it is possible to verify that in Portugal, Groër worked in 16 urban plans, including 15 in Portugal mainland (Abrantes, Almada, Aveiro, Beja, Braga, Cascais, Chamusca, Coimbra, Évora, Figueira da Foz, Lisbon, Oeiras, Pombal, Sintra, Vila Franca de Xira) and one in Luanda, the capital of Angola, a former Portuguese colony (Marat-Mendes and Sampayo 2015). Groër planning proposals included several scales of approach, including the regional, municipal but also the local one, including the building scale.[12]

---

[6] Thus, despite having received such orders, Sousa Mendes granted visas to Portugal to several refugees, including Jews, fleeing from Germany, Austria, France, and other countries already occupied by German armies, but also other individuals with Portuguese, British, American citizenship, who were trying to return to their homelands.

[7] Information available in a postcard written by Gabrielle de Groër to architect Moreira da Silva in 1943, archived in the Fundação Instituto Marques da Silva with the following identification FIMS/MSMS/42370252 and FIMS/MSMS/42370253.

[8] Information stated in Groër's contract with the Municipality of Lisbon for the development of the *Plano Director de Urbanização de Lisboa.*

[9] The west region of Lisbon was designated by Costa do Sol in 1935 by Law 1 909 of 22 May.

[10] According to Camarinhas, in 1937 Faria da Costa wrote a letter to Groër inquiring about the possibility to come and work in Portugal.

[11] According to Law 24 802 from 1934, many municipalities were obliged to develop General Urban Plans. Thus, settlements with 25 000 inhabitants or which expected to grow into that number within a ten-year period, were forced to prepare their urban plans to modernize existing infrastructures and assign the extension areas where urbanization would be allowed.

[12] That was the case of the urban remodulation proposal for downtown Lisbon, Baixa.

Table 2 Urban Plans elaborated by Groër

Date	Plan/Study	Architect-Urbanist	Country
1928–1930	Plano de Extensão-Remodelação-Embellezamento do Rio de Janeiro	Aldred Agache Étienne de Groer W. Palachon	Brazil
s.d	Plan d'ámenagement et d'extension de la regions Creilloise (***)	Étienne de Groër	France
1922	Plan General pour da reorganization et l'agrandsissement de la Ville de Belgrade (2nd prize for the competition)	Étienne de Groër	France
1924	Plan d'ámenagement for Bezons, Val-d'Oise	Étienne de Groër Alfred D. Agache	France
1924	Plan d'ámenagement et d'embellissement et l'extension de Courneuve	Étienne de Groër	France
1925	Plan de Poitier	Étienne de Groër	France
1927–1928	Project for the Societé des bains de Mer Monte-Carlo	Étienne de Groër Charles Letrosne	France
s.d	Aménagement des quartiers archéologiques; assainissement des parties insalu-bres' for Orléans/Loiret	Étienne de Groër	France
s.d	Plan d'ámenagement et d'embellissement et l'extension de la Ville' for Tours, In-dre-et-Loire	Étienne de Groër	France
1931	Plan d'ámenagement et d'embellissement et l'extension de la commune de Penmarc'h	Étienne de Groër	France
1931	Plan d'aménagement, d'extension et d'embellissement de Tours, Indre-et-Loire	Étienne de Groër	France
1931	Démarches pour les 'Plans d'ámenagement for Brunoy/Seine-et-Oise	Étienne de Groër	France
1932	Plan d'ámenagement et d'extension for Douarnerez, Finistère	Étienne de Groër	France
1935	Plan d'ámenagement et d'extension for Deuil, Seine-et-Oise (***)	Alfred D. Agache and Étienne de Groër	France
1933–1936	Lisbonne: Urbanisation de la Region Ouest (Plano da Costa do Sol)	Alfred D. Agache Étienne de Groër Faria da Costa	Portugal
1938	Plan Régional d'urbanisme de la Côte Basque	Étienne de Groër	France
1940	Plano de Urbanização de Coimbra	Étienne de Groër Moreira da Silva	Portugal
1942	Plano de Urbanização de Luanda	Étienne de Groër Moreira da Silva	Portugal
1942	Plano de Urbanização de Évora	Étienne de Groër	Portugal

(continued)

Table 2 (continued)

Date	Plan/Study	Architect-Urbanist	Country
1943	Plano de Urbanização de Beja	Étienne de Groër	Portugal
1944–1946	Plano de Urbanização da Costa do Sol	Étienne de Groër	Portugal
1941–1946	Plano de urbanização, alargamento e embelezamento da cidade de Braga	Étienne de Groër	Portugal
1946	Plano de Urbanização do Concelho de Almada	Étienne de Groër Faria da Costa	Portugal
1947	Plano de Urbanização de Abrantes	Étienne de Groër	Portugal
1946–1950	Plano Regional Moscavide-Vila Franca de Xira	Étienne de Groër Nikita de Groër	Portugal
1938–1948	Plano Director de Urbanização da Cidade de Lisboa	Étienne de Groër	Portugal
1948	Plano de Urbanização da Chamusca	Étienne de Groër Faria da Costa	Portugal
1948	Plano de Urbanização da Figueira da Foz	Étienne de Groër Faria da Costa	Portugal
1948	Plano de Urbanização de Pombal	Étienne de Groër Faria da Costa	Portugal
1948	Plano de Remodelação da Baixa Pombalina	Étienne de Groër Faria da Costa	Portugal
1949	Plano de Urbanização de Sintra	Étienne de Groër	Portugal
1950	Anteplano de Urbanização da Parte Leste do Concelho de Almada	Étienne de Groër	Portugal

In Portugal, Groër continued to write about urban planning issues, published in Portuguese scientific journals, including *Técnica*, the Journal belonging to the *Instituto Superior Técnico* of the University of Lisbon and *Boletim de Obras Públicas*, the Journal of the Ministry of Public Works (Groër 1945, 1945–1946). These manuscripts reveal Groër's pedagogic concerns about the importance of the discipline of *Urbanisme*, according to the IUUP tradition, while underlying Howard's Garden City theory in a very passionate manner. In Portugal, Groër also participated in scientific conferences (Groër 1945). It was in this same period that Law34:607 of 1945 decreed the obligation of the two Portuguese Schools of Architecture, in Lisbon and Porto, to include a course in *Urbanism*e (the 15th and 16th course *Urbanologia*) in their curriculum. Groër was nominated the professor for those courses, in Lisbon and Porto, but because of his intense planning activity, he refused such position (Lebre 2017). Regarding his personality, few personal documents, including telegrams, postcards and letters were identified, revealing Groër as a reserved person.[13]

[13] Some of Groër's personal documents are archived in the University of Coimbra but also in the files of other architects, as is the case with Moreira da Silva (Pires, 2012), whom Groër worked with in the Urban Plan for Luanda.

Apart from the written elements included in Table 2, several reports and graphical materials were also produced by Groër, during his stay in Portugal. These pertained to the urban plan proposals he had been commissioned to complete for several different municipalities.

The fact that his plans, even though approved, were never published led to his reduced visibility, as was the case for Costa do Sol. Yet, other contemporary plans, e.g., The Rio de Janeiro plan by Agache (1930), The Regional Plan of New York and its surroundings (1928) or the *Plan Directeur d'Organisation Générale De La Region Parisienne* (1960) were published by their municipalities which retained their visibility, contrasting with the Lisbon approach.

4 Étienne de Groër's Vision of a Garden City

The Garden City was very much venerated by Groër, as registered in his own writings (Groër 1945–1946). A vision was acknowledged by him since his formative years in architecture at the Russian Academy of Arts in Saint Petersburg, former Petrograd. In 1911, when the first course in Urbanism was created in Russia, several international texts, including those of Camillo Sitte (1843–1903) and Ebenezer Howard were translated into Russian by professors from the Russian University Institutions, inspiring a generation of architect-urbanists to develop the construction of Russian Garden Cities and support the urban development of municipalities. In 1914, the Garden City Russian society was founded with the support of Protopovov, who had contributed to the Russian Journal *Gorodskoye Delo* (The Urban Question) since 1909, disseminating the concept of garden cities, as well as Vladimir Semyonov, the author of the first Russian Garden City and of several other regional plans (1874–1960) (Groër 1921; Miller 1998).

Thus, Groër is part of the generation of Russian architects who were trained under the influence of the International Garden City Movement. Due to the circumstances of the war, he fled Russia but took his knowledge with him, applying this later in his practice, writings and academic experience. Consequently, when Groër was contracted by the Portuguese government to assist in the planning of the urban extension of the city of Lisbon and Costa do Sol, he was a mature 56-year-old architect-urbanist with considerable planning experience, who was not searching for an urban theory but rather wanted to apply a specific acknowledged urban theory—Howard's Garden City. Yet, his attention to the contributions of transportation, water, sanitation and energy, took him to complement Howard's theory with existing scientific techniques to solve contemporary urban problems.

Groër's thoughts on the importance of developing a plan for a city and how to do it were published in three manuscripts, after the studies for the Costa do Sol initiated with Agache were concluded, in the French journal Travaux (Groër 1936a, b, 1937a). The first article focused on the roots of urbanism while detailing the scales of approach: national, regional and municipal, its fundamental laws and the necessary data to prepare the survey analysis prior to the plan proposal. The second article identified how a master plan for a city should be organized and which elements must be considered. While comparing the advantages of the monocentric or polycentric schemes to organize the city, Groër advocates for the latter as the most advantaged one. Its scheme goes in line with the Garden City vision, favoring cleaner air into the city and linking rural areas closer to urban areas (see Fig. 4). His 1937 article focused on more practical and administrative issues, including the importance of municipalities publishing the urban plan proposals, to be better understood and accepted by the general population (Groër 1937a: 38).

While following this line of thought, Groër sought to sensibilize Portuguese planning authorities and the general population on his urban plans, guaranteeing their acceptance. This denotes a pedagogical sensibility which is in line with the IUUP's school of thought. A text published by Groër in 1946, titled 'Introduction to Urbanism', clearly suggests how he aimed to guarantee such pedagogical sensibility through further guidance for the municipal planning technicians (Groër 1945–146; s.d.). Interestingly, the municipality of Lisbon later published short booklets disseminating the *Plano Director de Lisboa,* elaborated by Groër in English, French and Portuguese, even before its final approval by the municipal authorities (CML s.da; s.db. 1948).

The Garden City vision perceived by Groër aimed to cover the overall region of a city, focusing on its development as well as the development of other main settlements, or satellite towns, and the rural areas which aimed to be productive (generate fresh air and agriculture products). This was exemplified in his proposal for the Region of Luanda, completed in 1942 (Groër 1945–1946: 25), but also by successful examples from French, German, English, American and Norwegian Regional Plans, and their implications on a national scale are also noticed by Groër (1946b).

Groër's understanding of the Garden City vision defines the geographical and economic conditions of the region as the main strategic resources to guarantee the viability of the city. Furthermore, Groër acknowledges the importance of the rural area due to its role in economically developing the region, as well as the urban life of the municipally where it is located (Groër 1946b). Groër set a maximum number of inhabitants for the city and for its satellite towns for the convenience of such socio-economic and environmental programs. Rapid transport systems were also determinant to connect all satellite towns with the city. Each of those satellite towns was required to be surrounded by its own individual rural areas, independent of the rural areas of nearby satellite towns (Groër 1945–1946: 24). Groër's concept of the garden city was not merely of spatial order but also included administrative and fiscal concerns. For example, Groër considered the acquisition of the necessary land to develop the city and its surroundings by the municipality of the utmost importance, allowing the land to be rented for a maximum period of 90 years, but never the private acquisition of land, allowing the municipality to protect the interests of the city against speculation (Groër 1945–1946: 24).

Groër's urban planning theory was not exclusively based on Howard' Garden City, but on three main foundations: First, on Ebenezer Howard's Garden City theory; secondly on zoning, considered by him as an important tool that had been assimilated by the Garden City itself to rationally distribute all the different functions in the territory, and therefore not novel at all; and finally the application of the techniques allowed by all sciences (sanitary, water, energy supply, waste treatment and eradication, and transports). This third foundation, for Groër, constituted an important strategy to guarantee the healthiest living conditions for the cities and its inhabitants, to promote work productivity (Groër 1945–1946: 25).

5 Étienne de Groër's Garden City Vision Applied to Lisbon and Its Surrounding Region

Planning the Lisbon Region was regarded as a paramount task by Étienne de Groër, as there was a need to acknowledge the historical evolution of the social, economic and geographic conditions of a given territory and support its future planning. Such task for the Lisbon Region was initiated in 1933, through his collaboration with Alfred Agache, for the Costa do Sol, the coastal region to the west of Lisbon (Groër 1936).

Interestingly, Groër was contracted to work on the plans for the Costa do Sol and Lisbon simultaneously, in 1938, as already refereed. However, the contracts for these two works were only signed in 1944, the same year that Law 24,802 of 21st December of 1934 was substituted by Law 339,121 of 5th December of 1944. The latter established that the former Urbanization Plans should be renamed as the General Plans of Urbanization and Expansion to allow the organization of partial/detailed urbanization plans as well. This new legal framework obliged all municipalities and towns with more than 2500 inhabitants to proceed with the necessary surveys to support the development of their General Plans of Urbanization. Thus, just before the mid-twentieth century, Portugal witnessed an intensive survey and planning activity.

Such socio-economic and geographic analysis or surveying procedure constituted a requisite set by the municipality when he was called in 1938 to assist in developing the necessary studies to support the urbanization plan for the city of Lisbon. Indeed, in his introduction to the Master Plan for Lisbon, *Plano Director de Lisboa*, concluded in 1948, Groër states he delineated the survey conducted for the preparation of the plan for the engineer, Anónio Emídio Abrantes in 1938, who was working for the Municipality of Lisbon (Groër 1948: 2). Such survey together with the Agache report for the Costa do Sol from 1936 allowed Groër to retrieve viable information about Lisbon and its region. That is exactly what he states in the introduction of the *Plano de Urbanização do Concelho de Almada* (Groër 1946b; Flores 2006) entitled Lisbon and its Region, which contained updated information of what he had previously prepared for chapter I of the general report for the *Plano de Urbanização da Costa do Sol*, in 1946 (Groër 1946a, b). Thus, while initially working for Costa do Sol and Lisbon, between 1938 and 1948, he produced the foundations for the analysis of other municipalities within the Lisbon region (Almada, Sintra and Moscavide-Vila Franca de Xira), and which he successively updated between 1946 and 1950. Furthermore,

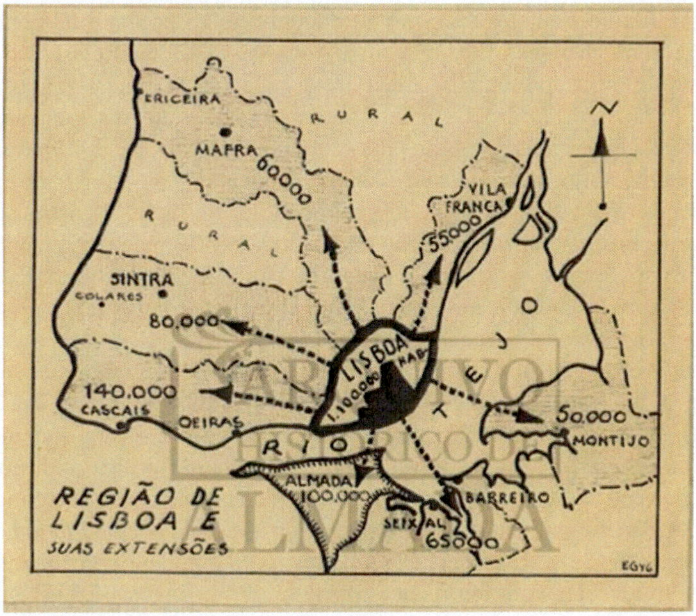

Fig. 1 Lisbon region and its extensions, by Étienne de Groër. *Source* Groër (1946b), Archive of Câmara Municipal de Almada

all the reports for these plans include a common contextualization of Lisbon and its region in their first chapter.

Lisbon and its region is identified by Groër as integrating: (i) the capital of the Portuguese Empire, Lisbon; (ii) several suburban settlements distributed in a tentacular manner, through different distribution routes; and (iii) several extensions of large agricultural spaces between suburban extensions. Furthermore, Groër identifies three specific categories for those extensions, (1) residential and touristic suburbs; (2) industrial suburbs; and (3) mixed-used suburbs (see Fig. 1) (Groër 1946b: 157).

Regarding the residential and touristic extensions, in 1946 Groër identified four specific areas in the Lisbon Region. Three located on the north bank of the Tagus River: (i) Costa do Sol (the extension between Lisbon, Oeiras and Cascais); (ii) the extension between Lisbon and Sintra; and (iii) the extension between Lisbon, Mafra and Ericeira; and on the south bank of Tagus River, he identified Caparica, located in the municipality of Almada. For the industrial extensions, Groër identified three specific suburbs areas, two of them located on the south bank of Tagus River: (i) the Barreiro and Moita group and (ii) Montijo and Alcochete group. Located on the north bank of Tagus River was the Vila Franca de Xira group, composed by Alhandra, Alverca, Póvoa de Sta Iria, Vialonga and Vila Franca de Xira settlements. Finally, for the mixed-use category, Groër identified the Almada suburb. Interestingly this appears as an updated revision of the Costa Sol Plan, where Almada was formerly classified as a residential suburb and now recognized as mixed with other uses (Groër

1946a). Furthermore, in 1948, in the introduction to Lisbon and its region, Groër identifies certain changes in this territory. Almada is no longer classified as mixed use, but a residential and touristic area (Groër 1948a. 81).

Figure 1 systematizes Groër's understanding of Lisbon and its region, delimitating Lisbon, and its extensions, including the suburban areas of Cascais, Sintra, Ericeira, Vila Franca, Montijo, Barreiro and Almada. Depending on their main uses, such extensions were classified by Groër as residential, touristic, industrial or mixed. Moreover, the links established by such settlements and the main city, Lisbon, through transport routes were essential to connect the entire region and establish its main structure. Figure 1 also indicates the maximum number of inhabitants allowed in the planning of Lisbon (or 1946) and its extensions.

The study of the Lisbon region implied a historic, demographic, economic and geographical analysis, allowing to perceive how this territory had evolved between 1900 and 1940. It was precisely the observed demographic changes, namely in the growth of the suburban extensions in terms of population, which allowed Groër to perceive the need to contain Lisbon and its suburban areas, while protecting the rural areas. In 1948 in the analysis of the Lisbon region, presented in the *Plano Director de Lisboa* (Groër 1948a), Groër increases the maximum number of inhabitants allowed for the suburban extensions but reduces the number for Lisbon (see Fig. 3).[14]

The several urban plans prepared by Groër for the Lisbon region, which are available in the municipalities, will now be analyzed in a comparative manner. This includes the *Plano de Urbanização da Costa do Sol* (PUCS) (Groër 1946), the *Plano de Urbanização do Concelho de Almada* (PUCA) (Groër 1946), the *Plano de Urbanização de Sintra-Anteplano* (PUS) (Groër 1949) and the *Plano Diretor de Urbanização da Cidade de Lisbo*a (Groër 1948a). These plans contain graphical material and written reports which served as the main source of our analysis.

The four urban plans under analysis were delineated by Groër for the Lisbon Region. The three first ones (PUCS, PUCA and PUS) comprehend the urbanization of specific suburban areas, between Lisbon and its extension (Cascais, Almada and Sintra). While the PUCS covers three municipalities (Lisbon, Oeiras and Cascais), the PUCA, the PUS and the Master Plan for Lisbon were prepared for three specific isolated municipalities, namely Almada, Sintra and Lisbon, respectively.

The analysis of all these plans allows to identify a common structure, as delineated by Groër, including: (i) an introduction to Lisbon and its Region; (ii) a survey and analysis of the municipality and region under analysis; (iii) Forecast of the proposed urbanization plan; (iv) detailed analysis (including graphical materials) for specific areas or settlements of the plan; and (iv) urban legislation. Furthermore, one should notice that all plans are introduced with a common analysis of Lisbon and its region, despite their specific geographical area. Relating any municipality

[14] The maximum number of inhabitants for Lisbon had been agreed between Groër and the Ministry of Public Works in 1938, while predicting that in 40 years' time Lisbon, the capital city of Portugal, would not exceed 10% of the total amount of the country (Groër, 1948a: 3). Interestingly, Groër was mostly correct, seven decades after the completion of his study, Portugal presents nearly 10 million inhabitants. Nevertheless, Lisbon has lost much of its population and its suburban areas have surpassed the estimated numbers.

within the region of Lisbon with Lisbon and its surroundings was considered important by Groër, to guarantee the preservation of the generous agricultural areas (rural land) (Groër 1946a, b, 1948). Rural areas could be organized as green belts or green cords (see Figs. 2 and 3). Green belts were mainly applied around the municipality of Lisbon (approximately 3 kms wide). Furthermore, smaller urban settlements, as those planned for the Costa do Sol foresaw the existence of narrower green belts, which when connected, formed the Lisbon-Cascais green wedge or cord (Marat-Mendes 2009). Thicker Green cords separate the Sintra and Marfa extension and the Mafra from Vila Franca de Xira extension, respectively.

Groër applied an adaptation of the Garden City vision to Lisbon and its surroundings, remarkably through the development of several studies for isolated suburban areas, considering them as satellite towns, including Almada, Seixal, Barreiro,

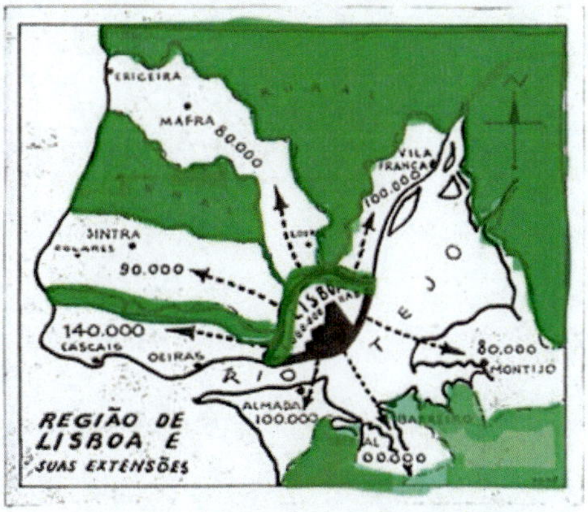

Fig. 2 Rural Areas within Lisbon Region (in green), as proposed by Groër. *Source* By author, adapted from Groër (1948a) archived at Gabinete de Estudos Olisiponenses—Câmara Municipal de Lisboa

Fig. 3 Lisbon Green Belt of 3 km's width (in green), as proposed by Groër in 1948. *Source* By author, adapted from Groër (1948) archived at Gabinete de Estudos Olisiponenses—Câmara Municipal de Lisboa

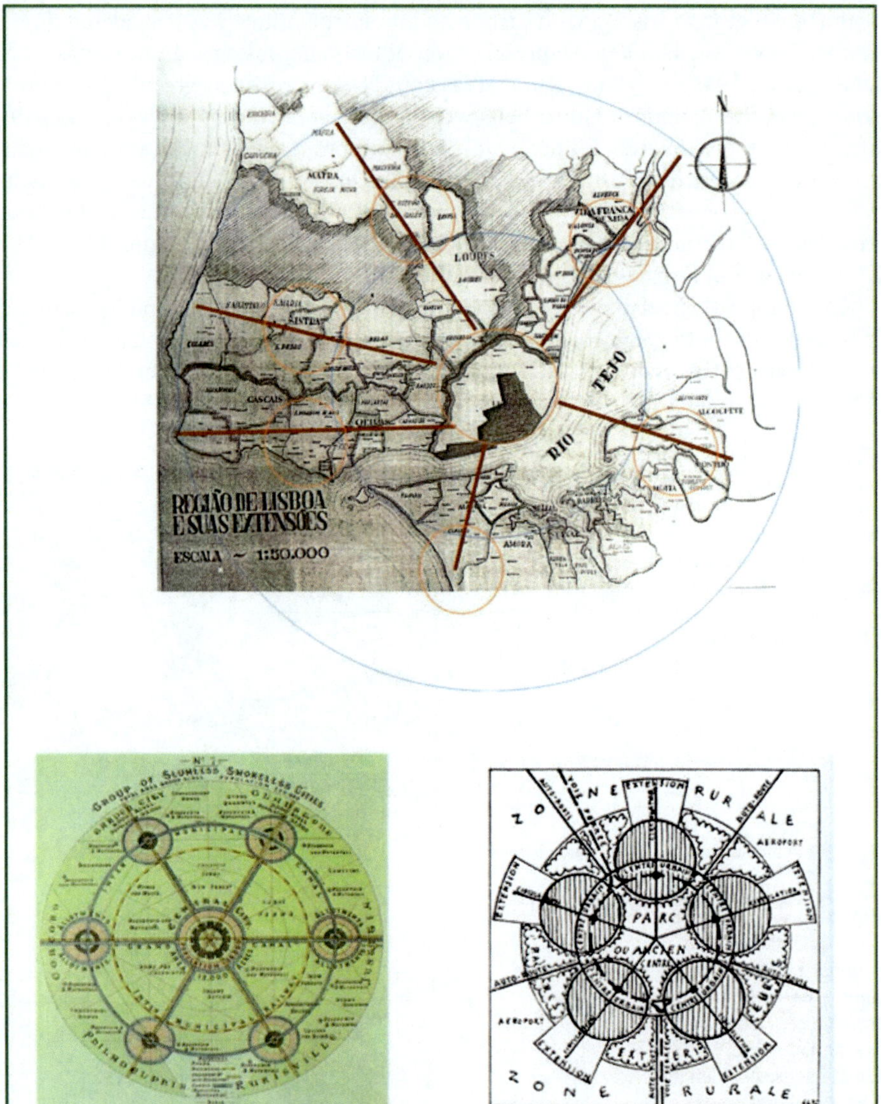

Fig. 4 Groër garden city application in Lisbon Region (top). *Source* By author, adapted from Groër (1946a). Bottom left, Ebenezer Howard for the Garden Cities diagram (Howard (1898) in public domain. Bottom right, scheme by Groër representing a polycentric good example to organize the garden city (Groër (1921) with reserved rights by revue Travaux)

Montijo, Mafra, Sintra, Cascais and Vila Franca de Xira. Satellite towns that depended on their nearby suburbs and rural lands. Figure 4 presents Groër's adaptation of the Garden City concept for the Lisbon Region, as proposed by the author of this chapter. Additionally, a legislative and administrative framework was previewed by Groër for each one of the studied urbanization plans, to support its further implementation.

Notably, Groër placed strong concerns on productivity in his plans, both for agricultural land but also for other uses, such as industrial and economical ones (Marat-Mendes et al. 2020). To protect such areas from further inappropriate urbanization Groër determined that it was completely forbidden to develop rural areas with urban infrastructures (sewers, electric, water or gas infrastructures) or buildings that did not serve farmer's needs (Groër 1946a: 92). This would according to Groër inhibit the reduction of rural areas. Furthermore, rural areas could also be included in urban plans, named as rural and forest zone, without costs implications to the municipality (Groër 1946a). Industrial areas were also covered by specific legislation, forbidding any construction not related to industrial or commercial purposes within their perimeter. The rural area was subject to regulation which was also included in urban legislation (Groër 1948a: 87). According to Groër, urban agriculture was already a reality in the Lisbon region, wherein the rural area was reserved for agriculture and silviculture and protected the city, and the urbanization plans of further urban extensions (Groër 1948a: 90–92). Moreover, Groër indicated several urban design restrictions for the delimitation of agricultural parcels, including maximum height of fences and materials that could be used.

For the *Plano Director de Urbanização de Lisboa,* Groër allowed the rural area to accommodate industrial constructions if needed to be kept apart from residential urban areas. Such rural area would take the form of a green belt, 3 kms in width, surrounding the entire perimeter of the city. The other rural areas in the Lisbon region did not follow this green belt concept, but rather a green wedge structure (see Figs. 1 and 2). This was the case of the green wedge between the west extension of the Costa do Sol and the Lisbon Sintra extension. According to Groër (Marat-Mendes 2009), this constituted a continuous rural area that should be preserved, while linking all the existing productive agricultural areas, green spaces and 'open'[15] areas. Some agricultural areas, such as the Carcavelos vineyards, were identified by Groër as having an interesting economic value. Furthermore, he also highlighted the need to increase soil productivity in the municipality of Cascais due to the poor soil, and the strategic importance of the Oeiras municipality within the Lisbon region due to its very fertile soils (Groër 1946a). The production of horticultural products from these areas would be directly supplied to the Costa do Sol coastline settlements (Groër 1946a). Groër recognizes the proximity of the rural areas to the urban areas in Costa do Sol and Almada, as a very important issue to make dairy products accessible to all, without being subject to long transportation modes and therefore provided at a more reduced price than in the city (Groër 1946a, b). This already denotes Groër's

[15] The concept of 'open' areas included all parks, public gardens, open fields and playing fields of a city. All of them regarded as green areas or the lungs of a city.

perception of short routes or distances, while maintaining a liveable and healthy region in mid-twentieth century.

In the Almada plan, Groër identifies agricultural production and industry as determinant factors to sustain the economy of the municipality as well as appropriate transportation modes. Therefore, in his survey for Almada, he provides a detailed analysis of the existing activities related to local agriculture, silviculture, piscatory and mineral production. For commercial proposes he identifies wine, olive oil, cereal and cod as the main products for exportation, but also materials for construction (such as sand, gravel and clay). Furthermore, while identifying the excessive sub-division of land in the municipality, limiting land productivity, he recognizes the high level of quality of the local products (Groër 1946b).

The plans for Almada, Lisbon and Sintra aimed to limit urbanization, protect existing settlements against further urbanizations and guarantee easy access to fresh agricultural products. All agricultural businesses could be freely developed. Yet, built structures in such rural area were only possible if they were built to support agricultural purposes, and could not exceed 1% of the agricultural allotment (Groër 1948a: 91). Other green areas included parks, playgrounds, sports grounds and public gardens, which in the case of PUCS were projected to be around 30 m^2 per person (Groër 1948a: 91). Finally, rural areas could flow into the urban limits, as was the case in the Costa do Sol planned settlements (Marat-Mendes 2009). All these were designated rural and forest areas. Apart from these areas, Groër also identified existing agriculture settlement areas, which should be protected and delimitated. They could be improved with better built conditions and could not be extended outside their limited zone. Outside their perimeter, the rural and forest area regulations would be applied.

6 Conclusions

Étienne de Groër regarded the planning of the Lisbon Region as an opportunity to apply Ebenezer Howard's Garden City vision in a specific territory yet updated with his experience and knowledge of the territory under analysis.

In Lisbon, not only did Groër apply a survey process to analyze the overall territory and the individual municipalities, but he also promoted the dissemination of his thoughts on the importance of a National Plan of Urbanism subdivided into Regional Plans to assure the development and proper organization of a country's economy, through a correct distribution of its industry (Groër 1948a: 2). This was further accomplished through specific publications and communications in academic and public events.

Groër gives continuity to Agache's Regional Planning perspective, which had identified several extensions for the city of Lisbon, including its west extension, Costa do Sol. Yet his longer stay in Portugal, due to World War II, allowed him to better know the territory and work for several municipalities at the same time on urbanization proposals.

Unlike in England, where specific garden cities associations were formed to support the venture of garden cities construction, such as Letchworth and Welwyn Garden City, in Portugal, there was no indication of such associations. Yet, it would be the ministry of Public Works which supported the development of Portuguese urban planning through the work of architect-urbanists devoted to the Garden City vision, as was the case of Groër.

Groër's involvement with the Costa do Sol Urbanization Plan as well as the urban plans for Lisbon, Sintra and Almada granted him the opportunity to urbanize almost all of the Lisbon region, based on a common systematization of the analysis of Lisbon's surroundings. Not all these plans were approved and implemented as foreseen by Groër. This was the case with Lisbon and Vila Franca de Xira (east regional extension). Yet, Sintra, Almada and Costa do Sol were approved and secured by law until the1990s when a new planning instrument was approved by the Portuguese law, the *Plano Director Municipal.* These plans have allowed the municipalities of Oeiras, Cascais, Sintra and Almada to preserve some of its former settlements and green areas. Yet the rigid regulation which forbade any further activities in the rural areas and farming settlements failed to promote an economic framework for the region, placing great pressure on the informal urbanization process which affected the Lisbon region during the 1960s (Cabrita and Marat-Mendes 2019). Immediately, rural land and former agricultural settlements were the first areas to witness great changes, through the rise of illegal constructions (Rodrigues 2014). These were areas without urbanization plans, without communication roads, water, sanitization and electric infrastructures. Nevertheless, in such emerging illegal situations, agricultural land kept providing the basic agricultural products for the families which occupied them (Marat-Mendes et al. 2021a, b).

As already described, Groër's vision for the Lisbon region was contemporary to many other planning proposals for capital cities, such as Paris, New York and Rio de Janeiro. Yet, while the full publication of these latter plans granted them their weight in urban history, alongside the recognition of their authors work, allowing future generations to better understand these cities and the ideas behind their plans, that is not the case of Groër's work for Lisbon. This is a challenge which should guide Portuguese authorities to consider the urban plans delineated by Groër as key elements to better perceive the heritage of the Lisbon region as an opportunity to reimagine its present condition from a sustainability perspective.

Acknowledgments The author gratefully acknowledges the support of the French journal Travaux and of the Portuguese archives of Câmara Municipal de Almada and Gabinete de Estudos Olisiponenses - Câmara Municipal de Lisboa for their permission to include the figures included in this chapter.

References

Almeida S (2009) O País a Régua e Esquadro. Urbanismo, Arquitectura e Memória na Obra Pública de Duarte Pacheco. PhD Thesis. Universidade Nova de Lisboa, Lisboa

André P, Marat-Mendes T, Rodrigues P (2012) Alfred-Donat Agache urban proposal for Costa do Sol. From the territory to the city. In: Proceedings of the 15th International Planning History Society: Cities Nations & Regions in Planning History, Universidade de São Paulo, Brazil, 15–18 July 2012, pp 315–316

Agache A, Prefeitura do Districto Federal, (1930) Cidade do Rio de Janeiro: Extensão-Remodelação-Embellezamento. Foyer Brésilien, Paris

Agache A (1936) Lisbonne. Urbanization de la Region Ouest. Mimeo, Paris

Anon (1930) Societé française des urbanistes – Bureau 1930. Supplement a L'architecture Usuelle 259:11

Bardet G (1947) L'Urbanisme. Puf (Que sais-je?), Paris

Boulfroy J (1939) Le Problème de la ville moderne: la cité-jardin. Institut the Urbanism Universita de Paris, Paris

Cabrita M, Marat-Mendes T (2019) A identidade cultural saloia e os novos usos para o território. In: d'Encarnação, J, (Coord) Dos patrimónios de Cascais. Associação Cultural de Cascais, Cascais, pp 163–178

Camara Municipal de Lisboa (s.da) Master Plan for Lisbon. Câmara Municipal de Lisboa, Lisboa

Camara Municipal de Lisboa (s.db) Plano Director de Lisboa. Câmara Municipal de Lisboa, Lisboa

Camara Municipal de Lisboa (1948) Quelques notes sur le dévelopment de Lisbonne. Câmara Municipal de Lisboa, Lisboa

Camarinhas C (2009) L'urbanisme de Lisbonne. Elements de Théorie urbaine appliquée. Thése pour obtenir le grade de Docteur en Urbanisme et aménegement. Universye Paris IV – Sorbone, Paris

Camarinhas C, Brito V (2007) Elementos para o estudo do Plano de Urbanização da cidade de Lisboa (1938). Cadernos Do Arquivo Municipal 9:182–183

Carrasco D (2018) Diplômes et diplômés en urbanisme: Paris 1919–1969. Architecture, aménagement de l'espace. Université de Nanterre, Paris X

d'Almeida P (2015) Bairro(s) do Restelo Panorama Urbanístico e Arquitectónico. Caleidoscópio, Lisboa

d'Almeida P, Marat-Mendes T (2019) Forma Urbana e Sustentabilidade. O Bairro do Restelo. Revista de Morfologia Urbana 7(2). http://hdl.handle.net/10071/20275

d'Almeida P, Marat-Mendes T, Toussaint M (2020) Portugal's rising research in architecture and urbanism: the influence of international research centers and authors. J Urban Hist. https://doi.org/10.1177/0096144220968078

Direcção Geral do Ordenamento do Território e Desenvolvimento Urbano (DGOTDU) (2005) Urbanistas Estrangeiros em Portugal 1930–1960. [CD-RoOM], DGOTDU, Lisboa

Dinis M (2015) Arquivo histórico do Ordenamento do Território e Desenvolvimento Urbano – proposta de ponto único de acesso online. MSc dissertation. Universidade Nova de Lisboa, Lisboa

Flores A (2006) Urbanização do Concelho de Almada (1946): Análise e Programa - Relatório. Anais De Almada, Revista Cultural 7(8):151–236

Galvão M (2020) Étienne De Gröer e o Plano de Almada. Msc Dissertation. Universidade de Coimbra, Coimbra

Geertse MA (2012) The International Garden City Campaign: Transnational Negotiations on Town Planning Methods 1913–1926. J Urban Hist 42(4):733–752

Guerrand RH (2002) Sobre los Orígenes del movimento de las ciudades-jardines em Europa. Ciudades 6:17–20

Groër É (1921) De l'Urbanisme en Russie. La Vie Urbaine 10:417–426

Groër É (1929) La Circulation et les transports à Rio de Janeiro. Le Maître D'oeuvre 4(33):41–48

Groër É (1932) Le Gratte- Ciel est-il Nécessaire? La Vie Urbaine 25:59–67

Groër É (1932) Urbanisme Au Canada. Urbanisme 2:65–70

Groër É (1932) Urbanisme Au Canada. Urbanisme 3:102–104
Groër É (1932) L'habitation Haute En Amérique. Urbanisme 16:220–223
Groër É (1933) Urbanisme Au Canada. Urbanisme 2:61–67
Groër É (1933) Le Zoning et la Législation Urbaine de la Région Parisienne. Urbanisme 16:220–223
Groër É (1933) L'habitation Haute Em Amérique. Urbanisme 16:220–223
Groër É (1934) Le Urbanisme Au Canada. Urbanisme 30:331–335
Groër É (1936) Porquoi et comment un plan d'amenagement doit se faire. Travaux 43:343–348
Groër É (1936) Porquoi et comment un plan d'amenagement doit se faire. Travaux 45:436–440
Groër É (1937) Porquoi et comment un plan d'amenagement doit se faire. Travaux 49:34–38
Groër É (1937b) L'urbanisme et la defense antiariénne. Le Monde Souterrain 11
Groër É (1938) Nécessité D'un Zoning Dans Paris. Travaux 63:144–147
Groër É (1945) Le Tracé D'un Plan D'urbanisation. Técnica 157:463–472
Groër É (1946a) Plano de Urbanização da Costa do Sol. Ministério das Obras Publicas, Lisboa, Relatório Geral
Groër É (1946b) Relatório do Plano de Urbanização de Almada, Etienne de Groër. In Anais Arquivo Histórico Municipal de Almada nº 2474, Câmara Municipal de Almada
Groër É (1949) Plano de urbanização de Sintra. Câmara Municipal de Sintra, Sintra
Groër É (1945–1946) Introdução ao Urbanismo. Boletim da Direcção Geral dos Serviços de Urbanização 1:17–86
Groër É (1948a) Plano Director de Lisboa, Câmara Municipal de Lisboa, Lisboa
Groër É (1948b) Anteprojecto de Urbanização e Embelezamento e de Extensão da Cidade de Coimbra. Câmara Municipal de Coimbra, Coimbra
Groër É (s.d.) Introdution à l'urbanisme (sl)
Lebre R (2017) From the organization of space to the organization of society: a study of the political commitments in post-war Portuguese architecture, 1945–69. PhD Thesis. Universidade de Coimbra, Coimbra
Lobo M (1995) Planos de Urbanização à época de Duarte Pacheco. FAUP, Porto
Marat-Mendes T (2009) O PUCS e os vazios planeados. Novas oportunidades para o ordenamento sustentado da Costa do Sol. In: Pereira M (ed) O Plano de Urbanização da Costa do Sol. Uma visão inovadora para o Território. Editora do Município de Oeiras, Oeiras, pp 91–122
Marat-Mendes T, Sampaio M (2015) The Plano de Urbanização da Cidade de Luanda by Étienne de Groër and David Moreira da Silva (1941–1943). In: Nunes da Silva C (ed) Urban Planning in Lusophone African Countries. Routledge, United Kingdom, pp 57–77
Marat-Mendes T., Sampayo, M (2010) Étienne de Groër: The different scales of the urban intervention in Lisbon territory. In: Proceedings of the 1st International Conference of the EAHN, Guimarães, pp. 32–43
Marat-Mendes T, Cabrita M, Oliveira V (2014) Teaching urban morphology in Portugal. In: Cavallo R, Komossa S, Marzot N, Pont M, Kuijper J (eds) New Urban Configurations. IOS Press, Amsterdam, pp 131–136
Marat-Mendes T, Oliveira V (2013) Urban planners in Portugal in the middle of the twentieth century: Étienne de Groër and Antão Almeida Garrett. Plan Perspect 28(1):91–111
Marat-Mendes T, Borges J, Dias A, Lopes R (2021) Planning for a sustainable food system. The Potential Role of Urban Agriculture in Lisbon Metropolitan Area Journal of Urbanism: International Research on Placemaking and Urban Sustainability. https://doi.org/10.1080/17549175.2021.1880960
Marat-Mendes T, Isidoro I, Catela J, Pereira M, Borges J, Lopes S and Henriques C (2021b) Drivers of change: how the food system of the Lisbon Metropolitan Area is being shaped by activities, initiatives and citizens needs towards a sustainable transition. Cidades, Comunidades e Territórios. http:// journals.openedition.org/cidades/3343
Marat-Mendes T, Borges J, Lopes S, Pereira M (2020) Where the fields have no name: Urban-rural transitions in the Lisbon Region planning history. Cidades, Comunidades e Territórios 41:105–132
Marques B (2015) João Guilherme Faria da Costa. O caso unico da Figueira da Foz. MSc dissertation. Universidade de Coimbra, Coimbra.

Marques E (2017) O início do planeamento urbano em Almada: 1946. Plano de Urbanização do Concelho de Almada. Étienne de Groër e (Faria da Costa). MSc dissertation. Universidade Nova de Lisboa, Lisboa.
Marques B (2018) Faria da Costa e o(s) Plano(s) de Urbanização da Figueria da Foz. Caleidoscópio, Casal de Cambra
Meller H (1997) Towns, plans and society in modern Britain. Cambridge University Press, Cambridge
Miller, (1998) Semyonov, Parker and Unwin, and the iconography of plans: A commentary. Planning History 20(2):25–33
Pereira M (1994) O processo de decisão na política urbana. O exemplo da Costa do Sol. PhD Thesis. Universidade Nova de Lisboa, Lisboa
Pereira M (2009) O Plano de Urbanização da Costa o Sol. O pioneirismo de um plano sub-regional. In: Pereira M (ed) O Plano de Urbanização da Costa o Sol. Uma visão inovadora para o Território. Editora do município de Oeiras, Oeiras, pp 120–126
Pires M (2012) O Ateliê de Arquitetura/urbanismo de David Moreira da Silva e Maria José Marques da Silva Martins. Visibilidade da Memória. PhD thesis. Universidade do Porto, Porto
Poëte M (1929) Introduction à l'Urbanisme. Boivin, Paris
Rodrigues G (2014) Qualificação de áreas urbanas de génese ilegal. O caso da Zona Oriental do Concelho de Loures. MSc dissertation. Universidade de Lisboa, Lisboa
Rioux G (2013) Le milieu de l'urbanisme à Montréal (1897–1941), Histoire d'une «refondatio». PhD Thesis. Université Du Québec À Montréal, Montreal
Silva CN (1994) Política Urbana em Lisboa 1926–1974. Livros Horizonte, Lisboa
Silva S (2018) O Plano de Urbanização da Amadora de Faria da Costa: o projecto e a sua implementação. MSc Dissertation. Universidade Lusíada, Lisboa.

City and Countryside: A Historical Landscape System to Enhance. The Case of the Territorial Plan of Sorrento-Amalfi Peninsula by Roberto Pane and Luigi Piccinato, 1968–1987

Paola Branduini and Andrea Pane

Abstract In the cultural context of the 1960s the Plan for the Sorrento and Amalfi peninsula, drafted by Roberto Pane and Luigi Piccinato, highlights an advanced vision of landscape protection and enhancement as a system of tangible and intangible heritage, overcoming the preservation of single monuments at that time. This vision embeds the role of the agricultural landscape, not only as a crown to the urban center, but as a place of co-construction and co-evolution with the city. The Plan offers a unique perspective through the lens of heritage. Starting from the Neapolitan cultural context of the 60' in which the Plan took shape, we present the two main protagonists of the Plan's construction with the other key figures that contributed to its contents. Then we describe the tentative drafts approaching the approved version in 1987 and the Plan's fundamental features, focusing on three aspects considered significant and innovative: the landscape system, the role of farmers in landscape management and the landscape restoration interventions. The conclusion reflects on the contemporary lesson provided by the Plan in terms of landscape safeguarding, dynamic conservation and co-construction.

Keywords Landscape · Cultural heritage · Preservation · Landscape system · Rural architecture

1 Introduction

The rural landscape is cultural heritage (UNESCO 1972, 2019; ICOMOS-IFLA 2017; Scazzosi 2018; ICOMOS 2019) in that it is a complex system composed of material elements of agricultural landscape, bonded by intangible social, symbolic and economic relations (Branduini 2017; Scazzosi 2018) in continuous dynamic

P. Branduini (✉)
ABC Department, Politecnico di Milano, Piazza Leonardo da Vinci 32, 20133 Milano, Italy
e-mail: paola.branduini@polimi.it

A. Pane
Department of Architecture, University of Naples Federico II, via Toledo 402, 80134 Napoli, Italy
e-mail: a.pane@unina.it

C. Brisotto and F. Lemes de Oliveira (eds.), *Re-Imagining Resilient Productive Landscapes*, Cities and Nature, https://doi.org/10.1007/978-3-030-90445-6_8

transformation (Scazzosi 2020). These elements of rural landscape are made heritage by their physical permanence through time, their historical authenticity (ICOMOS 1994), as well as the interpretation and significance attributed by people to places, techniques and skills that have enabled landscape creation and to features dictated by economic and behavioral factors (UNESCO 2003; Laviscio and Scazzosi 2015; Branduini et al. 2016; Scazzosi 2016).

The countryside has always been indispensable to the city, but its importance is rarely recognized, it is rather considered a place of residential expansion and related services. Even in the preservation of ancient buildings, urban planning of the twentieth century preferred the historical centers over the rural *edificato diffuso.*

From the 1950s, some Italian architects—such as Giovanni Astengo in the Assisi plan (1955–59) (Astengo 1958); Roberto Pane and Luigi Piccinato in the Sorrento and Amalfi peninsula plan (1973–77); and Alberto Samonà in the Volterra plan (1991) (Canestrari 1991)—understood the importance of the agricultural landscape, not only as a crown to the urban center, but as a place of co-construction and co-evolution with the city.

In this regard, the present contribution will analyze the Plan for the Sorrento and Amalfi peninsula, drafted by Pane and Piccinato, to highlight the advanced vision of the architects in overcoming the protection of single monuments in favor of the enhancement of a landscape system. We claim that the architects anticipated time when applying the "landscape as a system" vision, coming from the historical and geographical reading of Marc Bloch, Lucio Gambi and Emilio Sereni (Scazzosi 2018) that found contemporary agrarian landscape preservation (Di Bene et al. 2019).

This contribution is part of urban planning studies that, since the 1990s, have turned the gaze of planning from the countryside toward the city, exploring what kind of integrated management is needed in the territory. This territory, generally referred to as peri-urban, widespread city, or fringe area, is defined only in relation to the city. There is a need, therefore, to describe this territory as other than the totally urban or agricultural one (Lanzani 2003).

When themes of peri-urban and urban agriculture entered the cultural debate, many urban planners, sociologists, agronomists and ecologists began developing complementary points of view on peri-urban areas. The ecological approach imagines the area as ecosystems in tension between the urban and rural systems (Mininni 2012); the *territorialista* approach attributes great importance to the farmers' role as rural landscape architects (Magnaghi 2000; Ferraresi 2009; Fanfani 2015); while other urban planners consider agriculture material for the construction of new landscape scenarios (Viganò 2001).

This contribution offers a unique perspective through the lens of heritage: of its protection and enhancement and of material permanencies and intangible knowledge that have forged the landscape and continue to modify it, as the Sorrento and Amalfi Plan clearly highlights. The contribution first illustrates the Neapolitan cultural context in which the Plan took shape, then presents the protagonists of the Plan's construction, finally explains the Plan's fundamental features, focusing on three aspects considered significant and innovative: the landscape system, the role of farmers in landscape management and the landscape restoration interventions.

The conclusion reflects on the contemporary lesson provided by the Plan in terms of landscape safeguarding, dynamic conservation and co-construction.

2 The Neapolitan Context and the Plan Construction

2.1 Plan Premises

In the early 1960s, the urban area of Naples appears as one of the most densely populated in Italy, the result of a chaotic building development that had begun immediately after the Second World War (Belfiore 1994). The city, one of the most bombed in Italy (Casiello 2011), underwent a convulsive and disorderly reconstruction, while any attempts to plan its development gradually failed with the collapse of the master plan coordinated by Luigi Cosenza (1945–46) and the rise of a monarchical-conservative council headed by the mayor Achille Lauro: during the 1950s he kept the city in check, allowing for intensive construction, green spaces subtraction and monuments destruction. The complaints of this situation were underlined by aforementioned Cosenza and Roberto Pane, who will later be one of the authors of the Territorial Urban Planning Plan of the Sorrento-Amalfi Area (PUT). It is Pane, in particular, who raised the Neapolitan events to national attention, presenting a public complaint by his "Document on Naples", published during political elections in 1958, where he was a candidate on the Community Movement list led by Adriano Olivetti (Pane R. 1958). Even the second author of the PUT, Luigi Piccinato, although distant from Naples in those years, was indirectly involved in these struggles: the master plan he drafted for the city in 1936–39, during the teaching period in Naples, remained in force after the rejection of the Cosenza plan, but was even tampered with by the fraudulent transformation of the large agricultural areas into building areas (De Lucia and Iannello 1976). The Neapolitan building robbery—symbolized by the Cattolica Assicurazioni sky-scraper, built in defiance of any local and national regulation—was so intense and unpunished as to even inspire a famous film, "Hands on the city", directed by Francesco Rosi and awarded the Gold Lion at the Venice International Film Festival in 1963 (Pane A. 2015b).

At the very beginning of the 1960s, however, Naples experienced a first significant turning point, with the fall of the Lauro council (November 1961) and the emergence of timid hopes at the national level with the first center-left government and the land law carried out by the deputy Fiorentino Sullo, unfortunately destined to be wrecked shortly thereafter (Salzano 1997). In this context, a first study for a regional plan for Campania was launched, entrusted by the Ministry of Public Works to a commission led by Nino Novacco and Manlio Rossi Doria, which developed an in-depth reading of the territorial dynamics of the region. At the same time, however, the plan for the Industrial Development Area (ASI) of Naples also took shape, sanctioned by the Interministerial Committee for Economic Planning on 31 July 1961

(Piaz 1985). This last instrument, which had the legal effect of a territorial coordination plan, proposed a concentration of new industrial districts with numerous roads and motorways even toward the Sorrento peninsula, until then excluded from the tumultuous urban expansion of the regional capital, thanks to natural defenses of its orography. In fact, among the planned new infrastructures, a superhighway stood up—soon known as the "Gava-Tocchetti project", from Castellammare di Stabia to Massa Lubrense, climbing up the hilly slopes of the municipalities of Vico Equense, Meta, Piano, Sant'Agnello and Sorrento. Faced with this proposal—which also went beyond the very competences of the ASI plan—associations and cultural forces were mobilized, explaining the risks of «a project that directly attacked the Sorrento hills landscape and, indirectly, threatened the entire peninsula, allowing a glimpse of extensive urbanization programs with the prospect of a substantial "absorption" of the Sorrento plain into the expanding conurbation» (Piaz 2010).

2.2 *First Plan Drafts*

In this context, the awareness of the need for a specific urban planning tool for the Sorrento peninsula landscape protection and sustainable development would have begun to develop, and the aforementioned friends and colleagues Pane and Piccinato would have soon been the protagonists. They had been involved in the commission for the Naples city plan, appointed in December 1962 by the Christian Democrat mayor Vincenzo Palmieri, composed of numerous Neapolitan professionals and academics, including Corrado Beguinot, Amadeo Bordiga, Luigi Cosenza, Giulio De Luca and Roberto Pane himself (De Lucia and Iannello 1976; Dal Piaz 1985). The commission outcome would have been only partial, but rather significant at the territorial level: in 1964, in fact, a first plan of the Neapolitan area was presented: it included some theses of the Novacco-Rossi Doria group regional plan but, above all, denied the freeway proposed by ASI plan and proposed on the contrary a new motorway axis east of Vesuvius (Pane A. 2017a). A few years later, in 1966, Pane and Piccinato worked together to draft a territorial plan for Campania entrusted by the Ministry of Public Works: it will be presented in 1969, it proposed a settlement reorganization of the regional territory (Dal Piaz 1985; Pane 2017a).

These were the prerequisites for appointing a commission, by the regional Provveditorato alle Opere Pubbliche, in charge of preparing a "Scheme of territorial planning of the Sorrento-Amalfi Area", intended as an excerpt of the territorial coordination plan: Piccinato would have been the coordinator, Pane would have taken part as over twenty years' expert together with the younger Angerio Filangieri, Giuseppe Muzzillo and Alessandro Dal Piaz. The final papers, delivered between April and May 1968, constituted a fundamental premise for the subsequent territorial planning (Fig. 1). Among the most evident aspects, the clear rejection of the Gava-Tocchetti highway proposal which was replaced by a new ordinary ridge road network—defined as the "first route of penetration" into the Sorrento peninsula—aimed at decongesting the traffic on the coast and connecting the mountain centers

Fig. 1 L. Piccinato, R. Pane, G. Muzzillo, A. Filangieri, A. Dal Piaz, Scheme of territorial planning of the Sorrento-Amalfi Area. The road network, April 1968 (Luigi Piccinato Archive, Rome)

starting from Pompeii-Scafati, through Gragnano, reaching the plain of Sorrento in the locality of Trinità (Pane A. 2017a). Besides the usual morphology and land use boards, the report testifies to the interest in territorial reorganization with active landscape protection. In fact, the chapter entitled "Historical, artistic and environmental values" reads clearly: «The fundamental objective of this study is the defense of the landscape, historical and environmental heritage of the Area, […] this basic choice is not driven only by a generically cultural concern for a heritage of nature and history, but also from the conviction that the only concrete and consistent economic prospect for the Area is tourism, and from the awareness that this landscape, at all scales, is among the main resources, if not the main, for a qualified tourist economy». Once the documents were delivered, Piccinato was asked to work on the building indexes by superintendent Giovanni Travaglini, that would have been defined in the subsequent municipal level instruments (Pane A. 2017a).

2.3 *First Landscape Plan*

In the early 1970s, meanwhile, the Ministry of Education also moved to finally draw up a landscape plan of the Sorrento peninsula, an optional instrument provided by law n°1497 in 1939 for the protection of "natural beauties". For this task the peninsula was divided in two sides, entrusted to two working groups: the Sorrento one was coordinated by Renato Bonelli and the Amalfi one by Mario Coppa. The initiative appeared absolutely valid and the Bonelli group's work of 1975 was remarkable (D'Orta 2011), however, the resulting frame was rather fragmentary. On the one hand, territorial planning emerged based on the scheme drawn up in 1968 by the group coordinated by Piccinato and on the other hand landscape protection, without reaching an appropriate integration between the two levels. The split between protection and urban planning had been already denounced, for over twenty years, by many of the protagonists of Italian architectural culture, starting from Carlo Ludovico Ragghianti, who already in 1945 had called for an integration, up to the same Pane and Piccinato, who in the 1950s had focused on the topic several times.

It will be precisely the subsequent events of the Sorrento Peninsula Territorial Coordination Plan to sanction the possibility of this integration, clearly defined in the work carried out by Pane and Piccinato. The fundamental premise for a significant change of pace is provided, at the beginning of the 1970s, by the implementation of the delegation to the Regions, which induced the new regional councilor for urban planning of Campania—the socialist Francesco Porcelli—to propose, during the course of 1972, the approval of the "structure scheme" delivered in 1968 and never made operational. All mayors took position against that except for the mayor of Vietri, already engaged in the battle against the building speculation started in 1968 in the locality of Fuenti. Such a decisive opposition forced the Region to decide to appoint the same study group to draft a real territorial coordination plan. That very group who had already worked at the "Structure scheme", but to which new members had been added, including Coppa himself, with Pane and Piccinato as co-presidents. Convened in its first session in March 1973, the commission began to work intensely, producing already at the end of the same year a first guideline document, drafted in substance by Alessandro Dal Piaz, to which Roberto Pane attached his own report denouncing the tampering in progress on the coast, including the "Amalfitana Hotel" near Fuenti, later published (Pane R. 1973).

However, the situation evolved further during 1974, when the Region—chaired by Vittorio Cascetta—decided to officially entrust to the commission, in addition to the territorial coordination plan, also the landscape one. With a specific agreement of 24 September 1974, the commission—which in the meantime registered the resignation of Coppa—was therefore in charge of drawing up the "Territorial coordination plan and territorial landscape plan of the Sorrento-Amalfi area". For Pane and Piccinato it was finally possible to integrate the territorial and protection vision, according to the requests they had expressed several times in the past. Starting from the following months, with a very tight sequence of meetings, particularly concentrated in the spring–summer 1975, the plan took shape, to be delivered in its first version in the course of 1976 and then definitively in 1977.

In addition to the conceptual and technical work, carried out between meetings and drawing tables, the preliminary cognitive process makes use above all of inspections in the peninsula. Here the photographic eye of Pane proves to be fundamental, providing copious documentation not only of the landscape values, but above all of the assaults carried out by recent construction. Part of this documentation will be published by Pane himself, a few years later, on his «Napoli nobilissima» and still today constitutes a precious collection of images that allows us to draw a complex vision of the territorial and landscape dynamics of the peninsula in the mid-1970s (Pane R. 1977a). It had to constitute a fundamental accompaniment to the plan, motivating its implementation to avoid the serious tampering that was then underway, among which that of Fuenti appeared only as the most emerging.

3 The Planners

As mentioned above, the Territorial Urban Plan of the Sorrentino-Amalfi Area is due to two leading figures of the Italian architectural and urban planning culture of the twentieth century: Roberto Pane (1897–1987) and Luigi Piccinato (1899–1983), united by a common formation and decades of friendship. Both figures have been the subject of extensive literature in recent years, which has also crossed national borders (Casiello et al. 2010; Belli and Maglio 2015). Almost the same age, they met at the School of Architecture in Rome founded by Gustavo Giovannoni, where they graduated one year on, Pane in 1922 and Piccinato in 1923 (Pane A. 2015a). Both of them obtained free teaching in December 1930; since the academic year 1930–31, they reached, at the newly established School of Architecture in Naples, directed by Alberto Calza Bini, where Piccinato, arrived from Rome due to contrasts with Piacentini and Giovannoni, taught «City building and art of gardens» (later called «Urbanistica») while Pane «Scenography» (Mangone and Telese 2001; Mangone 2008). From this date, the contacts between Pane and Piccinato would have been continuous, despite the divergence of their respective academic paths starting from 1947. Before highlighting their points of contact, however, it seems appropriate to briefly outline their biographical profiles.

Having become full professor of «Stylistic and constructive characters of monuments» in 1942, Pane maintained a variety of interests throughout his life, adding to his main research line in history of architecture—testified in those years by the monographs on Andrea Palladio (1948), Gian Lorenzo Bernini (1953), Ferdinando Fuga (1956)—studies aimed at deepening the environmental characteristics of his city (Napoli imprevista, 1949) and the Sorrento peninsula (Sorrento e la costa, 1955). It is precisely this last volume, entirely illustrated by his photographs, which constitutes an important premise for the future work of the PUT. Starting from the study of Sorrento—intensely attended since 1939 as director of the local school of art for inlay and cabinet-making and from 1946 as designer of the master plan (D'Orta 2010)—Pane developed a peninsula analysis based on a rigorous historical research and a direct reading of environmental and landscape values through images, which underline the very close relationship between natural and anthropic factors (Pane R. 1955). The versatility of his approach pushed him to integrate the photographic tool with the cinematographic one, believing that the representation of an environment was possible only through the fourth dimension. This is how, in close relationship with the aforementioned volume, the two documentaries "Miti e paesaggi della penisola sorrentina" and "Architettura della penisola sorrentina" (Myths and landscapes of the Sorrento peninsula and Architecture of the Sorrento peninsula) were born, both directed by Pane in the same 1955, the second being awarded at the Venice film exhibition in the documentary section (Mangone 2014). A few years later, in 1962, Pane would have returned to deal with the environmental and landscape values of the peninsula, writing a few pages for the "Tuttitalia" encyclopedia, illustrated by many of his photographs, where he also focused on the «urban chaos», alluding to the difficulties of a correct planning that he personally experienced for Sorrento (Pane R. 1962).

This specific interest in the Sorrento peninsula is accompanied by a similar attention to other Campania landscapes, such as the Capri island, to which he dedicated a monograph in 1954, then reissued in 1965 and 1982 (Pane R. 1954), or the vesuviana coast (Pane et al. 1959). These works are followed by a comprehensive look at the entire region, carried out on the "Italia '61" exhibition, organized in Turin for the centenary of Unity, where he curated an exhibition with the significant title "Campania. La casa e l'albero" (The house and the tree), also illustrated by a catalog (Pane 1961). On all these occasions, Pane did not fail to express his lucid awareness about the need to overcome the separation between urban planning and protection, which led him to emphasize on several occasions the close relationship between the landscape plans (governed by law 1497 of 1939, but then very little practiced in Italy) and regulatory plans (Scazzosi 2010).

For his part, Piccinato, trained after graduating from the Technische Universität of Munich (1926–27) and in the Roman studio of Marcello Piacentini, was involved from an early age in concrete planning experiences. These range from self-commissioned proposals (the "counter-plan" of Padua, 1926), to more or less successful results in public competitions, among which the first prize for Sabaudia plan, the most outstanding new town of the regime (1934, with G. Cancellotti, E. Montuori, A. Scalpelli), later built, which constituted a turning point in his career (Merlini 1992). At the same time, since the early 1930s, Piccinato had been involved in local and national commissions for the formation of urban planning tools and regulations, as well as in the activities of the newly established National Institute of Urban Planning (INU).

In Naples, where he spent seventeen years as a teacher at the Faculty of Architecture (1930–47), Piccinato became the protagonist of a master plan (1936–39), which, although in partial continuity with a preliminary scheme that his teacher Giovannoni had coordinated in 1926, constitutes the best product of the Neapolitan urban planning of the twentieth century, unfortunately unapproved (Pane A. 2015a). In this plan, the idea of "rehabilitating" the old city through punctual operations from Giovannoni's thinning out coexists with a modern vision of urban expansion, based on an open stellar scheme that envisages new neighborhoods separated from the existing city by large green agricultural areas, considered strategic by Piccinato to avoid the "wildfire" expansion of the city.

Starting from the early 1940s, a series of growing national and international commitments distanced Piccinato from Naples; among these, the coordination of the commission for the first Italian urban planning law (Law August 17, 1942 n. 1150), as well as the intense work on the themes of reconstruction carried out in the hectic years of the immediate postwar period, during which he also founded the magazine «Metron» (1945) and collaborated with Bruno Zevi in the Associazione per l'architettura organica (organic architecture) (Pane A. 2017b). In 1947—when he also published his Urban Planning manual with the Sandroni publisher of Rome—Piccinato left the University of Naples and taught for a few years in Argentina, where he also participated in the Buenos Aires master plan, returning to Italy in 1950, as full professor of Urban Planning at the University Institute of Architecture in Venice (IUAV). However, contact with the Neapolitan and Campania events never failed and ended up accompanying Piccinato throughout his life: after the unfortunate outcome of the 1936–39 plan, Piccinato drew up the inter-municipal plan of Sorrento (1940) and the arrangement plan of Monte Faito (1941), experiences that certainly anticipated the relationship with the Sorrento peninsula that would have developed later on the occasion of the PUT. For the city of Naples, he is then called upon to chair a series of subsequent commissions for the drafting of urban planning instruments, often resulting in only partial results, such as the one already mentioned for the General Regulatory Plan, appointed in 1962 and arrived at a first draft of plan of the Neapolitan district in 1964 (Dal Piaz 1985; Pane A. 2017a).

In these years the frequentations between Pane and Piccinato continued and intensified, consolidating a relationship started since 1930 and marked by affinity of views on the topic of territorial protection and planning (Piccinato 1955; Pane R. 1956). After several occasions of meeting and exchange during the 1950s—among which certainly the 4th INU conference, held in Venice in October 1952, where Pane presented a report on Landscape and environment, particularly appreciated by Piccinato (Pane R. 1953; Piccinato 1953)—the relationship became more assiduous precisely through the participation of both in the aforementioned planning commissions for the city and the region. In all these experiences, not always successful, Pane and Piccinato often share common positions, fighting "no plan" urbanism in favor of defining rules for an orderly development of the city and the territory.

To the figures of Pane and Piccinato, who, as already highlighted, covered the role of co-chairmen of the commission for the drafting of the PUT between 1973 and 1977, it must however be added the fundamental contribution of two other technicians of considerable stature who participated since the end of the 1960s to the genesis of the plan: Alessandro Dal Piaz (1939) and Angerio Filangieri (1925–2012). Dal Piaz and Filangieri were already present in the first commission charged with drawing up the "Scheme of territorial planning of the Sorrento-Amalfi Area", delivered in the spring of 1968 and already identified as the initial nucleus of the PUT: in fact, they had an important role in the plan drafting. The first of the two—urban planner trained with Marcello Vittorini (1927–2011), who also dedicated a direct testimony to the exciting experience of the PUT (Dal Piaz 2010)—was certainly responsible for the drafting of a first guideline document in 1973 (Pane A. 2017a), as well as the drafting of the implementation rules of the plan. The second—economist and geographer with a strong interdisciplinary vocation, trained with Manlio Rossi Doria (1905–1988) and then professor of agricultural planning and territorial organization at the University of Naples—is responsible for studying the rural landscape of the peninsula, carried out through an extensive documentation begun in the 1960s, and the definition of the relative standards also in relation to the sustainability of tourism development, the subject of an initial study since 1965 (Filangieri 1965).

The coordination of the entire working group of the PUT was entrusted to the architect Giuseppe Muzzillo, who was also present since the commission for the drafting of the general "Scheme" in 1968. Finally, at least the other components of the study group should be mentioned, whose specific contribution is not always easy to identify and which in some cases only participated in the first phase of elaboration of the plan. Among these the urban planner architect Mario Coppa (1923–1999), already in charge in 1970 of the drafting of the landscape plan of the Amalfi side of the peninsula, who resigned in July 1974, Dr. Edoardo Del Gado, who also resigned during the works and other four representatives appointed, respectively, by the Province of Naples (Giulio Vitolo, engineer), by the Province of Salerno (Giuseppe Gallo, engineer), by the Municipalities included in the perimeter of the plan and falling within the Province of Naples (Arrigo Marsiglia, architect) and by those falling within the Province of Salerno (Gaetano Francese, engineer) (Pane A. 2017a).

4 The Plan

The Territorial Urban Plan of the Sorrento-Amalfi Area (PUT) was drafted between 1973 and 1977 and was not approved until ten years later, following the so-called "Galasso" law on the landscape protection, namely L. 431 of 1985. The Plan aimed to regulate a territory of over 350 square kilometers which were divided into 33 municipalities. The version presented in 1977 included many of the analyses and some forecasts of the 1968 structural scheme but expanded them in an updated perspective. The peninsula was analyzed for its geomorphological, agrarian, landscape, environmental, architectural and economic aspects as well as their reciprocal interrelationships. The territory was divided into six macro-areas and was regulated into 15 prescription zones which had to be considered in the drafting of the urban plans: (1) protection of the natural environment of first and second degree; (2) protection of centralized ancient settlements; (3) protection of scattered ancient settlements; (4) first degree settlement and environmental redevelopment; (5) second degree settlement and environmental redevelopment; (6) completed urban areas; (7) settlement rationalization and protection of agricultural resources; (8) territorial parks; (9) special parks; (10) urban parks; (11) tourist facilities; (12) integrated sports equipment; (13) integrated natural reserves; (14) existing tourist settlements; (15) interchangeable equipment for transport (Fig. 2).

The Plan was divided into seven parts: (1) introduction, territorial organization, finalization, definition, training and implementation; (2) geomorphology and soil defense; (3) historical, artistic and environmental heritage; (4) population, economic

Fig. 2 Territorial Urban Planning Plan of the Sorrento-Amalfi Area approved by Regional Law 35 of 22 June 1987. The general zoning of the territory (Luigi Piccinato Archive, Rome)

activities and built heritage; (5) communication systems; (6) implementation and management regulations; and (7) territorial planning.

For the urban issue, the Plan aimed to strongly contain the new development, reducing new residential settlements and instead concentrating on redevelopment, rationalization and adaptation of the existing buildings. It also limited the expansion of tourist and sports facilities (public necessity construction allowed at most 15% of the surface, maximum height 8 m with a construction index not exceeding 0.6 / 1 mc/m^2).

Tourism was considered a fundamental resource for the area, as mentioned in paragraph two. It was argued that the success of tourism depended on the conservation of the landscape quality and on the enhancement of agricultural production activities that allow not only its maintenance but also its economic liveliness. The Plan seeked to address the economic crisis that had struck the canning manufacturing sector linked to agriculture in the Nocera and Sarno by defining an organic framework of qualitatively valid choices for the entire territory and quantitatively defined for the Sorrento area, spatially and functionally integrating the marginal areas within it. Following an updated ecological vision, the proposed territorial structure aimed at the integrated safeguarding of what was «before [being] aesthetic values, were biological and social values» (Scazzosi 2010). Hence, the conviction that it cannot be protected without active interventions to support agricultural and artisan activities, both fundamental elements for maintaining a balanced structure of the territory and, therefore, of the landscape.

Regarding transportation infrastructure, in addition to the construction of the roads already proposed in 1968, the Plan encouraged the use of public transport, providing for both the construction of funiculars and cable cars in specific and limited locations, as well as strengthening connections via sea. The Plan criticized the existing "model", essentially based on a perimeter road ring along the coast (the "Sorrento" and "Amalfi" state roads) which required crossing all the inhabited centers in sequence, congesting them. Conversely, it proposed an integrated system of two new hilly arteries for internal penetration, tunnel junctions to save coastal centers from crossing traffic, mechanical carriers (funiculars or cable cars) from suitable internal hilly locations toward coastal locations to enable parking (Dal Piaz 2010).

As far as the city-countryside relationship is concerned, the Plan went beyond the vision of the countryside as a boundary of the city toward a place of social interaction and environmental benefit for citizens and established the link with the city on a historical basis. It did not intend to bring history back, but to re-evaluate «the pre-existing historical, cultural and landscape features of the area linked to its ancient urban commercial production vocation». In particular, the Sorrento peninsula had a historical commercial link with Naples «as the predominant consumer market» so much so that «it established its mortgage on the coastal unifying organization» (PUT 1977/2007, p. 68).

From this starting point, three key aspects emerge and are discussed below: (1) a careful reading of the landscape as heritage values, within the framework concept of landscape system; (2) the role of farmers, as bearers of tradition and steward of the landscape; (3) the landscape restoration approach.

4.1 The Idea of "Landscape as Heritage" System

The Plan shows an attention to reading the landscape not as a sum of elements but as a complex system of resources passed down through history and adapted to contemporary commercial needs. The landscape system is composed by the historical viability, by the cultivation of terraces and by the handicraft manufacturing activities, that "can find new life only if connected to the agricultural productivity of the inland areas, re-evaluating the genuine and the natural" (PUT 1977/2007, p.16).

A considerable value is recognized in traditional practices and the need to safeguard them is declared because they guarantee the balance of a landscape that has shaped itself over the centuries. Regarding construction, regulations are envisaged for guaranteeing an appropriate use of building materials and construction techniques, banning—in the case of restorations—plastic plasters, synthetic paints, metal and plastic fixtures, the joints fulfill in the reconstruction of dry stone walls, favoring instead the use of traditional materials (Art. 26). Similarly, regarding agriculture, in the proposed plan, the conservation of traditional "straw canopy" is encouraged, not only as a qualifying element of the landscape, but also for strictly agricultural reasons (Fig. 3). The numerous traditional uses of the chestnut are described as evidence of agricultural multifunctionality and in favor of their conservation: the chestnut poles are used to support the vine rows and the pergolas for the winter covering of citrus groves (for 1 ha of citrus grove the preparation of a pergola requires about 4,000 poles from 6 to 8 m, which must be renewed on average every 18 years), while the dry chestnut leaves, collected on branches during the summer season and administered during the winter season, supplement the nutrition of the cattle, especially in the Agerola area (PUT 1977/2007, p. 39).

Attention is therefore turned not only to the conservation of the tangible heritage of rural artifacts, but also of the intangible ones. Consolidated human relationships, knowledge and practices keep the landscape alive and allow it to reproduce by responding to the needs of contemporary society (Branduini 2017; Scazzosi 2018). Regarding the hydrological structure, the Plan stated that in the case of streets and squares widening, it is not necessary to cover the terminal sections of the streams, but to leave the areas free for flooding. These are traditional practices of "common sense" in water management, which are recognized for their effectiveness and sustainability and therefore preferred over huge engineering works of hydraulic regulation.

The PUT is therefore founded on «an idea of a system that is based on the quality of the environment in the first place, from which psychological well-being derives, resulting in a tourist potential», says Antonio di Gennaro (PUT 1977/2007, p. 15–18). Moreover, Pane himself, as noted by Lionella Scazzosi, stated in the same years: «Therefore we seriously defend environmental assets only if, to the defense of the formal aspects, we add that of the local ecological conditions» (Pane 1977b; Scazzosi 2010).

Fig. 3 Sorrento, traditional crops protected by the «pagliarelle» (straw canopy) in a photo by Roberto Pane (Roberto Pane Photo Archive, University of Naples Federico II)

4.2 Farmer's Role in Landscape Management

In maintaining the environmental balance and human well-being, the active role of farmers is declared fundamental by the authors of the Plan: it is the «surplus of manpower that allowed the construction of such an elaborate landscape» (PUT, Plan Proposal, p. 45). Therefore the conservation of the landscape depends on the maintenance of production structures.

In this sense, the farmer must be recognized for the fundamental role he plays in rural and urban society, because with his work he maintains the landscapes that benefit the citizens of Naples and produces food that they consume. In the case of abandoned land, the Plan proposes that farmers, individual or associated, with the public authorities, could take charge of safeguarding the heritage. The survival of the terraces must first include inter-sectoral protection that compensates for the relatively high costs of traditional agricultural systems, taking into account the advantages that this form of landscape and land use determines for tourism. Secondly, it must address the implementation of technical interventions to consolidate them. Finally, and above all, it must offer incentives to improve the overall conditions of agricultural production. The actions that are considered fundamental for maintaining the landscape as a cultural asset are the on-site training of a qualified group of professionals (farmers, artisans, architects, archaeologists, geologists, engineers, historians, doctors of letters and philosophy and law); and guaranteeing employment, productivity and the appropriate agricultural and tourist use of the territory. The Plan also proposes exchange meetings between local publics, private operators and local stakeholders directly involved, since public institutions must have an active role in requalification and not only in economic support (PUT 1977/2007, p. 146).

Multiple and diversified actions are proposed to strengthen agriculture in the relationship to the city, such as assistance and research centers, to improve the quality of products; better accessibility to land and cultivated areas (roads for tricycles); creation of producers' cooperatives and direct sales centers, which would include the vertical integration of cooperative management from production to sale to enhance the network of direct relationships between city and countryside; creation of a brand of local products to improve the visibility of agricultural products; creation of food supply agreements with tour operators; hotels and campsites; and the improvement of the living conditions of farmers, through the implementation of toilets and telephone connections, what for us today would be internet connections.

These proposals essentially aim at strengthening the multifunctionality of the agricultural and forestry space, rationalizing the production chains, by supporting common management and marketing in an overall rural development strategy.

4.3 *Landscape Restoration Actions*

As anticipated, in the plan rules (Regional Law 35 of 22 June 1987), the attention to landscape as a heritage is translated into precise indications for each type of intervention. For new buildings construction the «building typology must take into account, in a cultural sense, the ancient construction logic, trying to interpret the relationship that was established between the building and the surrounding environment as a result of construction technology, materials and human needs in respect of the morphology and agricultural resources» (Art. 26) (Fig. 4). «The materials must be used in a natural sense and must re-evaluate the high craftsmanship tradition of local workers» (Art. 26) (Fig. 5); therefore, materials of «industrial standardization» must be banned. The dry stone walls of the agricultural terraces (Art. 34) must also be built in accordance with the construction techniques of the existing ones, i.e., exposed stones without jointing (Fig. 6).

Further indications are also provided for the landscape restoration: «to remove the causes that have altered the natural environment through the demolition of the previous works» and «to restore the primitive structure, both in regards to soil morphology and vegetation» (Art. 22). In this sense, it is proposed to eliminate alterations, which consist of volumes or profiles (buildings, advertising posters, masonry works, soil excavations, pylons and electrical networks, roads, etc.) that «in relation

Fig. 4 Massa Lubrense, the marina of Fontanelle in a photo by Roberto Pane (Roberto Pane Photo Archive, University of Naples Federico II)

to the context are harmful to the environment». As Lionella Scazzosi (2010) points out, we speak of environmental alteration in an imprecise sense with respect to the current notion of the term, since in reality the alteration due to improper volumes does affect the landscape, or rather the cultural construction of the natural environment e.g., an alteration of the environment could be represented by planting a conifer in a spontaneous broadleaf forest, in order to alter its ecological balance. Where it is not possible to eliminate the alterations, it is suggested to plant green diagrams, or cover concrete masonry with *opus incertum* stones without filling joints, in order to mask the external appearance of the places, not to act on the substance.

However, it should be noted that Pane—like much of the architectural culture of the second half of the twentieth century—uses the term «environment» (in Italian *ambiente*) in a broader sense, which also includes the built heritage and can therefore sometimes be used ambiguously to identify the landscape. The dynamism of the landscape is observed by Scazzosi in previous writings (Pane 1967), even if Pane calls it environment: «The environment is not only a context of aesthetic forms to be handed down to the future, but something alive, which, as such, cannot fail to be subordinated to the evolution of associated life» (Scazzosi 2010). However, in the

Fig. 5 The typical vaults that cover the traditional houses in the Sorrento-Amalfi peninsula in a photo by Roberto Pane (Roberto Pane Photo Archive, University of Naples Federico II)

Plan (and therefore in practice), while showing respect for traditional construction techniques and recognizing the role of agricultural workers in the transmission of knowledge, surface maintenance interventions are still suggested.

The restoration of the landscape must preserve the historical characters but keep the use alive. As the landscape is rapidly changing, it is important to keep active the practice that allows the perception and understanding of the historical character. When a landscape has evolved while maintaining its historical character, it has found a balance between permanence and change and is resilient to temporal transformations (Scazzosi 2020).

Fig. 6 Pogerola, the integration between agricultural landscape and built heritage in a photo by Roberto Pane (Roberto Pane Photo Archive, University of Naples Federico II)

5 Conclusions

The article analyzed the genesis and contents of the Territorial Urban Plan of the Sorrento and Amalfi Area, drafted by Roberto Pane and Luigi Piccinato and their working group, in particular, Alessandro Dal Piaz and Angerio Filangieri. It highlighted the preservation of the agricultural landscape as a fundamental condition for safeguarding the local identity and enhancing territorial development, through a series of stringent rules to guide the orderly development of the city and its territory.

The preliminary understanding of the territory began with rigorous historical research and from a direct reading in the field of environmental and landscape values through images that underline the very close relationship between natural and anthropic factors.

The interpretation of the landscape by the Plan is systemic and takes into consideration not only the spatial components of the agricultural landscape, but also the economic, productive and identity components. The strengthening of agriculture and associated infrastructures is fundamental to guarantee the quality of the agricultural landscape, which is the home and workplace for coastal inhabitants and a place of entertainment for Neapolitan citizens and tourists from all over the world. To this end, the Plan strengthens the multifunctional role of agriculture, rationalizing the production chains, by supporting common management and marketing with modern overall rural development strategy.

To connect the peninsula to the urban centers of Naples and Salerno, the Plan rejects the solution of a highway to penetrate the peninsula and proposes a new motorway axis east of Vesuvius with light internal penetration routes that follow the historical and environmentally sustainable network.

5.1 A Guarantee for Safeguarding a Precious Landscape

Through a decisive protection action, the Plan has guaranteed the safeguarding of a landscape that can still be appreciated today. Compared to the coastal territory between Naples and Salerno, the one affected by the Sorrento and Amalfi Plan is the only one spared from the uncontrolled construction of the 1980s. The PUT still has the effectiveness of a landscape and territorial-urban plan and allows the landscape indications to have a prescriptive validity, implementing both a territorial and building scale control, as «the documents come to provide very precise indications as in a hand-book», as Salzano points out (PUT, p. 12).

With the approval of the Plan in 1987, Dal Piaz argues, the Campania region had been far-sighted, but then ran aground, so much so that «almost all the commitments regarding support for sustainable production activities were canceled, thereby making a decisive step that then facilitated the task of the detractors, interested in presenting the approved plan only as a cage of constraints». Although «the plan has been experienced in the area as a strictly binding instrument, contrary to its vocation and the logic that presided over its elaboration, innovative for the times and extraordinarily current today», affirms Raffaella Di Leo, it had the merit of delivering to the new millennium an agricultural landscape in which man's building hand is still legible.

The Plan confirms that the positive effects of prescriptions, care and respect can be seen over time. These may be appreciated by reversing the current vision on preservation as a constraint and considering what that landscape would have become if limits to transformations had not been applied. New construction and infrastructure offer novelty and apparent solutions to critical issues that have an immediate and easily perceivable impact on the population, soliciting votes for politicians.

5.2 *The Dynamic Landscape Conservation for Urban Agriculture*

The dominant idea of Pane and Piccinato was not binding and passive protection but proposed what we now call dynamic conservation. The rural landscape created by urban agriculture is subject to changes which, within certain limits, must be accepted on the condition that they can be guided and oriented to be "appropriate" (Scazzosi 2020). Dynamic conservation does not place "a priori norms but (makes us) aware of the need to achieve a harmonious coexistence between old and new" (Pane in Scazzosi 2010). For Pane, this is achieved by listening to people, to their contemporary needs and to the current uses that the ancient matter can have. This idea agrees, adds Dal Piaz, with «[Pane's] rigorous attention to the ethical dimension of spatial planning, with its task of making collective needs, citizenship rights, and interests prevail in the event of a conflict» (Dal Piaz 2010). Dynamic conservation of the landscape heritage considers the needs of the population through a continuous process of updating and integrating the ancient technical knowledge experienced over the centuries (collective genius) with current technology, manual work with mechanization and renewable resources with fossil ones (Branduini and Scazzosi 2020). It can be a resource for the management of urban agricultural landscapes, susceptible to rapid urban transformations, promoting self-sustenance food (food gardening), and outdoor uses, which can support the balance between tradition and sustainable innovation.

5.3 *The Seeds of Engagement and Landscape Co-Construction*

Listening to communities, in particular rural ones, provides the seed of co-construction between farmers and citizens of agri-urban territory. This is the vision that André Fleury developed in twenty years of studies, extending the concept from peri-urban to urban agriculture. This includes the economic and social links that agriculture creates with the city beyond the physical proximity of the city (Donadieu and Fleury 1997) and the recognition of the multifunctional role of agriculture in urban planning as providing fresh food, environmental quality and cultural good citizens' supplier (Fleury 2020). All these components are already present in the PUT, thanks to the attention paid to farmers in defining their requests. These farmers are now prepared to engage in the participatory process with citizens (Branduini, 2020). Ultimately, even if «in Pane's writings it does not seem to emerge a clear stance in favor of the participation of populations in the places' knowledge and in the actions' definition, he would probably have supported this idea and would have explored its tools and contradictions» (Scazzosi 2010).

Acknowledgments We acknowledge the Roberto Pane Photo Archive, University of Naples Federico II and Luigi Piccinato Archive, Rome for allowing the reproduction of their images.

References

Astengo G (1958) Assisi: Piano generale e Piani particolareggiati di primo intervento. Urbanistica 24–25:1–124

Belfiore P (1994) Dal dopoguerra ad oggi. In: Belfiore P, Gravagnuolo B (eds) Napoli. Architettura e urbanistica del Novecento. Laterza, Roma-Bari, pp 67–123

Belli G, Maglio A (eds) (2015) Luigi Piccinato (1899–1983). Architetto e urbanista. Aracne, Roma

Branduini P (2016) Il patrimonio rurale nutre la città. Economia e Società Regionale 2:44–54

Branduini P, Laviscio R, Scazzosi L, Supuka J, Toth A (2016) Urban agriculture and cultural heritage: an historical and spatial relationship. In: Lohrberg F, Licka L, Scazzosi L, Timpe A (eds), Urban Agriculture Europe, Jovis, Berlin

Branduini P (2017) La trasmissione della conoscenza del paesaggio agrario: esperienze multimediali dinamiche. Scienze Del Territorio 5:172–180. https://doi.org/10.13128/Scienze_Territorio-22244

Branduini P (2020) Engagement, participation and governance of urban agricultural heritage. In: Scazzosi L, Branduini P (eds) AgriCultura. Urban agriculture and the heritage potential of agrarian landscape, Urban Agriculture series, Springer

Branduini P, Scazzosi L, Pratesi C, Meregalli D (2020) Paesaggi rurali, Politica Agricola Comunitaria e pandemia. Opportunità da cogliere. Ri-Vista 2

Canestrari M (ed.) (1991) Il Piano di Volterra, Officina

Casiello S, Pane A, Russo V (eds.) (2010) Roberto Pane tra storia e restauro. Architettura, città, paesaggio. Atti del convegno, Napoli, 27–28 ottobre 2008, Marsilio, Venezia

Casiello S (ed.) (2011) Offese di guerra. Ricostruzione e restauro nel Mezzogiorno d'Italia. Alinea, Firenze

D'Orta L (2010) Il progetto di PRG per Sorrento di Roberto Pane (1946–1963). In: Casiello S, Pane A, Russo V (eds.) Roberto Pane tra storia e restauro. Architettura, città, paesaggio. Atti del convegno, Napoli, 27–28 ottobre 2008, Marsilio, Venezia, pp. 517–522

D'Orta L (2011) Lo studio di Renato Bonelli per il piano territoriale-paesistico della penisola sorrentina (1970–75). In: Fiengo G, Guerriero L (eds.) Monumenti e documenti. Restauri e restauratori del secondo Novecento, Atti del Seminario Nazionale, Arte Tipografica, Napoli, pp. 201–212

Dal Piaz A (1985) Napoli 1945–1985. Quarant'anni di urbanistica, Franco Angeli, Milano

Dal Piaz A (2010) L'esperienza innovativa del piano territoriale e paesistico dell'area sorrentino-amalfitana 1974–1977. In: Casiello S, Pane A, Russo V (eds.) Roberto Pane tra storia e restauro. Architettura, città, paesaggio. Atti del convegno, Napoli, 27–28 ottobre 2008, Marsilio, Venezia, pp. 523–525

De Lucia VE, Iannello A (1976) L'urbanistica a Napoli dal dopoguerra ad oggi: note e documenti, Urbanistica 65

Di Bene A, Scazzosi L, Laviscio R, Valentini A (2019) Indirizzi metodologici per l'adeguamento e la conformazione della pianificazione urbanistica comunale al PIT-PP della Regione Toscana, Maggioli editore

Donadieu P, Fleury A (1997) De l'agriculture péri-urbaine à l'agriculture urbaine. Le Courrier De L'environnement 31:45–61

Fanfani D (2015) Il parco agricolo periurbano. Un percorso condiviso di governance e progettazione del territorio. In: Gisotti MR (ed.) Progettare parchi agricoli nei territori intermedi. Cinque scenari per la piana fiorentina, Firenze University Press, Firenze

Ferraresi G (2009) Produrre e scambiare valore territoriale. Dalla città diffusa allo scenario di forma urbis et agri, Alinea Editrice, Firenze
Filangieri A (1965) Piano di massima per lo sviluppo del comprensorio turistico della penisola sorrentina, Italia Nostra, Portici
Fleury A (2020) Agricultural and urban policies in Europe: the co-construction of periurban agricultural landscape. Experiences, problems, perspectives. In: Scazzosi L, Branduini P, (eds) Agri-Cultura. Urban agriculture and the heritage potential of agrarian landscape. Urban Agriculture series, Springer
ICOMOS (1994) Nara Document on Authenticity, Nara, Japan, https://www.icomos.org/charters/nara-e.pdf. Accessed 9 July 2021
ICOMOS-IFLA (2017) Principles concerning Rural Landscapes as Heritage. ICOMOS Doctrinal text, Delhi https://www.icomos.org/images/DOCUMENTS/General_Assemblies/19th_Delhi_2017/Working_Documents-First_Batch-August_2017/GA2017_6-3-1_RuralLandscapesPrinciples_EN_final20170730.pdf. Accessed 9 July 2021
ICOMOS (2019) The Future of Our Pasts: Engaging Cultural Heritage in Climate Action. https://indd.adobe.com/view/a9a551e3-3b23-4127-99fd-a7a80d91a29e. Accessed 9 July 2021
Lanzani A (2003) I paesaggi italiani, Meltemi editore, Roma, p. 20
Laviscio R, Scazzosi L (2015) Cultural approaches toward a world rural landscape recommendation. In: Di Stefano M (ed) Heritage and landscape as human values. Proceedings of the symposium, 18th ICOMOS General Assembly, 9–14 November 2014, Florence/Naples, Edizioni Scientifiche Italiane, pp. 233–240
Magnaghi A (2000) Il progetto locale. Verso la coscienza di luogo, Bollati Boringhieri, Torino
Mangone F, Telese R (2001) Dall'Accademia alla Facoltà. L'insegnamento dell'architettura a Napoli, 1802–1941. Hevelius edizioni, Benevento
Mangone F (2008) La nascita della Scuola superiore di Architettura a Napoli. In: Gravagnuolo B, Grimellini C, Mangone F, Picone R, Villari R (eds.) La Facoltà di Architettura dell'Ateneo fridericiano di Napoli, 1928/2008. Clean, Napoli.
Mangone F (2014) Roberto Pane. In: Trione V (ed) Il cinema degli architetti. Johan&Levi, Milano, pp. 162–165
Merlini C (1992) Luigi Piccinato. Una professione per la città e la società. In: Di Biagi P, Gabellini P (eds.) Urbanisti italiani. Piccinato, Marconi, Samonà, Quaroni, De Carlo, Astengo, Campos Venuti. Laterza, Roma-Bari, pp. 23–95
Mininni M (2012) Approssimazioni alla città: urbano, rurale, ecologia. Donzelli, Roma
Pane A (2015a) Diradamento e risanamento delle "vecchie città". L'opera di Piccinato tra continuità e rottura con Giovannoni da Padova a Napoli. In: Belli G, Maglio A (eds.) Luigi Piccinato (1899–1983). Architetto e urbanista. Aracne, Roma, pp. 53–77
Pane A (2015b) Napoli: Francesco Rosi e Le mani sulla città, 50 anni dopo. 'ΑΝΑΓΚΗ 75: 75–84
Pane A (2017a) Il paesaggio tra urbanistica e tutela: la genesi e il destino del PUT della penisola sorrentino-amalfitana (1973–1987). In: Aveta A, Marino BG, Amore R (eds.) La Baia di Napoli. Strategie integrate per la conservazione e la fruizione del paesaggio culturale. Artstudiopaparo, Napoli, vol. I, pp. 150–156
Pane A (2017) Da vecchie città a centri storici: il contributo di Luigi Piccinato alla conservazione urbana tra ricostruzione e primo boom economico. Storia Urbana XL 156–157:97–123
Pane R (1953) Paesaggio ed ambiente. In: La pianificazione regionale, Atti del IV Congresso Nazionale di Urbanistica, Venezia, 18–21 ottobre 1952, Roma, pp. 89–95
Pane R (1954) Capri. Neri Pozza, Venezia
Pane R (1955) Sorrento e la costa. Edizioni Scientifiche Italiane, Napoli
Pane R (1956) Appello alla cultura. In: Lettere al direttore. L'architettura. Cronache e storia, I, 5: 653–654
Pane R (1958) Napoli d'oggi. In: Documento su Napoli. Edilizia e urbanistica, Edizione di Comunità, Napoli, pp. 5–12
Pane R, Alisio G, Di Monda P, Santoro L, Venditti A (1959) Ville vesuviane del Settecento. Edizioni Scientifiche Italiane, Napoli

Pane R (1961) Campania. La casa e l'albero. Montanino, Napoli
Pane R (1962) Bellezze naturali e caos urbanistico. In: Tuttitalia. Enciclopedia dell'Italia antica e moderna. Campania, vol. I, Sadea, Milano, pp. 326–328
Pane R (1967) Italia da salvare: tattica e strategia. Lettere a "Casabella" di Rosario Assunto e Ro-berto Pane, «Casabella» 320: 6
Pane R (1973) Documento presentato da R. Pane e sottoscritto dalla Commissione della Regione Campania per il piano territoriale della penisola sorrentina e della costiera amalfitana, aprile 1974. Restauro, II 10:71–74
Pane R (1977) Documentazione ambientale della Costiera amalfitana. Napoli Nobilissima, XVI 1:3–41
Pane R (1977) Il restauro dei beni ambientali, la Carta di Venezia e l'illusione tecnologica. Restauro 33–34:22–23
Piccinato L (1953) Relazione conclusiva sulle discussioni e interventi di carattere generale. In: La pianificazione regionale, Atti del IV Congresso Nazionale di Urbanistica, Venezia, 18–21 ottobre 1952, Roma
Piccinato L (1955) Appello ai conservatori. L'architettura. Cronache e Storia I(1):35
PUT (1977) Piano Urbanistico Territoriale dell'Area Sorrentino-Amalfitana. Italia Nostra, Istituto Italiano di studi filosofici, ristampa 2007, Napoli
Regione Campania (1987) Legge Regionale 27 giugno 1987, n. 35 (GU n. 037 SERIE SPECIALE N. 3 del 26/09/1987—BU Campania n. 040 del 20/07/1987) Piano Urbanistico Territoriale dell'area Soerrentino-Amalfitana.
Salzano E (1997) Leggi e istituzioni. In: Dal Co F (ed.) Storia dell'architettura italiana. Il secondo Novecento, Electa, Milano, pp. 344–367
Scazzosi L (2010) Roberto Pane e il paesaggio: attualità del pensiero. In: Casiello S, Pane A, Rus-so V (eds.) (2010) Roberto Pane tra storia e restauro. Architettura, città, paesaggio. Atti del convegno, Napoli, 27–28 ottobre 2008, Marsilio, Venezia, pp. 465–469
Scazzosi L (2018) Landscapes as systems of tangible and intangible relationships. Small theoretical and methodological introduction to read and evaluate rural landscape as heritage. In: Rosina E, Scazzosi L (eds) The conservation and enhancement of built and landscape heritage. A new life for the ghost village of Mondonico on Lake Como. PoliScript, Milan, pp. 19–40
Scazzosi L (2020) Urban Agriculture as Heritage: methodological issues and perspectives. In: Scazzosi L, Branduini P (eds) AgriCultura. Urban agriculture and the heritage potential of agrarian landscape, Urban Agriculture series, Springer
Scazzosi L, Branduini P (2020) Toward a linkage between urban agriculture and cultural heritage. In: Scazzosi L, Branduini P (eds) AgriCultura. Urban agriculture and the heritage potential of agrarian landscape, Urban Agriculture series, Springer
UNESCO (1972) World Heritage Convention, WHC, Paris. https://whc.unesco.org/archive/convention-en.pdf. Accessed 9 July 2021
UNESCO (2003) Convention for the Safeguarding of the intangible Cultural Heritage, Paris. https://ich.unesco.org/en/convention. Accessed 9 July 2021
UNESCO, World Heritage Committee. (2019) The operational guidelines for the implementation of the world heritage convention. Paris. https://whc.unesco.org/en/guidelines/. Accessed 9 July 2021
Vejre H et al (2016) Can agriculture be urban? In: Lohrberg F, Licka L, Scazzosi L, Timpe A (eds.) Urban Agriculture Europe. Jovis, Berlin
Viganò, P (2001) Spostamenti concettuali. In: Viganò P (ed.), Territori della nuova modernità. Provincia di Lecce, Assessorato alla gestione territoriale, Piano territoriale di coordinamento, Electa, Napoli
Zasada I (2011) Multifunctional Peri-Urban Agriculture. A Review of Societal Demands and the Provision of Goods and Services by Farming. Land Use Policy 28 (4): 639–48

Productive Urban Landscapes

Dynamic Usonia: The Evolution of Wrightian Organic Principles for Community Sustainability

Cecilia Muzika-Minteer

Abstract Frank Lloyd Wright's vision of an organically formed, organized, and functioning agrarian community, as it evolved in the early to mid-twentieth century, is outlined by principles intended to be ever-changing. In considering change, I propose it is essential to evolve these principles to corollate with contemporary necessities, as they are not stagnant ideas, but ones intended to be reactionary to time and place. Wright's principles of *plasticity*, *continuity*, *simplicity*, *horizontality,* and *harmony* support the organic goals of *reintegration* and *communal individuality*. This organic community was first proposed as Broadacre City and evolved into the farther-reaching Usonia through the reiterations of the text accompanying Wright's organic community plan, reinforcing the action of decentralization and values of traditional agrarianism. Though Broadacre City has been studied endlessly in architecture and planning circles, since its conception, an understanding of the dynamic principles Wright infuses into the form, function, organization, governing systems, and land use have yet to be evolved holistically, as the organic theory requires, to apply to self- and communally-reliant, sustainable, and reactive communities in the twenty-first century. By focusing on the aspects of these principles that make a community and its population both communally and self-reliant and sustainable, we may better understand how Usonia may still be the organic community of the future.

Keywords Organic architecture · Community agrarianism · Community sustainability · Urban agrarianism · Organic planning

1 From Broadacre City to Usonia

Criticism of the city, its operations, underlying systems of inequity, and its stifling, stagnant architecture, did not begin, nor did it end with Frank Lloyd Wright. Solutions, deemed to be organic, that respond to the unsustainable processes and forms within the city have been labeled utopian for centuries due to a fracturing of norms

C. Muzika-Minteer (✉)
The University of Texas at Austin, Texas, TX, USA
e-mail: cecilia.klein@austin.utexas.edu

C. Brisotto and F. Lemes de Oliveira (eds.), *Re-Imagining Resilient Productive Landscapes*, Cities and Nature, https://doi.org/10.1007/978-3-030-90445-6_9

and values that influence a human settlement's operation. Wright's organic theories are some of the many used as inspiration to reorder human settlement patterns. Solutions that aim to reevaluate and reform the ideological norms and values that govern the city, and since the mid-twentieth century, the megalopolis have been proposed as radical ways of reshaping a deteriorating and suffering society. In this sense, Wright's Broadacre City, and many of its underlying design principles, aligns with the trend of decentralization and the values of traditional agrarianism seen across the globe in the nineteenth and twentieth centuries. The contained diversity of Wright's organic settlement plan evolves over the latter part of his career in the publications *The Disappearing City* (1932), *When Democracy Builds* (1945), and *The Living City* (1958). The settlement's form relies on natural processes, cycles, and the inherent formal and functional dynamism necessary for a reactionary settlement to serve humanity rather than unhuman systems. By exploring Wright's community design through the theorized site's textual accompaniments, this research proposes that the hopefulness of Broadacres City, and Usonia, lies in the purpose imbued in spaces through formal, functional, and organizational means. This chapter will explore Wright's dynamic principles around which he forms his Agrarian Utopia—principles that are reactionary and promote individual and community lifestyles that make sustainability, reliance, and flexibility more possible. By exploring and evolving these principles, Wright's Usonia may provide an outline for an Anthropocene community founded on equity in the twenty-first century rather than the polarizing and often dehumanizing attempts toward equality in Wright's time.

The values that govern space and design in Wright's Broadacres, which evolves through the mid-twentieth century into Usonia, are romanticist and conflicting, yet they provide an outline for a settlement pattern that prioritizes spatiality, design, and organization ideal to the human condition. Broadacre City is the original decentralized community plan outlined by Wright, but as the ideas of this community evolve to be farther-reaching, and both more specialized and generalized to fit into the North American landscape, it transforms into Usonia. Wright relies on formalizing his organic ideals of *communal individuality* and *reintegration* through arterial-like highways, equitable distribution of land and resources, design and use of communal and public space, and both standardization and variety of architectural forms. This helps to create equal opportunity for all citizens to reach their potentialities and ultimately leads to the betterment of society. The ideals of his perspective that influence the design, scale, use, and organization of a Usonian community and the landscape rely on a community's ability to feed, shelter, and clothe its population. As the agrarian worker provides two of these three necessities within a community, this increases the need for productive use of landscape to occur at large and small scales. The various productive landscapes within a community are to produce as much variety as possible in response to local climate, and for people of all ages and abilities to participate in food cultivation and production. Wright's organic principles that work in tandem toward and with *reintegration* and *communal individuality* are *plasticity*, *continuity*, *simplicity*, *horizontality*, and *harmony*. When these organic principles are formalized in a community, Wright's urban agrarian utopia begins to take shape.

1.1 Outline of Wright's Organic Principles

The organic principles Wright outlines in his publications about Broadacre City, and eventually Usonia, all depend on one another to create an organic community and life for humanity. Without the application of all of these principles, in an organic system, none will come to fruition, as they are all interwoven to create a whole. *Reintegration*, though applicable in the short-term societal and community changes that would occur in the systems governing space in Usonia, would appear in the long-term as an organic pattern of human occupation across the landscape. It is an action and process that lends itself to change (Wright 1932, 1945, 1958). *Communal individuality* is opposed to the toxic and rugged individualism that is the norm of the United States. It is mirrored in the systems that govern a community and intends to outline a holistic community created by the application of organic design of spaces and systems (Wright 1932, 1945, 1958). *Plasticity* is seen in Wright's forms and ornament through a pattern that provides flexibility and depth to space and design. It also allows for change to occur in spaces as architecture, planning, and design adapts. In ornament at a building's scale and individual architectural forms at a community scale, *plasticity* engages with *continuity* to create a holistic and organic design, making form and function synchronous. (Wright 1932, 1945, 1958 and Kaufmann 1978). *Continuity* is, as Edgar Kaufmann Jr. extrapolates, the holistic and dynamic relationship in which "architecture, mankind, and nature" are engaged. As this organic lens views, it is a natural process, *continuity* must, therefore, be instilled in a building for it to be considered organic (Kaufmann 1978).

Simplicity is the opposite of excess and is ideal for both spatial and system design. In architecture, *simplicity* should be adhered to with form, function, material, and pattern (Wright 1932, 1945, 1958). It also helps to create a path for *plasticity* and *continuity* to be built into spaces and systems. *Horizontality* is the "earthline of human life" and the ideal of spatial freedom that promotes self and communal connection to the landscape and other individuals (Wright 1932, 1945, 1958). It also helps to balance community and private space. *Harmony* exists where formally and materially landscape and architecture become one and also when the values instilled in the architectural forms are human values (Wright 1932, 1945, 1958). Considering the original intent of these principles is central though their dynamic nature also requires them to be placed in an evolving context holistically, beyond Wright's early to mid-twentieth century vision, which will be explored later in this chapter.

The foundation of Wright's utopia makes life and prosperity a right to all citizens rather than something to earn through short-term monetary gains. Like other utopia's, Wright's Broadacre City, and eventually Usonia, use architecture and planning to create a better society, thus making it a social idea rather than strictly figurative. Land, air, water, and sunlight are a birthright, and thus Wright relies on the land-owning citizen-farmer to form the core of his "everywhere and nowhere" Utopia (Wright 1932, 1945, 1958). Both the forms within the settlement, and those who dwell within, provide multifunctional and multipurpose benefits to the community as a whole. The intent is to expand the individual's sense of identity to a community scale, beginning with the sustaining quality of those who provide agricultural products to the population. This necessitates a localized supply and demand for anything a farmer may provide, and for what the population may consume.

The blossoming of the *Self* in Usonia, much like the core of other modernist settlements conceptualized by other architectural leaders of the time, differentiates Wright's concept from other decentrists and modernists.[1] To Wright, organic norms and values must be predicated on holistic human betterment, or rather on community betterment, embodying his concept of an organic-Self. This stands in opposition to the inorganic norms and values formalized in the modern city, which according to Wright, prioritize the abstract systems and their architecture that perpetually reinforce the devaluing of human life (Wright 1953). The result intends to shrink the ecological footprint of industrialization and expand the cultural footprint through the same system of localized social and economic continuity. Through the formalization of Wright's organic principles, this works to reform communally centered ideals that benefit both the local community and the larger human community. To put it simply, the settlement becomes more self-sustaining as the value of all citizens grows; this is especially true for the agrarian worker, which is a mantel that should be dawned by each citizen in one form or another.

2 Land Distribution and Spatial Freedom

The philosophy of the land-owning citizen-farmer shows Wright's reliance on Jeffersonian simplicity which is formalized in the rigid yet fluid grid that stretches across the landscape.[2] Like Ebenezer Howard's Garden City,[3] Wright's Usonian project focuses on a localized economy, geographically layered planning concept, and a solution for those who had been most abused by a cruel economic system, which Wright refers to as a dehumanizing "uneconomic system" which creates the "slums."[4] This combines with the American colonial ideal of manifest destiny to create a settlement that, to Wright, embodies Romantic and democratic individuality. There is also an emancipatory strain of thought in making land available to those who have been most disenfranchised. Sherman's Field Order No. 15, released near the end of the American Civil War, intended to redistribute land to formerly enslaved African's

[1] Other decentrists of the early to mid-twentieth century include architects and theorists such as Lewis Mumford and Henry-Russell Hitchcock and industrialists such as Henry Ford.

[2] The 1785 Land Ordinance, promoted by Thomas Jefferson, was the framework through which white settlers began to settle and develop the western part of the United States.

[3] Initiated by the English urban planner in 1898, Ebenezer Howard's Garden City was responsorial to lacking urban welfare and a decreasing rural population due to industrialization. Howard's designs and publications continue to influence settlement designs across much of western Europe (Howard 1898, 1902).

[4] Wright's use of the word "slums," which is written with a racist undertone, is in his outline for Broadacre City and is used often in the text when criticizing centralization (Wright 1932, 1945, 1957, 1958).

that bore the injustices of chattel slavery.[5] Though this plan was halted, progressive politicians of the time, such as Charles Sumner and Thaddeus Stevens, who would have been well-known political leaders when Wright was a child, as he was born in 1867, promoted land redistribution as a step toward equalizing accessibility to landscape, resources, and opportunity.

However, Wright's acceptance of segregation and of Jim Crow era laws, practices, and ideals, as noted by Dianne Harris, shows an underlying tone of white supremacy and racism that has been continually propagated in the built environment, visible in the city, countryside, and in the curated and predominantly white suburb (Harris 2021). In his publications about Broadacre City and Usonia, as well as others, the dismissal of individuals of color, his consistent assumption of white clients, and his use of racist language all feed his conflicting vision of democracy. While Wright champions democracy, he also continues the same colonial ideals that lead to the genocide of peoples indigenous in the contemporary United States, whom he both romanticizes and culturally appropriates, and the brutally inhuman enslavement of Africans, rooted in imperialist rather than democratic norms and values. Wright's assumptions of equality, his focus on a single-class system, and reinforcement of the white supremacist status quo are clear in the accessibility and inaccessibility of Wright's Agrarian Utopia (Harris 2021). Wright's Usonian vision, spatial freedom and order, use of landscape, and pioneering spirit with an aim of decentralization are all to be led by white men (Wright 1932, 1945, 1958). Though Wright's principles and beliefs are in conflict with each other and harmful to the population, his utopia claims to serve, organic theory, by its nature, is meant to change. Accountability for underpinning racist structures is necessary to allow organic theory and principles to evolve.

Wright first introduces the concept of Broadacre City in a lecture at Princeton University in 1924, and a decade later, one of Wright's patrons, Edgar Kaufmann, whose commission of a vacation home led to the construction of Fallingwater, supported the design and construction of a model of the Broadacre City plan, completed at Taliesin in the Winter of 1935. Later, reiterations that display the contained diversity have been created, and recreated, by artists, historians, and theorists (see Fig. 1). The geometric aerial view brings Wright's hope of an Agrarian Revolution to life, with his reliance on the organic principles of *simplicity, continuity, plasticity, horizontality*, and *harmony* (Wright 1932, 1945, 1958). Wright relies on a new vision of space, one that is founded on early to mid-twentieth century reactionary yet regressive ideals of equality in addition to the influence of the automobile. As the economic collapse of the 1930s made clearer the disparities between industry and individual, as well as urban and rural, Wright saw the cure in a new land distribution policy. One acre, the new measure of space, rather than the dehumanizing

[5] General William T. Sherman issued Field Order No. 15 on January 16, 1865 intended to redistribute thousands of acres of land exclusively to formerly enslaved African's and their families, though it never came to fruition due to white supremacist social, economic and political backlash under the Johnson administration.

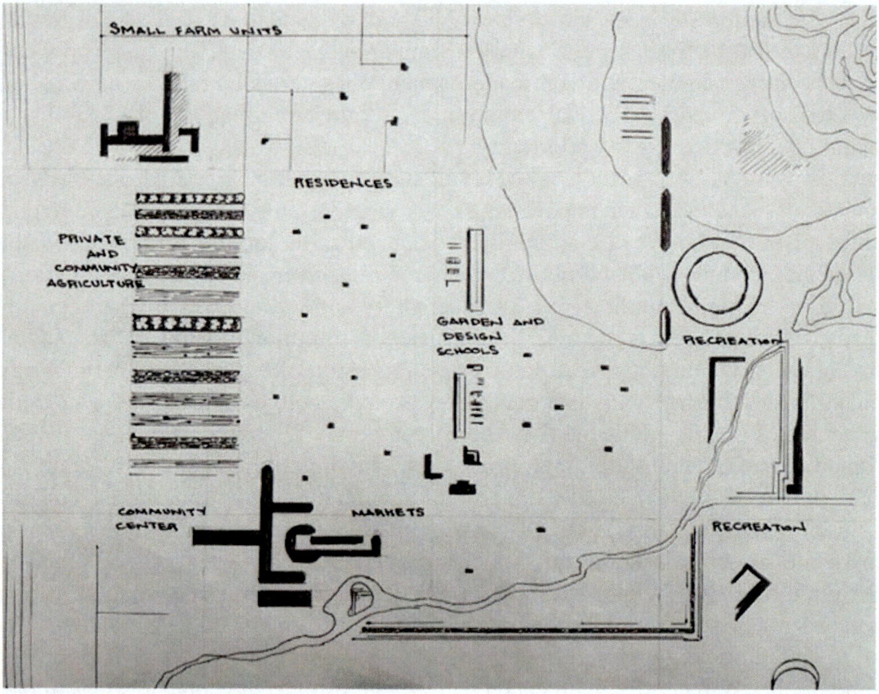

Fig. 1 Wright's Broadacre City plan was recreated by Bruce Pfeiffer to display the organization of space in the idealized community (Reproduced Illustration by Regina Carmine et al. 2021)

square-footage of the city, is a birthright in Broadacre City. Broadacres is a long-range dynamic system for human species survival, which, placed in a transformative context, is always changing to be balanced and harmonious with a multiplicity of social, economic, and political factors and advances.

Wright's vision of a balanced human community lends itself, through his perspective, to an agrarian landscape pattern in order to promote both communal- and self-reliance and sustainability. Just as Wright's perspective of social justice issues of the era are regressive, how a community becomes reliant and sustainable, and whom it is set to benefit, is limited. This affects the form, function, and organization of built spaces and landscapes. The contained diversity of Usonia would be woven into the landscape as small farms would be dispersed more frequently not to compete with early twentieth-century industrial farming, but to fill gaps in human need and create a strong localized economy. Land is a birthright, contingent on to whom Broadacres is accessible, yet Wright states the intent to make acreage available in accordance with one's ability (Wright 1932, 1945, 1958). Communities are sustained and become reliant by balancing the use of the "factories and fields," though it is clear that this organic vision of land distribution hinges on productive land use beyond the large and small farms, to change the lifestyle of the average resident of a Usonian community. Wright's intent with industry is to beautify it, make it serve the many rather than the few, and to cause minimal environmental damage (Wright 1932, 1945, 1958). The life-sustaining role of the agrarian worker lends itself to organic forms, which are ideal for both human individual and human species well-being. The farmer needs an organic architecture in the same way they need water and soil and all citizens need the farmer and should thus take some life-giving necessities upon themselves. Land distribution and community patterns help to place the responsibility of sustaining one's community and oneself in the hands of all citizens.

With the support of an organically designed human habitat, centered around access to space and ground, agrarianism is built-in, and the agrarian worker can thrive in a society that values their integral role in the continuation and evolution of the human species. Wright imagines that "dairying, fruit growing, truck gardening, raising the rarer meats, fowls, eggs, all in which freshness is the first consideration, will be [the farmers] direct contribution to the city in which he himself lives" (Wright 1932, 1945, 1958). The value of the agrarian worker and of one's individual productive and sustaining space is, to Wright, equivalent to the value of human life itself, for "the food master of humanity" decides the survivability of the species (Wright 1932, 1945, 1958). Due to its organically evolving nature, one might argue that Usonia is not a utopia, which relies on stagnant perfection rather than a dynamic, changing, and imperfect whole, yet the reliance on traditional agrarianism does give the settlement a stationary outline.

Allotments of three-, five- and ten-acre plots for use as "small farm units" allow the farmer to participate in more Self exploration and partake in the leisure time offered to other citizens by the farmer's continued sustaining efforts. The system of localized supply and demand, based around centralized markets and assured crop sales, is built into the formal, functional, and organizational scales of Usonia. Transportation and communication networks directly connect agrarian workers and all citizens to markets, community centers, industrial sites, and other multifunctional facilities. It is Wright's intent that larger markets be like "county fairs" and "highway attractions," and smaller markets be dispersed more frequently, just miles from farmers and all citizens (Wright 1932, 1945, 1958). In community spacing, Wright balances the connectivity to community spaces and large plots of private space by centralizing automobile transportation—within one hundred miles is everything a citizen might need. Yet, essential spaces, like markets, just a few miles away, are evenly dispersed through the landscape, among the individual acre allotments. These formal, functional, and organizational details of Usonia centralize individual freedom of movement—accessibility provided by community and landscape provided by decentralization.

3 Formal and Social Communal Individuality and Reintegration

The equal right to space in Usonia is founded on Wright's readings of William Blake in that "the organic relation between the welfare of one and the welfare of the whole can create a permanent and universal well-being" (Wright 1932, 1945, 1958). Spatial humanism on a communal scale is how Wright's land distribution policy translates to *communal individuality.* This is a primary objective of *communal individuality* as both a formal and social idea, for if one is to thrive, one's neighbor must also thrive. For settlements to form around the concept of *communal individuality*, Wright's organic principles are intended to promote the maximum amount of self- and communal-sustainability and reliance. This works with *reintegration* across the landscape to create beautifully functional spaces that from one generation to the next are intended to become more balanced, creative, and unified, connecting people with landscape. Wright's concept of *communal individuality* proposes a new system for community connection in which one values individuality over individualism, and thus forms and patterns within a settlement balance communal and private property, space, and landscape.

However, Wright genders space and constructs its functioning under the domestic family unit, just as his sense of spatial freedom is rooted in racialized and imperial motives. For an organic community to evolve and possess a true sense of *communal individuality*, it must formalize flexibility through *reintegration* to become responsorial. Overall, spatial, design and landscape flexibility, hybridity, and multiplicity are central priorities of built forms, and also governing systems, making a community's interconnectivity clear. Organicism, as defined by the goals of *communal individuality* and *reintegration*, is thus ideal for the evolution of humans, socially, individually, and as a species. Furthermore, there are integral organic principles and forms Wright notes that theorists and architects have further investigated into the twenty-first century which can help to outline the dynamic long-term changes necessitated in an organic community. There is a multitude of architectural theorists that have applied different aspects of Wright's organicism, creating a lineage that evolves the details of Wright's *communal individuality* and *reintegration* and the elements that buttress them.[6] These concepts can be best enacted and evolved through the integral community spaces, especially the communal, as well as private productive landscapes of Usonian settlements.

The organization of community and private spaces are central to instilling the goals of Wright's organic settlement. The integral community spaces and forms in Usonia are markets, community centers, education centers and landscapes, such as parks, gardens, and preserves. Each of these spaces, both specifically and generally, is productive, educational, and multifunctional, encouraging connection between

---

[6] A few examples: Principle of "Critical Regionalism," Kenneth Frampton (1998); principles of community, Serge Chermayeff, Alexander Tzonis and Christopher Alexander (1963, 1971); principle of miniaturization, Paolo Soleri (1969), principle of elasticity, Alvar Aalto (1945); many more.

the population and their surrounding environment. The market, and therefore, its form and location is essential for the interconnection of economy, transportation, communication, and resources, encouraging centripetal *reintegration*. The community centers act similarly as vessels for interconnectedness, stimulating communal and individual betterment and exploration. Community spaces also act as the center for technological support for agrarian workers and a lifeline to the community. In Usonia, education centers offer balance just as with other public areas that invite opposites and tension into the form and use of space. Garden schools and design schools are to provide spaces that enhance the exploration of the mind and others that encourage the invigoration of the body (see Fig. 2). Balance creates ingenuity, innovation, invention, and variety which allows for citizens to be active in their individual and community well-being. Though ultimately, every space in an organic community should be educational, citizens of a Usonian community are, idealistically, always learning because they are connected to the world and people around them, no longer solely tucked-away in the inner-lives—an extension of Wright's Organic-*Self*. This happens more through each generation by the action of *reintegration*, continually propagating and reinforcing a sense of *communal individuality*.

Landscapes, for both productive and recreational purposes, are vessels for community *continuity* and individual exploration; both community and private landscape sustains human life and is active even without human activity. Landscapes, as Wright proposes, would be synonymous with architecture, a oneness that lends itself to many evolutionary predecessors of an organic human habitat. Wright states, "architecture

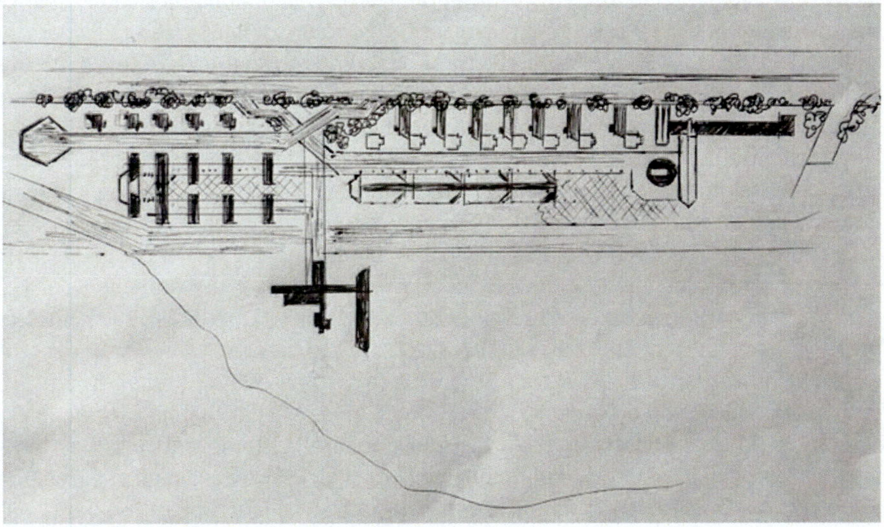

Fig. 2 Wright's design for a garden school in a Usonian community shows the balance of lifestyle promoted in the form, function, and organization of a variety of community spaces. (Reproduced Illustration from Wright's "Broadacre City. Education Center, Unbuilt Project" (1934) by Regina Carmine et al. 2021)

and acreage will now be seen together as landscape," and the landscape is humanity's home and lifeline (Wright 1932, 1945, 1958).

To survive and thrive, as some base goals of an organic settlement, reliance and sustainability on a communal and individual scale can only be accessible to all if the necessities of life are treated as such. Wright looks toward agrarianism to help create interrelation and promote individuality. Therefore, participatory landscape interaction is encouraged for all in a Usonian community, including those dwelling in skyscrapers, not solely farmers and those who maintain an individual acre. Gardens are to surround skyscraper complexes to balance the verticality of the architectonic form. Though this design promotes the balance of opposites, it is also reflective of the late nineteenth-century City Beautiful efforts that originated in the Chicago area. Wright's organicism necessitates a productive element and thus a productive landscape with other experiential opportunities is favored over a façade of green nature or beautified parks—life-giving rather than life-taking (Wright 1932, 1945, 1958). In contrast to Usonia, Wright paints a picture of the fight to survive in the early twentieth-century human settlement by short-term benefit, dehumanizing and ultimately cannibalistic, often mirrored in beautified spaces that mask underlying cruelty, such as with City Beautiful efforts. A settlement that values "capitalism run-amok", embodied by the rent-profiteering of space, rather than people, cannot be sustaining or allow for freedom of the individual and is, therefore, opposed to the goals of *communal individuality* and *reintegration* (Wright 1932, 1945, 1958).

Though privacy is necessary for individuality, in Wright's spatial explorations, it must be balanced with communal space in which Wright does not intend to enforce a hierarchical social structure. However, through a pattern of container and exchange stretched across the landscape, a stringent order controls interaction, a pattern of contained diversity which is furthered through *reintegration*. Much of this pattern is generalized by Wright in the hopes that the concepts may be adaptable to different places and times, but such generalization also invites control that becomes dehumanizing if the political and economic framework of the settlement is manipulated by an "architect-king" or another leader.[7] An organic community founded on a holistic productive and livable landscape comes about through the process of *reintegration* as Wright notes, "decentralization and reintegration will go to work together over the whole land to create a better livelihood by so recreating the framework of our modern life that our work, leisure, and culture will be our own and, as nearly as possible, one" (Wright 1932, 1945, 1958).

Reintegration is then, to Wright, the system through which humanity will act in their individual and communal interest, made possible through a holistic pattern across the landscape. Formalized in the transportation systems and spaces of overlapping use, such as markets, community centers, education centers, and landscapes, *reintegration* works toward community interest rather than self-interest, which "perpetually endangers" the human species (Wright 1932, 1945, 1958). The process of

[7] Wright describes the rule of a social and design-minded "architect-king" which is then further extrapolated by Wright historians, such as Anthony Alofsin, to describe and critique the governing system of a Usonian community.

reintegrating social systems, functions, and forms of the human settlement requires an active community and an architecture that lends itself to such pliability. Over time, Wright imagines, "smaller units might be again integrated or systematically 'chained' over large areas, to down costs and facilitate distribution; to add new economics to mass production and standardizing of gasoline" (Wright 1937, 1945, 1958). This result grows from an integral holism in which the citizen and their habitat rely on one another—organically interwoven in a dynamic balance. Through *reintegration*, the whole of the landscape becomes a park, cultivated for the betterment of humanity's social and health needs and overall human well-being. However, this also insinuates a stripping of resources from the land for human use, where, even in an idealistic community of greedless humans that care about their neighbor as much as they do themselves, this system doesn't heal the scars of industrialization as much as it reopens them. Wright's reliance on communal resource use is intended to equalize humanity but can also easily morph into extreme resource depletion.

"The food master of humanity" is the originator, that which clothes and feeds communities, and is thus a role that should be built into the functions and forms of communal and private spaces (Wright 1932, 1945, 1958). Without this built-in reliance and relationship, *reintegration* and *communal individuality* are not possible. Just as communal and private spaces all provide opportunities for education and exploration, all landscape provides an opportunity for connection to life itself. By using the ground to sustain oneself and one's community, citizens are brought into communion with the very thing that keeps them alive rather than living in a community where human needs are not being met, which Wright sees as a symptom of discontinuity between the habitat and human population.

4 The Evolution of Organic Principles

Wright's organic principles are outlined to create communities and spaces that evolve and adapt, two details that are essential for each individual and the public and private spaces that serve them. Wright's intent is to make these organic principles and agrarian principles interchangeable, benefiting first and foremost all who wish to pursue a productive connection with the landscape. Greenhouses, fireproof and sanitary structures, covered, accessible, and connected, all simplify the farmer's life and allow them to take part in the luxury provided to the rest of the citizens through their life sustaining work (Wright 1937, 1945, 1958). The forms mirror the industrial nature of the industrial worker's residences—functional in its *simplicity* and *plasticity*. Prefabricated units combine fluidly, in service of the farmer, to offer practical emancipation by way of organic architecture (Wright 1932, 1945, 1958). Though standardized, the units come in a variety with modifications available for both agricultural functions and geophysical features. Units would be used for diverse reasons, all making the farmers "life worthy of conservation" (Wright 1932, 1945, 1958). Similar to the industrial worker's residence, spaces can be purchased and added when they are needed or affordable. Since much machinery is communally owned and used, the tool shed, and clutter it enables, is no longer necessary. Multifunctionality in private space is just as important as it is in community space; the ability for the small farm, individual acres, and residences to adapt to technological changes, and a variety of functions that promote survivability and prosperity, is essential for a space's organic capacities.

4.1 Plasticity and Continuity

A Usonian community would normalize the cultivation of productive land so that the landscape would be seen aerially as a diverse patchwork, an intricate organism itself. The human habitat would, over generations, expand to become more fully a part of the landscape and of Nature at every scale. From the entire planet to the individual residence, the human organism is both collective and individual, embodying Wright's Organic-*Self,* which is made more possible through the application of these principles. Though it is necessary to strip the imperialist perception of landscape from Wright's theories, the active principles of the site provide a multiplicity of organic solutions to allow communities to be dynamic and resilient both now and in the future. The private and community spaces that create a meaningful connection between individuals and community, as well as community and landscape, can help to instill the organic *continuity* necessary for humanity's short- and long-range betterment and survival.

A citizen's active relationship with the landscape, whether for agrarian purposes, individual kitchen gardens, flower gardens, or even recreational purposes, is essential to understand the *continuity* between the human species and the planet, a balance that had been damaged significantly from industrialization during the nineteenth and early twentieth-centuries. Wright's criticism of the scars left by industrialization is contradicted by his ardent belief in capitalistic endeavors, though he does not disguise his disapproval of the neo-capitalistic system embodied by the metropolis, ultimately displaying his Romantic belief in Americanized democratic ideals. As Wright centralizes the small farms and the productive private acre of the citizen-farmer, the dynamic organization, function, and form, evolving around Wright's organic principles, are intended to adapt to different spaces and times. Wright's principles buttress *reintegration* and *communal individuality* and form every scale of design from governing systems, private space, communal space, transportation and communication systems, settlement pattern, resource distribution, and landscape use. Still, the core of Wright's organicism hinges on efforts toward decentralization.

The abandonment of cities, as Wright promotes in his decentralization efforts, is an unsustainable solution and irrational, however, the organic principles outlined in Wright's textual reiterations are by nature adaptable. The use of *plasticity, continuity, simplicity, horizontality,* and *harmony* in community organization, architectural forms, aesthetic details, and spatial functions exist to support and work with *reintegration* and *communal individuality*. Normalizing productive use and *continuity* with one's local environment through organic principles applied to landscape, private and community spaces can enhance self- and communal-reliance and sustainability. Productive landscapes and resilient communities will evolve from spaces whose reliance on these formal and functional organic principles meets twenty-first century demands for equity. This grows from formal and social *plasticity* and *continuity*, allowing a community, both its spaces and people, to be intertwined with one another and the environment through societal systems that promote individual and community well-being. Ultimately, Usonia's lack of stagnant timeless perfection makes the *plasticity* of the settlement's plan and pettern a reflection of Wright's vision of the *continuity* of the human species.

4.2 Simplicity, Horizontality, and Harmony

Centralizing *simplicity* at every scale is the ideal of *reintegration* and *communal individuality*, untangling the dispersal and transportation of goods from the web of capitalist enterprise that creates unnecessary complexity in the exchange of resources. A simplification of the use of space and interconnecting systems lends itself to small agriculture—by minimizing acreage to the three-, five-, and ten-acre plots, space, and landscape can be used more efficiently and effectively. The simplification of space has agricultural benefits, through both architectural forms and organization as well as landscape use. Wright states "a composite farm building… would be assembled of units consisting of a garage, a dwelling, a greenhouse, a packing and distribution house, a silo, a stable and a diversified animal shed" (Wright 1932, 1945, 1958). Organic architecture will benefit agriculture, and therefore, the community as a whole because ultimately Wright's organicism intends to create a link between the necessary elements for human survival and the citizens themselves.

Through *plasticity* and *continuity,* the principles of *simplicity, horizontality,* and *harmony* can evolve from Wright's romanticized and colonial perception. *Simplicity* must evolve beyond Wright's Romantic Agrarian Revolution in form and function to respond to issues of spatial inequity and access to land and resources to provide an opportunity to more wide-reaching diverse populations. Additionally, *simplicity* must stretch beyond a formal descriptor of organic spaces to also promote balance in resource use and accessibility. Arguably, luxury offered to a few at the expense of the many is a dehumanizing principle that is so deeply ingrained in human habitats that it is utopian to dream of its dissolution. However, this necessitates *simplicity* at the core of political and economic action though enforcing such a concept is inevitably eco-fascist, and thus the expansion of one's identity through *communal individuality* may provide the ontological shift necessary for a change in *simplicity* without political overreach. For a community and architecture to give the sense that all "life is worthy of conservation," *simplicity* must be an active component of everyday life, once again reflecting the lineage of organic determinism, to create a life balanced for all (Wright 1932, 1945, 1958). Dedication to *simplicity* must be driven beyond material, formal, and aesthetic details and be focused on the architectural and formal influences on lifestyle changes that promote *simplicity* in environmental impact. *Simplicity* is an essential avenue for balance that will promote an active twenty-first century version of the necessities of life that Wright outlines. In form, *simplicity* is also necessary for formal *plasticity* in future technological endeavors and changes—flexibility for future use and built-in catastrophe.

The *horizontality* that Wright visualizes as organic for human life stands in opposition to the vertical settlements envisioned by architects such as Le Corbusier and Paolo Soleri.[8] To Wright, "a sense of life as organic architecture or architecture as a form of organic life" grows from *horizontality* (Wright 1932, 1945, 1958). *Horizontality* is defined by a right to space and the necessities of human life that Wright defines. To Wright, the verticality of the modern city negates the freedom of the individual and thus the *horizontality* of Usonia brings humanity more into balance with the planetary processes and themselves. To deny this connection is inhuman and thus Usonian communities are to be founded on a horizontally expanded human habitat, creating a visible link that is meant to translate into a productive and holistic relationship with the landscape. Just as furthering a vision of *simplicity* is predicated on economic, political, and social action, *horizontality* requires a shift in the same actions and norms. To Wright, *horizontality* is the form through which *simplicity* in institutions will take shape and create a community in which the benefits are human benefits rather than dehumanizing system benefits. Thus, *horizontality* is more necessary for the systems that govern space than the use of space itself, though one, ideally, helps to bring about the other. To evolve, *horizontality* requires a community's active participation to sustain oneself and one's community—*continuity* and *harmony*.

Wright's organic principle of *harmony* is one that is deeply reflective of the architect's romantic beliefs that influence his deterministic view of organic architectural creations. Yet, *harmony* in a Usonian community simply reflects the underlying purpose of organicism—the architecture best suited to provide the necessities of life that allow communities of people to reach toward their potentialities. *Harmony* requires equality on every scale of a community and is the ideal of human freedom and Wright's democratic spirit. Wright's sense of *harmony* grows from his Sullivan-inspired concepts of *plasticity* and *continuity*. Environmental, communal, individual, and cosmological *harmony* are infused into spaces through the organic forms and aesthetics that promote balance on each scale of interaction.

Harmonious landscapes have various productive uses, have enhanced biodiversity, and require balance through a realization of *continuity* with life, which, by Wright's definition, is interchangeable with organic architecture. Community spaces of *harmony* require multifunctionalities such as in educational spaces where productive landscape activities provide balance for formal learning, enhancing creativity, and innovation. *Harmony* also necessitates a balance between private and community space to provide populations with a variety of experiences for both communal and individual exploration, helping to buttress Wright's Organic-*Self* as a central force of Usonia. Thus, *harmony* in form is made more possible through social means—self- and communal-sustainability and reliance, brought about by productive private and

[8] Both Le Corbusier's Radiant City and Paolo Soleri's Arcologies concentrate on condensed verticality as the ideal human settlement pattern (Le Corbusier 1933 and Soleri 1969).

community land use, bringing citizens closer to the environmental *continuity* that keeps them alive. Form and function become one when architecture is harmonious with human life. Organic architecture is founded on these principles beyond the form of buildings and settlements, but rather on the value of human life and equality from which *simplicity, horizontality,* and *harmony* grow. Wright's *plasticity* and *continuity* embody this interdependence and are thus prepared to allow Wright's other organic principles to evolve.

Contextualizing and evolving Wright's organic principles requires *reintegration* to be an action that does not necessitate decentralization but rather requires an acknowledgment of the fractured norms and values that help to create dehumanizing systems and spaces to this day. The ideal of *communal individuality* can be propagated by instilling existing spaces and future spaces with the organic principles of *plasticity, continuity, simplicity, horizontality,* and *harmony* to be dynamic and evolutionary. This begins with enhancing communal- and self- sustainability and reliance through productive land use built into private and community spaces. However, the agricultural space of interaction with community and landscape, as well as individual and community, must be something offered to all, especially those communities that have been most affected by systemic inequalities that are reinforced through the built environment. Additionally, elitist versions of *simplicity, horizontality*, and *harmony* work against any real change necessitated by organicism.

Wright states of Usonia "It is demanding organic foundations economic, ethical, social, and aesthetic: insisting upon beginning at the beginning to plan: planned revolution—by evolution" (Wright 1932, 1945, 1958). Though idealistic, Wright makes clear the dynamic nature of organicism goes beyond an architectural movement—it is a social, economic, political, and architectural movement wrapped into one, a movement to reform society. It is Wright's evolving Utopia, and though his early to mid-twentieth century perspective provides challenges due to dramatic technological changes, and must be stripped of the dehumanizing elements that were the norms of a century ago, the ideas and principles are continuous in themselves, giving us a roadmap for shaping and transforming communities and architecture to create a more equal world during ongoing and worsening crises rooted in issues of social and environmental justice. This equality stems from self- and communal-reliance and sustainability founded on urban agrarian actions in private and community spaces. The interwoven nature of these principles makes them inseparable in Wright's utopia and thus planning must take all of them into account as factors that create a whole.

5 Reenvisioning our Twenty-First Century Usonia

From Wright's organic principles, we can outline what an organic community does and doesn't contain; such a community would learn from the dehumanizing principles that govern space and communities historically and today, realizing the flaws of organicism's nineteenth and twentieth century reiterations. An organic community is one that opens paths toward communal- and self-reliance and sustainability, that is founded on spatial equity and access to the necessities of life for all people. Systemically abused communities are those whom organic principles must serve if long-term change is to come about through short-term action. Spaces that provide human-sustaining *clean* land, air, water, and sunlight must be accessible by all in order for organic communities to be actualized. Communities and spaces reintegrated based on equal access to space, the necessities of life, and private and community spaces that provide an opportunity to actively partake in sustaining oneself and one's community, is the ideal of a twenty-first century organic architecture. The entire population must have access to spaces that are life-giving rather than life-taking: organic principles reinvigorate human life, making all human life worthy of care and protection. Wright reminds us that ultimately "architecture values are human values, or they are not valuable" (Wright 1932, 1945, 1958).

When considering the evolution of these principles, a contemporary focus on organicism must be heavily intertwined with issues of social and environmental justice or it will be just as dehumanizing as Wright finds the cities of the early twentieth-century. Organic balance, defined by *simplicity*, *horizontality*, and *harmony*, evolves into the twenty-first century as one of diversity and variety rather than singularity. An *equal right to space for all*, as Broadacre City proposed almost a century ago, means something drastically different today than it did in the early to mid-twentieth century. This requires the creation of an honest narrative surrounding the western architectural canon, which has often perpetuated a dishonest story of shallow beauty rather than the underlying intent which often buttresses structures of racism. Wright's failure to acknowledge and fight social and environmental injustices of the time should not diminish the potential for these principles, especially when placed in the context of organic communities and spaces capable of flourishing today and in the future. The prioritization of self- and communal-reliance and sustainability based on diversity, and biodiversity, that isn't contained by the norms and values of colonization and white supremacy, can give these organic principles new life. Though idealistic, Wright's motives are malleable in the hopes that through generations, the human habitat will be more formally and ideologically centered around *communal individuality*, which would allow the human species to survive and thrive. Wright's organicism, though flawed and telling of both Romantic and colonial beliefs, is dynamic in its generalizations and specificities. Only through this evolution will the underlying agrarian reliance and principles that instill diversity and balance offer effective, ongoing, and evolving short- and long-term solutions to the social, economic, and environmental crises humanity is, and will be, faced with today and in the future.

References

Alofsin A (2011) Broadacre City—Ideal and Nemesis. American Art 25(2):21–25

Dehaene M (2002) Broadacre City: The City in the Eye of the Beholder. Journal of Architectural and Planning Research 19(2):91–109

Donovan J, Cuming P (2010) Permaculture Part A—Introduction and Urban Farming. *Environment Design Guide*: 1–7

Dougherty J (1981) Broadacre City: Frank Lloyd Wright's Utopia. The Centennial Review 25(3):239–256

Dunec JL (2011) Urban Farming. Land Use. Natural Resources & Environment 26(2):62–63

Harris D (2021) Where was Jim Crow?: Living in Wright's American. Garden and Landscape Studies Public Lecture, Dumbarton Oaks

Johnson DL (1988) Broadacres Geometry: 1934–35. Journal of Architectural and Planning Research 5(2):129–144

Kaufmann E Jr (1978) Frank Lloyd Wright: Plasticity, Continuity, and Ornament. J Soc Archit Hist 37(1):34–39

Lapping MB (1979) Toward A Social Theory of the Built Environment: Frank Lloyd Wright and Broadacre City. Environmental Review: ER 3(3):11–23

Leonard E (2014) Middle Ground: Uniting City, Farm, and Nature with Diverse Agroecosystems. The American Journal of Economics and Sociology 73(5):1007–1134

Printezis I, Grebitus C, Printezis A (2017) Importance of Perceived "Naturalness" to the Success of Urban Farming. Choices 32(1):1–7

Stedman RC, Larson LR, Tidball KG, Tidball M, Curtis PD (2017) Hunting and the Local Food Movement: Insights from Central New York State. Wildlife Society Bulletin (2011-) 41, no. 4: 720–728

Twombly RC (1968) Frank Lloyd Wright in Spring Green, 1911–1932. Wis Mag Hist 51(3):200–217

Twombly RC (1972) Undoing the City: Frank Lloyd Wright's Planned Communities. Am Q 24(4):538–549

Uechi N (2009) Evolving Transcendentalism: Thoreauvian Simplicity in Frank Lloyd Wright's Taliesin and Contemporary Ecological Architecture. The Concord Saunterer, New Series 17:73–98

Waldheim C (2013) Notes Toward a History of Agrarian Urbanism In Krause L (ed) Sustaining Cities: Urban Policies, Practices, and Perceptions. Rutgers University Press, New York, p 63–75

Wright FL (1935) Freedom Based on Form. New Masses 16:23–24

Wright FL (1932) The Disappearing City. William Farquhar Payson, New York

Wright FL (1937) When Democracy Builds. University of Chicago Press, Chicago

Wright FL (1958) The Living City. Horizon Press, New York

Wright FL (1953) The City. The Future of Architecture. Horizon Press Inc., New York, pp 181–199

Wright FL (1953) The Tyranny of the Skyscraper. The Future of Architecture. Horizon Press Inc., New York, pp 163–181

Wright FL (1953) The Cardboard House. The Future of Architecture. Horizon Press Inc., New York, pp 143–163

The Future Without Centralities, "New Settling" by Mikhail Okhitovich and His Proponents

Fedor Kudryavtsev

Abstract In 1922–1930, discussions on socialist settling crystallized three versions of the socialist built environment. Workers' garden city, "sotsgorod" or socialist city and network of residential stripes and production cells of "new settling" have become probable spatial patterns of the future with no division into city and countryside. The future settling was to become a single space in which industry and agriculture would be integrated, and the quality of life would be largely the same. All three concepts had the potential for the development of new socialist settlements as a kind of integral productive landscape. "New settling" of disurbanism was the most radical option of this future. Mikhail Okhitovich and his fellow architects developed a clear and logical model of a uniform spatial network of settling that would provide the most favourable living conditions in any cell regardless of its location. It was a response to the problem of inequality of centre–periphery relations inherent to the urbanized world. Currently, expansion of a uniformly accessible environment of "feeder-to-cell" communication and transportation well fits the basic model of "new settling". This hypothetical process of "post-urbanization" can lead to equal and homogenous settling conditions too but in contrast to the concept of disurbanism, in a non-uniform physical environment.

Keywords Disurbanism · Okhitovich · Deurbanization · Socialist settling · Post-urbanization

1 Socialism without Cities

In the middle of the 20th century, the socio-economic system of socialism was at the peak of its development and covered around half of the world. By that time, the principles of socialist urban planning have spread throughout the world. In this part of the globe, urbanization has taken a different path. The cities here have become

F. Kudryavtsev (✉)
Moscow Institute of Architecture (state academy), Department of Urban Planning, Moscow, Russia
e-mail: ulab@marhi.ru; fedor.kudryavtsev@archnova-lab.ru

C. Brisotto and F. Lemes de Oliveira (eds.), *Re-Imagining Resilient Productive Landscapes*, Cities and Nature, https://doi.org/10.1007/978-3-030-90445-6_10

very different from those in countries where the land market and private investment in real estate flourished.

However, the socialist world, perhaps, could look very different. Soviet "disurbanists"[1] led by Mikhail Alexandrovich Okhitovich suggested dispersal of residential, service and industrial units and integrating them with agricultural areas in one whole. Uniform, non-urban and non-rural spatial networks of socialist "new settling" could become an alternative to a system of socialist cities.

Well documented in Soviet architectural, art and political magazines and newspapers of 1929–1930, theoretical views and projects of Mikhail Okhitovich and his fellow architects soon after became the subject of official criticism in Soviet Russia. In the West, before the publication of Anatole Kopp's book "Ville et révolution" (1967), only the memory of their discussion with Le Corbusier saved the disurbanists from complete oblivion (Cornu 1967: 201–202). In the 1960s, the reactualization of Soviet avant-garde architecture around the world brought back the legacy of disurbanism as an object of research. The ideas and projects of disurbanists in the context of their discussions with opponents have become an integral part of key compendiums on the history of early Soviet planning and architecture (Kopp 1967, Khazanova 1980; Khan-Magomedov 1987, 2001; Kosenkova 2018). Many publications have been devoted to certain aspects of these debates and their historical context (Cohen 1984; Volchyok 2008; Meerovich et al. 2011; Meerovich 2013; Hudson 2016), including the biography of Mikhail Okhitovich (Hudson 1992).

There were few publications about disurbanism as such, its potential to become an alternative to Soviet urbanization which appeared later (Khan-Magomedov 2009, Hatherley 2020). It is assumed that this chapter will continue this line of research. The key views of Mikhail Okhitovich and his opponents will be presented, as well as one of the two projects developed by the disurbanists, and the general scheme of the new socialist settlement, which became the culmination of disurbanism concept. The purpose of this chapter is to prove that the "new settling" of disurbanism is not only a theory of its time but also a long-term forecast and strategy for the future.

2 Designing the Future: Three Solutions to the Problem of Socialist Settling

The ideological model of socialism was just a theory during the early Soviet era. There were neither ready-made models of people's daily life, work, leisure and living environment, nor prototypes of urban and rural settlements that could be turned into reality. This unique situation has given architectural design and urban planning a dual task of simultaneously inventing new social models and their corresponding spatial forms. The doctrine of socialism was not the only source of inspiration. The very

[1] Henceforward, disurbanism and disurbanists refer to particular concept and its proponents described in the chapter, following the translation in the book of S. O. Khan-Magomedov (2001: 711).

idea of art, which constructs the future and a new kind of human being, was deeply rooted in the cultural context already in the pre-revolutionary time (Golomshtok 1994). Social experiments on the real lives of millions of people have allowed this concept to come to life.

However, the Marxist theorists formulated the programme of spatial planning of socialism long before it became a practical task. They indicated a goal of “elimination of an opposition between the city and the countryside” which should be solved “only by evenly distributing the population throughout the country, as far as possible, only by a close internal link between industrial and agricultural production, along with the necessary expansion of communication facilities” (Engels, “The Housing Question”, quoted from Khan-Magomedov 2001: 46–47) and “transmission of electricity over distance” (Lenin, “The Agrarian Question and the "Critics of Marx”, quoted from Khan-Magomedov 2001: 47). When this goal is solved, in “a new settling of mankind” there would be neither the “backwardness” of the countryside nor “unnatural accumulation of giant masses in large cities” (Lenin 1914, “Karl Marx”, quoted from Khan-Magomedov 2001: 47).

This agenda prompted experimental projects of the 1917—the early 1930s, generally known as the Soviet architectural avant-garde, to focus on topics of reorganization of everyday life and a problem of socialist settling. The core issue that made architectural and planning routine problem-oriented and experimental was the ultimate and holistic role of the new built environment in the development of the socialist type of a human being, a new social system and mode of production which were just emerging and still remained unknown: “The construction of socialist enterprises—a plant, a factory, a state farm that create and organize new labour, the construction of house-communes that create a new person, a communist way of life—all this culminates in the construction of a more grandiose whole, called the socialist “city”. And if we sometimes struggle over the correct solution of a separate unit of the socialist type (the house-commune), then, of course, there are much greater difficulties in solving the whole complex…” (Stroitelstvo Moskvy 1930: 2). In the briefest summary, the problem of socialist settlement was to create a spatial structure without the opposition between the city and the countryside for a new society without social classes.

In 1919, the goal of the founders of Marxism about overcoming the opposites between the city and the countryside was included in the official programme of the Russian Communist Party of the Bolsheviks (Khan-Magomedov 2001: 47). In 1920, a vision of a nationwide electricity supply network became the State Plan for Electrification of Russia (GOELRO). In 1928, the prospects for the planned distribution of industry took the form of the First Five-Year Plan for the Development of the People’s Economy. The development of socialist settling and the future of the large cities have become a subject of professional and public discussions (Khan-Magomedov 2001: 56–63, 194–198). These disputes were open and real despite the background of ongoing political repressions and cruel policies, such as an elimination of the “kulaks”, the rich rural population, as a social class (TsK VKP(b) 1930a).

In the 1918–mid-1920s, the concept of a garden city became almost an official planning policy. Its proponents (Vladimir Semenov, Nikolay Markovnikov and many others) argued that Howard based his concept on “a workers’ settlement, formed in a capitalist society, but which is a kind of germ of a communist city" (Khan-Magomedov 2001: 59). Despite the support of some of the high-ranked officials, housing cooperatives and city Soviets (councils of peoples' deputies) of Moscow and Petrograd they faced opposition from the proponents of a socialist city. The first public discussion on socialist settling was a kind of culmination of these controversies. It was launched in 1922 due to the reestablishment of Obshchestvo gorodov sadov (Garden Cities Society). Originally founded in 1913, now it was going to promote the concept of a garden city as a scientifically based and verified tool for the socialist reorganization of the settlement system. These intentions have become the subject of harsh criticism. It was often ideologically motivated as opponents accused the garden city for "planting a petty-bourgeois ideology" by the preservation of the old lifestyle instead of developing a new one (Khan-Magomedov 2001: 61). At the last public discussion in 1923, most of the speakers opposed the application of the garden city concept in Soviet Russia. This event marked the decline of the garden city version of socialist settling.

An alternative model of socialist settling as a collective living in close connection with the functioning of the factory continued to develop gradually, until it reached a culmination in Leonid Sabsovich's books (1929, 1930) on the "sotsgorod"—a socialist city. Then it faced a new competitor—the concept of "new settling" of disurbanism developed by Marxist theorist Mikhail Okhitovich and a group of Constructivist architects supporting him, led by Moisei Ginzburg. That again caused a public discussion in 1929–1930.

The disurbanists were not followers of the ideas of the garden city. They generally denied the future existence of a city in a favour of deurbanization and integration of built-up and agricultural areas into a single integral landscape. However, compared to their rivals, these two concepts had some common or at least similar positions. They were both opposing concepts of a compact socialist city and a collective living as the highest socialist imperative. Therefore, a brief overview of the development of the garden city concept and its criticism in the early days of Soviet Russia can help better understand what made disurbanism concept of Mikhail Okhotivich different from earlier discussed ideas.

In the Russian Empire, the garden city was the subject of intense public discussions and experimental application (Semenov 1993). In the new social and cultural context of Soviet Russia, this concept has undergone a certain transformation. It was widely applied on the scale of relatively small residential settlements with a population of several thousand people. Master plans of these small communities followed the principles of the first garden cities in England and Germany.

In terms of their functional programme, Soviet garden settlements differed both from European analogues and Howard's theoretical model. They tended to be more self-sufficient and autonomous in providing everyday services. This was especially true for settlements located far from cities. This was fairly common, as many of them were planned for workers of newly built mines or power stations launched under the GOELRO plan. An example is a competition for the settlement of workers of the New Grozny oil field in the North Caucasus, held in 1923. All winning entries were projects of garden cities with schools, hospitals, pubic baths, sports facilities, garages, administration and other public amenities (Khan-Magomedov 2001: 79, 80). Contrary to Howard's concept, inhabitants of garden settlements were supposed to work in a certain factory or a larger city nearby.

If the lack of available resources limited the implementation of garden cities in practice, then the theoretical and conceptual schemes of this idea in Soviet Russia went far beyond the scale of the original model. The first conceptual plans for the reconstruction of Moscow suggested to "dissipate" the city with a population of more than one million people in a system of the reconstructed historical centre and adjacent garden cities and suburbs (e.g., Draft plan for re-planning and expansion of Moscow under the leadership of Ivan Zholtovskiy, 1918–1919). On the contrary, the planning schemes of Sergey Shestakov, 1921–1925, and Mikhail Dykanskiy, 1926, assumed an even greater spatial extension of Moscow into a complex structure of garden cities, industrial and green zones for 4 million or even 12 million people, respectively (Khan-Magomedov 2001: 64–67, 68–70).

Agricultural activity on individual plots was another feature of garden city projects developed in Soviet Russia. Among other goals, the charter of the Garden Cities Society stated its intention to support gardening, the development of small-scale horticulture and orchards (Khan-Magomedov 2001: 56). In practical terms, it was a measure to create a better and healthier living environment. Conceptually, the opportunity of individual small-scale farming was a tool for involving industrial workers in food production. This was to make a garden city able to integrate industry and agriculture, city and countryside and thus solve a key problem of social settling. While this approach was not always shared by workers (Khan-Magomedov 2001: 61), in other cases it met the wishes of clients. A quote from the terms of reference for a workers' settlement project, formulated in 1918, well presents the wide range of potential agricultural works in a garden suburb or settlement of that time: "(1) a worker without household (pure proletarian) (2) qualified worker that has a small household (a kitchen-garden, a small garden, a goat, a birdhouse, rabbits, etc.) (3) the same on a larger scale (e.g., a cow instead of a goat and so on) (4) a worker-peasant" (Meerovich 2007: 136). The bitter experience of food shortages during the Civil War was another reason for food production in garden settlements and suburbs. These difficult times contributed to the relocation of a significant part of urban dwellers back to the countryside and spontaneous growth of "kitchen-garden cities" namely suburbs with local household agriculture. The reconstruction of such areas in the periphery of cities was a solid part of the work of architects (Khan-Magomedov 2001: 53, 54). The need for food production and the outflow of the urban population were strong driving forces for the application of the garden city model in Russia at that time.

Proponents of alternative concepts criticized the use of garden cities as a requalification of the old capitalist city instead of its reconstruction. The first experiments on a new type of urban form appeared as competition entries of phalanstery-like residential buildings with shared facilites, such as a library or a cinema hall, and house–communes with built-in canteens. They later took the form of residential superblocks (housing complexes) built for workers by state or city Soviets (councils) in inner-city districts (Khan-Magomedov 2001). In the late 1920s, these experiments evolved into the concept of a "zhilkombinat". This new term literally meant "housing plant" and indicated an urban block integrated into one coherent superstructure which was supposed to provide all necessary amenities and services for the everyday life of its residents (Kuzmin 1930). A group of such structures associated with industrial enterprises was to constitute a new socialist city. "It is unlikely that it will be possible to create a socialist city and a socialist settlement from separate scattered buildings" (Vesnin and Vesnin 1929).

The concept of a garden city was also criticized for the discrepancy between the scale of the settlements it proposed and the realities and needs of large-scale industrial development. "The city of 58 thousand people cannot claim the significance of a Centre in a scale of our time. This is a city of handicraft manufacturing, its population is the population of the one large textile indusrty plant", that was an opinion of engineer Boris Sakulin, the author of one of the first regional planning schemes of the Moscow urban agglomeration (Khan-Magomedov 2001: 71–72). Sakulin also believed that the garden city is "outside the state system", pointing to the contradiction between centrally planned heavy industry and garden settlements, the construction of which was largely initiated in a decentralized manner by local governments and housing cooperatives (Meerovich 2007: 154).

The task of integrating industrial and agricultural labour was common to the supporters of garden cities and their opponents, but the latter denied individual farming in both urban and rural areas. In the project of the New village by V. Bykov, 1923, new agrarian settlement looked in many ways like a garden city. However, agricultural workers were supposed to live here in low-rise apartment buildings, and not in detached houses. The project envisaged collective agricultural areas planned as a peripheral belt, but, unlike in workers' garden settlements, residents were not supposed to have any individual plots, individual land cultivation, gardens, orchards, poultry, etc. If the supporters of the garden city advocated the dispersal of the city, then their opponents sought to urbanize the village (Khan-Magomedov 2001: 62–63).

The application of the garden city concept seemed to have reached its peak in the regional planning scheme of the Absheron Peninsula in Azerbaijan. It was designed in 1924-1925 by Alexander Ivanitsky, Alexander Vesnin and Viktor Vesnin. Largely in accordance with Howard's vision, this plan covered the development of Baku and its environs and suggested turning this area into an integral system of garden cities and suburbs. It was implemented until 1927 when further development shifted to the construction of multi-storey housing in the Armenikend district on the outskirts of Baku (Khan-Magomedov 2001). According to the TASS report,[2] quoted in N. Kuzmin's article (1928: 82), the reason was "the revealed uneconomical nature of the construction of individual detached houses".

In the late 1920s, the garden city gradually stepped off the scene of debate on the future of socialist settling. The reasons for this failure could be both economic (Khan-Magomedov 2001) and political (Meerovich 2007). The new concept of disurbanism had to take into account the experience of the first stage of planning debate and

[2] An abbreviation of Telegrafnoye agentstvo Sovetskogo Soyuza (Telegraph Agency of the Soviet Union).

propose a new system of arguments and views as comprehensive as those developed by the other two parties in mutual criticism and exchange of ideas.

3 Mikhail Okhitovich: An Amazing and Tragic Biography

Mikhail Okhitovich was born in 1896 or 1898 (Khan-Magomedov 2009: 177), joined the Bolshevik Party in 1917 and served in the Red Army during the Civil War (Hudson 1992: 453). Later, he studied the legacy of Marxism at the Marx-Lenin Institute in Moscow. Here, Friedrich Engels' thoughts on the future disappearance of large cities inspired him to develop his doctrine of socialist "new settling" (Khan-Magomedov 2009: 181).

Mikhail Okhitovich was neither a planner nor an architect, but a sociologist, economist or philosopher. He was unknown in the field of town planning until he turned to the group of Constructivist architects (M. Ginzburg, M. Barshch, V. Vladimirov, A. Pasternak, G. Sumshik) and convinced them of his ideas. Together they formed a small group of "disurbanists", and there were just over ten architects who have ever worked in it (Khan-Magomedov 2009: 36–37, 177). In 1929–1930, this group elaborated a concept of disurbanism from the stage of a theoretical model to the comprehensive plan of a new system of dispersed socialist settling (Sektsiya sotsialisticheskogo rasseleniya 1930).

In 1931–1934, when planning discussion ended shortly after the Resolution "On the work on the reorganization of everyday life" (TsK VKP(b) 1930b), he experimented with the processing of straw, sawdust, reeds, needles for the development of new construction materials in one of the laboratories of Narkomzem (People's Commissariat of Agriculture) (Khan-Magomedov 2009: 164–170).

The fate of Mikhail Okhitovich was tragic. According to the scanty data on his biography beyond two years of the glory of disurbanism, he was first arrested around 1924 because of his support of Trotskyist opposition and released in 1928 or 1929 according to the memories of M. Barshch, quoted by Khan-Magomedov (2009: 178). In 1935, he was arrested again as a result of a campaign of harassment organized against him after a speech "The National Form of Socialist Architecture" at a conference of non-party architects (Hudson 1992: 456, 462). The history of disurbanism was only part of a period about five or six years that Mikhail Okhitovich spent at large between two prison terms, the second of which ended fatally in 1937 (Hudson 1992: 462).

4 "The Network Will Win, and the Centre Will Die"

Okhtovich was criticizing the "builder", that is, planners and architects, for "pure empiricism" and "inability to consider his craft, his profession on a non-professional, historical and scientific basis" (Okhitovich 1929: 131). Looking from this point of view, he analysed opposition between a city and a country through the lens of long-term trends in human settling. These studies convinced him that process of settling is guided by changes in the mode of production, the social form of housing and the social division of labour. And that the vector of settling began to change from urbanization, "the phenomenon of universal gravitation to the centres", to deurbanization, "the process of universal centrifugality, repulsion" (Okhitovich 1929: 134).

From the point of view of Okhitovich, the gravitation of the place of residence to the place of production was the basic law of settling. "Housing strives to be as close to production as possible. The dwelling strives to be located not even close, not near, but directly in the production …but so as not to harm the location of the production itself" (Okhitovich 1930a: 7). In its turn, production sought places convenient for transporting fuel, which, according to Mikhail Okhitovich, was a more important factor in the development of the city than the convenience for trade and exchange of products. The more production in the city grew, the less was the ability of local resources to meet the needs for fuel and raw materials, and the more important it became to be able to quickly deliver resources from afar. This interaction of production forces and transport created a form of settling in which vast and expanding rural areas as suppliers of raw materials were opposed to the uneven distribution of cities as centres of concentration of manufacturing. It was these different roles in the production chain that Okhitovich emphasized in his theoretical studies as a fundamental basis of the city to countryside contradiction. Based on this conclusion, he considered the integration of extracting and manufacturing raw materials as a solution to integrate a city and a country into a new evenly distributed settling system. "The process of eliminating the contradiction between town and country is not a process of urbanization of the countryside, as some supporters of socialist cities think, nor the agrarisation of the city. This process destroys the dependence of the city on the countryside as a source of raw materials—the city prepares it itself, just as it destroys the dependence of the countryside on the city as an industrial centre, for the countryside itself becomes the 'centre' of industry" (Okhitovich 1929: 134).

In his theoretical works, Mikhail Okhitovich outlined several trends in technological progress that will lead to these radically new conditions for the location of industry and the settlement of people. The first of them is the development of power supply networks. The new infrastructure will not only deliver but also collect energy from any connected point, including even the smallest suppliers and recipients. This will minimize the cost of energy transportation through the use of local energy sources. The proliferation of both delivery networks and power suppliers will make it possible to immediately process local raw materials anywhere in the world instead of transporting them to specific production centres. Advances in cultivation technologies, greenhouses, breeding, hybridization and other agricultural innovations will lead to the possibility of growing any agricultural product in any geographical location and at the same time eliminate the agricultural specialization of specific areas. The production and consumption of food, as a rule, will also occur in the same area (Okhitovich 1930a).

Altogether, these changes in technology will increase the amount and variety of local resources available for the production of food, energy or further manufacturing into finished goods or semi-products. Unlike many of his contemporaries, Mikhail Okhitovich was sure that new synthetic materials would be widely used in the construction industry. Agricultural waste will be the key feedstock for this technological chain.

In these new conditions, reducing of costs of transporting semi-products within one technological chain will become a decisive factor in production efficiency. Mikhail Okhitovich envisaged that a new production method would solve this problem through technologies that integrate all stages of manufacturing into one. With a touch of humour, he illustrated this as follows: " ...even now it is not difficult to imagine a combine harvester that would perform, say, one more additional operation—turning grain into flour" (Okhitovich 1930a: 10). According to Mikhail Okhitovich, networks of such dispersed production units would minimize costs of transactions between intermediate processing steps and would be more efficient than large-scale industries located in a specific centre. "Thus, the greatest concentration of production in the economy leads to the greatest decentration of production in space" (Okhitovich 1930a: 10).

In the concept of disurbanism of Mikhail Okhitovich, the problem of concentration of production and population and well-being of people were closely interrelated: "... to be a city in socialism or not to be, to be a personality in socialism or not to be, to be an individual dwelling, individual consumption in socialism or not to be..." (Okhitovich 1930b: 13). Securing the best possible conditions for an individual at any location became the main focus of his theory of "new settling".

Okhitovich argued that the uneven distribution of production led to an uneven distribution of people, forced proximity and congestion of urbanites and forced remoteness, almost isolation of villagers. If the choice between Howard's magnets of a city and a country was not good enough, then in a theoretical model of Okhitovich settling of people was more a fate than a free choice.

Since the time of the first "burg", a small city of crafts, settling has been driven by the desire to get time which has become a product and a commodity. Okhitovich has expressed this struggle to minimize daily commuting time at the expense of the quality of location or housing as the problem of homelessness. In his opinion, it was not just the lack of housing but the location of one's accommodation in a place different from the location of one's job. Uneven distribution of production and population made people suffer from this problem both in the immediate vicinity and in the distance from urban centres. (Okhitovich 1930a: 8).

The quality of housing was one of the main subjects of consideration in the concept of disurbanism. Mikhail Okhitovich developed various approaches to study this issue, and his argumentation challenged common points of view. He denied physiological needs as the driving force behind improving living conditions. In his opinion, sanitary norms were social norms that depended on the basic social conditions of a society and its mode of production. From his point of view, any norm applied to housing was indeed a socially acceptable limitation. This approach allowed him to suggest the ability to secure a full range of conditions for "the maximum functioning of a physiological, natural human being" as a criterion for the highest level of development of society and, consequently, favourable living conditions (Okhitovich 1929: 131).

Okhitovich developed his original methodology for analysing the evolution of housing typology to identify its future basic type, collective or individual. Simplified, his model includes several key postulates. The housing has two forms namely external, physical, and internal, non-material social. In this respect, barracks or hostels are different external shells of the same "primitive-communist" social form of housing, and an apartment and a cottage are casing of a bourgeois-family form. Whatever the size or location of housing, the distinctive socio-spatial pattern of its internal form remains unchanged.

Settling is a grouping of housing in space, so a city, village or apartment building are examples of settling. Settling may influence the physical form of housing but not the social one.

The social or internal form of housing is the result of the social division of labour inherent in a particular mode of production. When the mode of production changes from capitalist to socialist, a new internal form of socialist housing will appear. It will be the basis of the future typology of housing.

Following this concept, Okhitovich traced the evolution of the social form of housing from the time of prehistoric communist communities and agrarian societies to the capitalist society of his time. This study convinced him of two long-term historical trends that he used to formulate the future housing model in its entirety. He noted that both a basic social unit of the society and the physical size of its housing have gradually decreased throughout history. The second trend was an increase in the share of services and production that housing transferred to settling, especially to the city. These conclusions allowed him to substantiate a single housing unit for one person, connected to a well-developed system of public services as the future housing of socialism. (Okhitovich 1930a).

The need for individual personality and consumption in socialism was not obvious at that time and also had to be justified. Okhitovich used two types of arguments to prove it. Some of them were purely physiological: people are born individually and cannot collectively carry out such basic natural processes as swallowing while eating and sipping when drinking. Others were based on the Marxist approach to look for arguments in socio-economic relations. In this case, Okhitovich made a sequence of conclusions. Personality is the result of the emergence of property rights, joint work and the technical division of labour, "the need of the socialist mode of production, which has already been created in the depths of the capitalist one" (Okhitovich 1930a: 13). In his opinion, the elimination of personality would make the technical division of labour impossible, and this would undermine the efficiency of production. Finally, he came to the conclusion that the development of the socialist personality is one of the most important goals of a socialist society and any one-sided development of an individual should be avoided (Okchitovich 1930a: 12–13).

The recognition of the human personality as a key value of socialism pushed disurbanists towards the concept of permanently modifying and adaptive housing. The inevitable concentration of the population, congestion and unchanging form of urban dwellings made the city unable to meet these requirements for housing and, consequently, for favourable living conditions.

Okhitovich viewed urbanization as a historical process that has its natural end due to the exhaustion of the driving forces that set it in motion. He acknowledged that the city could also disappear in a capitalist society (Okhitovich 1930b: 11–13). However, neither the absence of a city under capitalism nor the new socialist city was a solution to the problem of "homelessness" and the task of providing maximum favourable conditions for biophysical and personal development for each member of society. The city not only can but also must be replaced by a new socialist deurbanized settling to fulfill this "moral imperative" of the doctrine of disurbanism.

5 A New Settling of Mankind

In 1929, the concept of disurbanism was presented in a series of articles and public discussions. These efforts eventually led to the development of a general scheme for a new settling system in early 1930 and a plan for its gradual implementation in the middle of that year. On September 16, 1930, the Division of Socialist Settling of the Construction Sector of the Gosplan of the RSFSR,[3] where disurbanists (architects K. Afanasiyev, M. Barshch, V. Vladimirov, M. Ginzburg, G. Zundblat, I. Milinis, A. Pasternak, G. Savinov, N.Sokolov, M. Okhitovich as sociologist, engineer S. Orlovskiy) worked at that time, submitted the results in the form of the report to the Presidium of Gosplan RSFSR (Khan-Magomedov 2009).

The scheme was aimed to the gradual transformation of the current two-polar settlement system into a new one, in which the city and the countryside would be integrated into one whole (Sektsiya sotsialisticheskogo rasseleniya 1930: 1). This planning document suggested a triangular cell network of settling stripes stretching along the lines of transport communications, combined with power supply lines (Fig. 1).

The disurbanists took into account a need for intermediate steps to deconcentrate the uneven settling system of their time into a new, uniformly organized one. They developed four variants, or principles, of a new socialist settling for application in different regions (Fig. 1). Each of them corresponded to a specific economic profile of the territory and was designed to deal with the specific type of settling formed around certain key elements: a city (scheme 1 in Fig. 1), large scale industrial enterprise (scheme 2 in Fig. 1), agriculture or mining industry (scheme 3 in Fig. 1). The fourth scheme demonstrated an ideal settling of disurbanism (scheme 4 in Fig. 1). The first, the second and the third options were aimed to reorganizing a specific pre-socialist form of settling. The "disurban" option targeted areas around existing cities. The "dicentric" variant was a tool for the redevelopment of "concentric" settling, namely mature industrial areas. In the third case, the subject of the new "acentric" planning was the "anticentric" settlement system where the concentration of agriculture or

[3] State Planning Committee of the Russian Soviet Federal Socialist Republic.

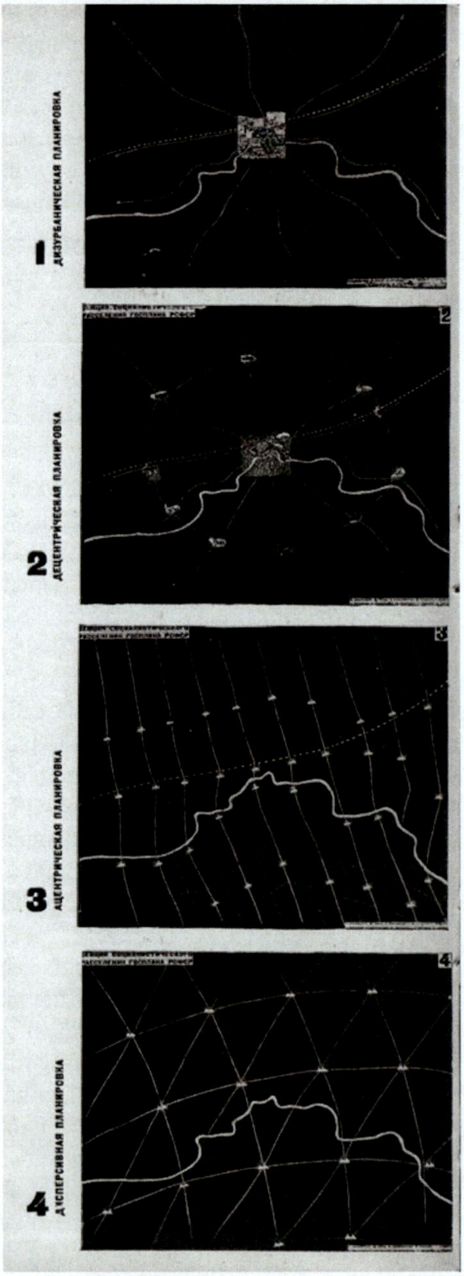

Fig. 1 Theses of the report "On the socialist planning of settling". Principles of socialist planning of settlement areas. 1—"disurban", 2—"dicentric", 3—"acentric", 4—"dispersive" (Sektsiya sotsialisticheskogo rasseleniya 1930: 2). No copyright illustration. A scanned copy is a courtesy of the library of the Moscow Institute of Architecture (state academy)

mineral extracting industries used to go separately from the location of manufacturing enterprises ("anticentrality"). The fourth layout, namely "dispersive", was the true form of disurbanism (Fig. 1) (Sektsiya sotsialisticheskogo rasseleniya 1930: 1–2).

Each planning scheme was supplemented with a summary of the key strategies for putting the plan into action. Agricultural strategies were an integral part of them. Measures to reorganize already urbanized areas presumed replanning and partial relocation of agricultural areas in such a way as to minimize the use of refrigerators and other "means of long-distance transportation". Planned actions also included stretching urban areas in favour of expanding areas of agriculture, a ban on the construction of new railways and prohibition of developing areas suitable for cultivation of vegetables and horticulture. The measures of other planning options followed the same general directions and included the development of local energy sources, food production enterprises, local transportation systems, including aviation and waterways. A special feature of option 3, which was the most related to agriculture, was a requirement to settle workers in the immediate vicinity of the location of jobs to avoid using transport for daily commuting to work. (Sektsiya sotsialisticheskogo rasseleniya 1930: 1–2).

The nodes of the grid were the locations of the manufacturing industry. The transport network connected them with residential areas stretching along car roads and at a distance of 50–150 m from them (Khan-Magomedov 2001:199) (Fig. 2).

This easement was the site of parkland and everyday social services. The triangular grid cells were left intact inside to make room for agriculture or mining. Their workers could settle in the inner areas of residential stripes. Thanks to this, agricultural workers lived within walking distance both from services located along the transport routes and from their workplaces in fields or farms. Residents who lived closer to the road had the same opportunity to travel quickly to work due to their proximity to public transport lines within the settling stripes (Fig. 2). Some fragments of the stripes were left for the location of sports and entertainment facilities, exhibitions, etc. (Khan-Magomedov 2001: 199). Average commute time was limited to 20 min by bus, so the maximum length of the settling stripe of a particular enterpise had to be not more than 20 km (Sektsiya sotsialisticheskogo rasseleniya 1930: 5).

Housing units could be chosen and customized according to the wishes of the residents. They varied from a single residential unit for one person or a couple with a total gross area of 16 m^2 (Fig. 3) to a block of several units staying apart or assembled into one structure. The dwellings were supposed to be constructed from prefabricated standard parts, mainly timber frames and slabs from local materials (Sektsiya sotsialisticheskogo rasseleniya 1930: 3,5).

The total number of housing variations was 61 to match various options for individual, family or collective living and different combinations of communities and communes (Khan-Magomedov 2001: 211). A distance of 50 m was set as a minimum spacing between dwellings to ensure the privacy of residents (Khan-Magomedov 2009: 100). The living environment had to eliminate all sources of negative influences: auditory, visual, olfactory, psychological and others. It was supposed to provide the maximum connection between people and nature. So "each dwelling

Fig. 2 Theses of the report “On the socialist planning of settling”. Axonometric view of a fragment of a settling stripe (Sektsiya sotsialisticheskogo rasseleniya 1930: 16). No copyright illustration. A scanned copy is a courtesy of the library of the Moscow Institute of Architecture (state academy)

turns into a permanent rest house, a sanatorium” (Sektsiya sotsialisticheskogo rasseleniya 1930: 15). The population density was limited to 200–400 adult inhabitants per kilometre of stripe length, depending on the capacity of nearest industries (Sektsiya sotsialisticheskogo rasseleniya 1930: 5).

Residential areas were to be served by local utilities, except for electricity supply. Instead of using centralized water supply, sewerage and heating networks, it was planned to use local water supply systems, dry closets with periodic removal of sewage and its further processing into fertilizers. In contrast to the decentralization of utility systems, common networks of public services were planned very well developed and easily accessible. They included housing maintenance services, transportation, catering, delivery of goods, sanitary and hygienic services, culture activities, communications, childcare (Sektsiya sotsialisticheskogo rasseleniya 1930: 3–4).

The new patterns of settlement were supposed to equalize the accessibility of workplaces, living conditions and leisure for any person, regardless of whether his or her occupation was related to agriculture, industrial production or services. Open and built-up areas, industrial and agricultural territories were installed into a complete network of infrastructures and residential stripes. In fact, they constituted a kind of

Fig. 3 Theses of the report "On the socialist planning of settling". Individual dwelling, type #30. Housing for individual non-family and non-collective living. Single residential unit for individual or couple without household (Sektsiya sotsialisticheskogo rasseleniya 1930: 14). No copyright illustration. A scanned copy is a courtesy of the library of the Moscow Institute of Architecture (state academy)

a uniform productive habitable landscape. The ideal scheme of "dispersive" option (scheme 4 in Fig. 1) would not have an antithesis between the city and the countryside since there would be no separate settlements in it at all. There was also no division between vast agricultural areas and cities as centres of production or culture. What could have been a densely built-up city or sprawling suburbs, now "rolled out" in stripes across what had previously been a randomly stretching countryside. The historically formed landscape was to become a regular grid of transport links and productive cells. Following the theoretical principles of Mikhail Okhitovich, extracting or cultivating raw materials and their processing could be integrated in each cell of the planned grid. Thus, any cell was a kind of local productive system that could be adapted to the local conditions of the landscape without destroying the integrity and universality of the settling system.

Before the final report, principles of disurbanism were tested in two competition designs. One of them was the contest for a master plan for the city of Magnitogorsk, which was to be built along with the construction of Magnitogorsk steel mill in the Ural Mountains industrial area. Another was the design of the "Green city", a recreational suburban area of Moscow. In both cases, the disurbanists suggested entries that went far beyond the initial goal of the competition briefs (Khan-Magomedov 2001).

In the contest for Magnitogorsk master plan, a team of disurbanists (M. Okhitovich, M. Barshch, V. Vladimirov, N. Sokolov) designed a system of settling stripes that linked the metallurgy plant and the surrounding territories (Fig. 4) instead of a compact city for about 30 000 people. Disurbanists introduced a special geographical term Magnitogorie to refer to the vast area they designed instead of the compact city of Magnitogorsk. This new term became the title of the completed entry and articulated a sharp difference between this proposal and the city master plan expected by the jury.

The length of each stripe was about 25 km. Residential stripes started from the plant and stretched along the roads (Fig. 4). The stripes connected a steel mill and existing cities, agricultural areas, stand-alone industrial enterprises such as a dairy farm and food production factory in the northern part of the area (Fig. 4). Smaller cities were the starting points of stripes of settling too, like the small town of Musak located west of the steel mill. Some of the stripes here were dead ends and broke off in a natural or agricultural landscape. Several roads between the stripes were only transport connections or left without settlement for further development (in Fig. 4 they are shown with dashed lines). Manufacturing zones were linked to railways for the delivery of semi-finished productsfor production and onward transportation of ready goods.

Each stripe had a cultural centre, a park of culture and recreation, placed close to its middle part in the most favourable natural landscape. For children 8–15 years old, various facilities were distributed across residential areas. The educational institutions also included three schools, evenly distributed along each stripe. In the social concept of "new settling", children were not supposed to be separated from their parents, unlike the proposals of the supporters of the "sotsgorod" (Khan-Magomedov 2009: 101–115).

Despite the overall focus on ensuring the same accessibility of services and quality of living conditions everywhere, the Magnitogorie project envisaged the creation of a central park of culture on the stripe between the local small town and the steel mill (Fig. 4). It was also supposed to create a single administrative and social centre, which was to be located closer to the factory, between two neighbouring settlement stripes (Fig. 4).

Three main agricultural areas with different production specializations were planned (Fig. 4). The two circles west and south of the mill marked agricultural

Fig. 4 Magnitogorie. Competition entry. Master plan (Barshch et al. 1930: 39). The inscription at the top of the drawing: “…a new settling of mankind…Lenin”. No copyright illustration. A scanned copy is a courtesy of the library of the Moscow Institute of Architecture (state academy)

areas specializing in field farming (in the north) and livestock farming (in the south). The boundaries of these agricultural areas were determined by the 15 km radius of accessibility of the services of the machine and tractor stations. Several stripes were specially located for the convenience of agricultural work in the third agrarian area of intensive irrigated horticulture and gardening along rivers and lakes. This arrangement of settling stripes was not only the result of the intention to provide better living conditions but also to maintain a balance between the demand for labour in agricultural production and the capacity of the nearby settling stripes (Fig. 4).

Each stripe had a clear structure of parallel linear zones (Fig. 5). A lane of protecting green space was located directly at the border of the road, behind it there

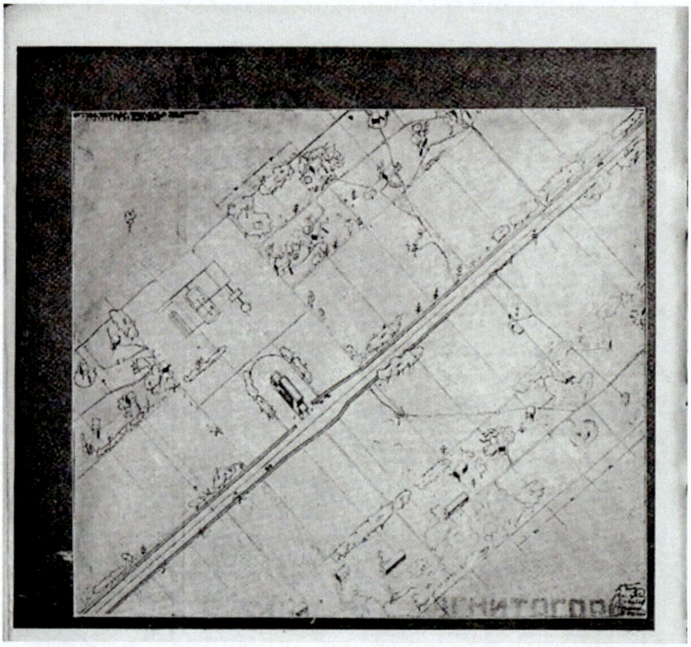

Fig. 5 Magnitogorie. Competition entry. A fragment of a settling stripe. (Barshch et al. 1930: 44). No copyright illustration. A scanned copy is a courtesy of the library of the Moscow Institute of Architecture (state academy)

were residential areas with freely placed dwellings of different types. On the inner side of the residential stripes, there were a variety of sports trails: for jogging, skiing, biking, walking and even riding on the sledges during winter times. The road in the middle of the stripe was served by cars, public transport lines and a system of delivery of ordered goods. To organize a system of daily services for residents, it was planned to locate two-storey service stations along the road at a distance of one kilometre from each other (Fig. 5). Public transport stops were not rigidly tied to station locations. Residents could get on or off public transport on demand along the entire stripe. The station itself was a local public culture and maintenance facility providing daily services like newspaper kiosks and barbershops, retail and order of goods, catering, storage of equipment for housing maintenance, a pantry of sports supplies, and the like. Kindergartens, hotels and sports facilities were also located near these stations. The ordered goods were to be delivered directly to one's home, as well as a ready meal in case of such an order (Barshch et al. 1930).

The project included plenty of types of housing, which were designed in detail: for one, two and more people. All of them were located at a distance of at least 30 m from each other to provide privacy for residents (Barshch et al. 1930: 40). The construction system was intended to allow these dwellings to be easily modified to accommodate the changing lifestyles of the residents. Moreover, each dwelling could be easily disassembled, moved to a new place and reassembled there again. Each resident could enjoy the freedom in choosing the type of housing and the degree of social closeness with neighbours. "An agricultural worker, and a worker or a workwoman of industry live here next to or equally close to the place of work, catering, culture, places of social and political life. But, in addition, to the natural benefits of nature, healthy living conditions and recreation" (Barshch et al. 1930: 42).

The new settling system was supposed to be better than the city not only in terms of quality of life but also in terms of economy. In the explanatory note to the Magnitogorie project, the authors argued that densely populated cities would require much more investments. Accommodating a large number of people on a small plot of land would need much more expensive durable building structures and materials, centralized engineering systems (Barshch et al. 1930: 40). In the long term, this will lead to additional maintenance and operation costs, especially for reconstruction. The new dispersed settling of population and production strived to be much more efficient through the use of local and cheap materials, local engineering systems and easier transformation and upgrading of the built environment (Sektsiya sotsialisticheskogo rasseleniya 1930: 6–7).

6 The Productive Landscape of the Future: Neither Urban nor Rural

In 1922–1930, discussions on socialist settling revealed three versions of the built environment of socialism.

All of them proceeded from the premise that in the future settling there will be no division into the city and the countryside, and there will be an integration of industrial and agricultural production, industrialization of agriculture and reorganization of socio-spatial patterns of everyday life. However, the models of the future they proposed turned out to be very different.

The concept of workers' garden settlements naturally had its roots in Howard's model. His initial concept was flexibly adapted for a variety of planning tasks, ranging from small suburbs to comprehensive schemes for the "dissipation" of Moscow into a complex metropolitan area.

The model of the socialist city, or "sotsgorod", had much in common with Fourier's phalanstery. It envisaged the formation of compact urban settlements, consisting of factories and associated group of "zhilkombinats", namely residential superblocks for collective living with a well-developed system of public services, but without housework and household chores.

"New settling" of Mikhail Okhitovich's disurbanism challenged the very idea of a city in the future. Its foundation was directly a Marxist theory and, as for many avant-garde artists of that time (Cohen 1984), the American experience of urban development. Disurbanists advocated a uniform network of settling stripes integrating residential, industrial and agricultural areas.

According to Okhitovich, the different comprehension of the socialist personality led to the fact, that each concept proposed distinguishable type of built environment and housing. In a summary of the debate on town planning (Khan-Magomedov 2009: 157), he argued that garden city proponents retained the bourgeois personality, the corresponding old forms of individual, family housing and a role of women in the domestic household, while the supporters of "sotsgorod" suggested to eliminate the personality, and ultimately ban it altogether. In the contrast, the social basis of the concept of disurbanism was the goal of the comprehensive development and flourishing of the socialist personality.

Each concept developed a particular settling pattern that integrated industrial and agricultural production. There were workers' garden settlements with subsidiary farming of industrial workers, a model of an agro-city and surrounding areas of collective industrialized agricultural production, and a large-scale network of settling stripes with land cultivation, processing of agricultural raw materials and manufacturing in any cell. The integration of agriculture into the basic settling patterns in all concepts lets presume that each of them had the potential to develop a new socialist settling into a kind of unified productive landscape.

In terms of the future relationship between city and countryside, the "new settling" of disurbanism was the most radical, most original in spatial form, long-term solution to the problem of socialist settling. It relied on solid theoretical studies of the history and future development of settlements. This concept targeted eliminating the inequality of centre–periphery relations inherent in the urbanizing world. And it was the only concept that proposed a large-scale spatial structure on a national and global scale to transform settlement systems that appeared in the past.

7 Long-Term Forecast for Post-urbanization

Since the debate on socialist settling, the antithesis between a city and a country has taken the form of an imbalance between the global North and the South, global cities and urban periphery, capitals and regional centres, city centres and outskirts.

These growing controversies could cast doubt on the prospects for a new evenly distributing settling of mankind and speak in favour of Le Corbusier, who criticized disurbanism and advocated concentration of people and urbanization as a natural desire of a human being (Khan-Magomedov 2001: 206, 207).

This would be true if the deconcentration of people and production in space was the primary goal of the disurbanists. In fact, the deconcentration of the population and industry was just a tool. The goal was a settlement that could provide the most favourable living conditions for a human being, regardless of his or her location. In this theoretical model, dependence on location means limiting the desired maximum of accessible possibilities for human being development, which the disurbanists saw as a criterion for a highly developed society. Thus, the fundamental problem was not the heterogeneity of physical space itself, but the heterogeneity of access to the best possible living conditions in this space.

From this "disurbanism" point of view, urbanization is the process of increasing the dependence of meeting people's needs on their location. On the contrary, deurbanization means a growing degree of independence.

The projects and theoretical studies of the disurbanists allowed them to develop a clear and logical model of such an ideal settling. It includes only two key elements. The first is the delivering and collecting links of communication and transportation systems, which can be called "feeders". The second is the universal units for living, development of raw materials, production and consumption, which can be called "cells" since they have a kind of metabolism. In the general scheme of the "new settling", roads and electricity supply lines were "feeders", and the triangles between them were "cells". Anyone living in any "cell" had equal access to the services and opportunities that the "cell" and "feeders" provided working together and supporting each other. Whatever the degree of homogeneity of the entire system, any person in any "cell" would have the same desired maximum level of living conditions. A non-uniform or disrupted grid would have satisfied that criterion too if it was capable to maintain this maximum in any "cell", which remained in a system. Easy scaling both in the scale of the whole and the elements was another opportunity of this model. It was ready to use for areas of different sizes.

This simplified model allows one to see emerging tendencies that fit well into this basic structure of the "new settling" of disurbanism. Communication networks, delivery services, collecting and distributing electric power smart grids, emerging personal or collective air transport systems could easily become "feeders" in this model. The coverage of these networks is not yet uniform but is growing rapidly to connect more "cells" whose service conditions are becoming almost the same across the network.

New technologies and new infrastructures tend to equally provide any location where they operate with the necessary services and goods. This does stimulate the development of "cells", first of all of a very small size. This process has been greatly accelerated by the COVID-19 pandemic. These very basic elements, envisaged by the model of "new settling", in today's realities are often squeezed up to the scale of one's individual space, namely an apartment or even a room. If it stood on pillars in a vast green site, such a structure could be a contemporary variant of the individual residential unit of the "new settling" in Fig. 3. Small family production enterprises connected with global delivery and communication networks, especially in remote rural areas, can be another example of new spatial and economic actors that combine production and consumption in one place, thus following the principles of disurbanism.

This very brief comparison of current development trends and the settling model of disurbanists found some analogies. The interpretation of this model in terms of "feeders" and "cells" can be explored more thoroughly as a tool for analysing contemporary development vectors of settling.

Presumably, today the development of networks of "feeders" equally increases the accessibility of external and internal resources in the entire area of their operation. This should push the development of new productive "cells" to use this new evenly distributed resource. This means that a single space of interactions between "cells" and "feeders" will evenly cover the spatially unevenly developed built-up areas.

Expanding this new unified space for production, communication and exchange will eliminate or significantly reduce the dependence of availability of jobs and essential services on location in the physical landscape. Any new "cell" will be able to connect tothe service area of the "feeder", for example, a mobile operator, at any point, both in the immediate proximity of the network physical infrastructure and in a more remote location.

The process of uniform expansion of "feeder-to-cell" communication environment over the existing built-up areas can be considered as a hypothetical "post-urbanization", which may lead to a new type of settling in the future. Unlike disurbanists layouts, fewer changes in the physical structure of the space may be required to ensure maximum living conditions. Under the conditions of "post-urbanization", scheme 4 in Fig. 1 of the "dispersive" "new settling" would have a less rigid grid of "feeders" and a much more complex set of much smaller "cells".

The potential of networks as an alternative to centralities is worth further development. The growing uniformity of the availability of Internet and networks of delivery services, platforms for distant work, 3D printing, vertical farms, telemedicine, passive houses and local renewable sources of energy opens up new opportunities for achieving the ideals of the theory of disurbanism: integrating work and housing, producing raw materials and their processing and, ultimately, providing the most favourable conditions for human development in any place.

References

Barshch M et al (1930) Magnitogorie. Sowremennaia Architektura (SA) 1–2:38–56

Cohen J-L (1984) America: a Soviet ideal. AA Files 5:32–40

Cornu M (1967) Anatole Kopp, Ville er révolution: architecture et urbanisme soviétiques des années 20,Éditions Anthropos, Paris [compte-rendu]. L'Homme et la société 5:201–209

Golomshtok IN (1994) Totalitarnoye iskusstvo (Totalitarian art). Galart, Moscow

Hatherley O (2020) The city without a centre: disurbanism and communism revisited. In: Bach J, Murawski M (eds) Re-centring the city: global mutations of socialist modernity. UCL Press, London, p 63–72

Hudson HD Jr (1992) Terror in Soviet architecture: the murder of Mikhail Okhitovich. Slav Rev 51(3):448–467

Hudson HD Jr (2016) Blueprints and blood: the Stalinization of Soviet architecture, 1917–1937 (Princeton Legacy Library,1746). Originally published in 1993. Princeton University Press, Princeton

Khan-Magomedov SO (1987) Pioneers of Soviet architecture. Rizzoli, New York

Khan-Magomedov SO (2001) Arkhitektura sovetskogo avangarda. Kniga 2. Sotsialniye problemy (Soviet avant-garde architecture. Book 2. Social problems). Stroyizdat, Moscow

Khan-Magomedov SO (2009) Mikhail Okhitovich. Fond "Russkiy avangard", Moscow

Khazanova VE (1980) Sovetskaya arkhitektura pervoy pyatiletky (Soviet architecture of the time of the First Five-Year Plan). Nauka, Moscow

Kopp A (1967) Ville et révolution: architecture et urbanisme Soviétiques des années vingt. Éditions Anthropos, Paris

Kosenkova YL (ed) (2018) Sovetskoye gradostroitelstvo. 1917–1941 (Soviet town planning. 1917–1941). Progress-Traditsya, Moscow

Kuzmin V (1928) O rabochem zhilishchnom stroitelstve (On construction of housing for workers). Sowremennaia Architektura (SA) 3:82–83

Kuzmin (1930) Problema nauchnoy organizatsiyi byta (The problem of scientific organization of everyday life). Sowremennaia Architektura (SA) 3:14–17

Meerovich MG (2007) Rozhdeniye i smert goroda sada: deystvuyushchiye litsa i motivy ubiystva (The birth and death of the garden city: actors and motives of the murder). Vestnik Evraziyi (Eurasia Gerald) 1:118–166

Meerovich MG, Konysheva EV, Khmelnitsky DS (2011) Kladbishche sotsgorodov: gradostroitelnaya politika v SSSR (1928–1932 gg.) (Cemetery of sotsgorods (socialist cities): town planning policy in the USSR (1928–1932)). Fond "Presidentskiy centre B.N.Yeltsina, ROSSPEN, Moscow

Meerovich MG (2013) Discussiya o sotsrasseleniyi. Noviye materialy (Discussion on socialist settling. New materials). Arkhitekturnoye nasledstvo 58:262–280

Okhitovich M (1929) K probleme goroda (To the problem of the city). Sowremennaia Architektura (SA) 4:130–134

Okhitovich M (1930a) Zametky po teoriyi rasseleniya (Notes on the theory of settling). Sowremennaia Architektura (SA) 1–2:7–16

Okhitovich M (1930b) "Marksistskaya" zashchita kommunalnogo sotsializma ("Marxist" defense of communal socialism). Sowremennaia Architektura (SA) 5:7–13

Sabsovich LM (1929) Goroda budushchego i organizatsia sotsialisticheskogo byta (Cities of the future and the organization of socialist everyday life). Gosudarstvennoye tekhnicheskoiye izdatelstvo, Moscow

Sabsovich L (1930) Sotsialisticheskiye goroda (Socialist cities). Gosizdat RSFSR "Moskovskiy rabochiy", Moscow.

Sektsiya sotsialisticheskogo rasseleniya stroysectora Gosplana RSFSR (Division of socialist settling of the construction sector of the State Planning Committee of the Russian Soviet Federal Socialist Republic) (1930) Tezisy doklada "O socialisticheskoy planirovke rasseleniya" (Theses of the report "On the socialist planning of settling"). Sowremennaia Architektura (SA) 6:1–17

Semenov V (1993) O blagoustroystve gorodov (On good city planning). Reprint edition of 1912. Klyuch, Moscow

Stroitelstvo Moskvy (1930) K diskussiyi (Introduction to the discussion). Stroitelstvo Moskvy 1:2

Tsentralny Komitet Vserossiyskoy Partiyi Bolshevikov (TsK VKP(b)) (Central Committee of the All-Russian Communist Party of the Bolsheviks) (1930a) Postanovleniye "O tempe kollekivizatsiyi i merah pomoshchi gosudarstva kolhoznomu stroitelstvu" ot 05 yanvariya 1930 goda (Resolution of 05 Jan 1930 "On the pace of collectivization and measures of state support to the building of collective farming"). Rossiyskoye istoricheskoye obshchestvo. Elektronnaya biblioteka istoricheskih dokumentov. http://docs.historyrussia.org/ru/nodes/64427-materialy-komissii-politbyuro-tsk-vkp-b-pod-predsedatelstvom-narkoma-zemledeliya-sssr-ya-a-yakovleva-po-podgotovke-proekta-postanovleniya-tsk-vkp-b-o-tempah-kollektivizatsii-i-merah-pomoschi-gosudarstva-kolhoznomu-stroitelstvu-postanovlenie-tsk-vkp-b#mode/inspect/page/1/zoom/4. Accessed 10 Jun 2021

Tsentralny Komitet Vserossiyskoy Partiyi Bolshevikov (TsK VKP(b)) (Central Committee of the All-Russian Communist Party of the Bolsheviks) (1930b) Postanovleniye "O rabote po perestroyke byta" ot 16 maya 1930 goda (Resolution of 16 May 1930 "On the work on the reorganization of everyday life"). Sowremennaia Architektura (SA) 1–2:3

Vesnin VA, Vesnin AA (1929) Predposylky stroitelstva novyh gorodov (Prerequisites for the construction of new cities). Ekonomicheskaya Zhizn 281(3302):2

Volchyok Y (2008) Urbanisti i desurbanisti – kulturologicheskiy experiment (Urbanists and disurbanists – culturological experiment). In: "IQ" science and education portal. National Research University "Higher School of Economics". https://iq.hse.ru/news/177680976.html. Accessed 10 Jun 2021

The City in the Landscape: Alfred Caldwell's Vision and Experiment for an Ecological City

Kristin Jones

Abstract Alfred Caldwell (1903–1998) was a twentieth century American architect, planner, landscape architect, and educator, whose work bears the influences of landscape architect, Jens Jensen, architects Frank Lloyd Wright and Ludwig Mies van der Rohe, and planner Ludwig Hilberseimer. "The city in the landscape" is a moniker he used to frame his conception of urban development within the broader context of landscape and to suggest a form of human settlement in harmony with nature. It is also a reference to his Plan of Chicago that was based on the planning principles of Hilberseimer for the incremental, planned decentralization of cities. This chapter explores the context out of which Caldwell's ideas and plans grew; his background and influences, assessment of persistent urban and rural problems and their interconnected roots, and arguments for structural changes that promised more livable and sustainable conditions. It then presents his own farm in Bristol, Wisconsin as an experiment in the application of these planning ideas. Key themes in Caldwell's work include private land ownership, stewardship, and a balanced approach to agriculture and industry, reflecting a more comprehensive and ecological perspective on architecture and planning that looks to a productive landscape to provide individual, social, and environmental health and resilience.

Keywords Urban farming · Ecological city · Resilient city planning · Societal transformation · Land and food

1 Introduction

At the midpoint of the twentieth century, the discipline of planning in the U.S. was still considered to be in its infancy. According to the U.S. Census, 1920 marked the turning point when more people were living in cities than in rural areas (USCB 1999). City Planning emerged as a profession to improve the conditions in urban

K. Jones (✉)
Illinois Institute of Technology, Illinois, USA
e-mail: kjones24@iit.edu

C. Brisotto and F. Lemes de Oliveira (eds.), *Re-Imagining Resilient Productive Landscapes*, Cities and Nature, https://doi.org/10.1007/978-3-030-90445-6_11

centers that were experiencing problems associated with rapid growth as a result of industrialization. In 1923, the first academic program in planning was established in North America, at Harvard (Harvard n.d.). The rising public awareness of urban problems led other American universities to begin programs in planning, including the Illinois Institute of Technology[1] and Massachusetts Institute of Technology in 1941, and University of California Berkeley and Columbia University in 1948.

With industrialization marching on, researchers in planning sought to understand how and why specific problems had emerged and to find solutions to the unhealthy urban conditions. In studying urban growth and structure, analyzing means of application for planning in practice, and formulating criteria for standards of urban development (Lillibridge 1953), research began to incorporate social and economic perspectives into existing disciplinary thought within architecture and landscape design, extending planning discourse beyond traditional formal considerations like functional relationships, circulation, and aesthetic qualities to include public health, policy, and civil rights issues. Environmental awareness also grew as the impacts of pollution and poor hygiene continued to worsen. By mid-century, a new development was also sweeping across America. The realization of the Fordian dream of a "universal car" for every family (Ford 2013) coupled with the advent of federally backed low-interest mortgage loans (Table 1) fueled the growth of and flight to the suburbs.

The City and Regional Planning program was founded at the Illinois Institute of Technology (IIT) by Professor Ludwig Hilberseimer within the Department of Architecture led by Ludwig Mies van der Rohe. The influential planning program continued through the end of the century. "The city in the landscape"[2] was a moniker adopted by both Hilberseimer and his colleague and student, Alfred Caldwell, to describe their planning perspective with respect to the broader natural context within which urban development was to be considered. It was also the title of Caldwell's (1948) IIT master's thesis which contains his plan for the greater Chicago metropolitan region.

This chapter makes a connection between Caldwell's work as planner and the contemporary concept of an "ecological city" for two primary reasons. First, to underscore his understanding of the relationship between nature and the built environment as interconnected living systems, and second, as a means to link his vision and experiment with contemporary discourse on healthy and sustainable urban planning. The following discussion on the state of planning in Chicago at the time serves to distinguish and ground Caldwell's contribution within the field.

[1] Planning courses were offered through Armour Institute of Technology as early as 1938, before the school merged with Lewis College of Science and Letters to form Illinois Institute of Technology in 1940.

[2] "The City in the Landscape" refers to both an idea and the title of Caldwell's master thesis, which were explored in depth in (Jones and García-Requejo 2020).

Table 1 A few important developments

1862	U.S. Department of Agriculture established
1903	U.S. Department of Commerce (and Labor) established
1909	Chicago Plan Commission established
1916	U.S. National Parks Service established
1917	American City Planning Institute (a predecessor of the American Planning Association) established
1920	U.S. Census first reports majority of Americans live in urban areas
1934	Establishment of federally backed FHA (Federal Housing Authority) mortgage loans to the average American (Housing Act of 1934)
1935	WPA (Works Progress Administration) established to put unemployed to work on public infrastructure projects
1935	Home Owners Loan Corporation "residential security" appraisal map (239 cities)
1935	Soil Conservation Service (Soil Conservation Act 1935) established
1937	Establishment of the U.S. Housing Authority to fund development of low-rent housing (Housing Act of 1937)
1940	U.S. Census reported more population growth in suburbs than cities for the first time
1947	Chicago Land Clearance Commission established to deal with urban slums
1954	Federal Housing Act of 1954 funded urban renewal and slum prevention
1956	Federal funding was authorized to create the U.S. Interstate Highway system

2 The Question: The Potential of Planning for Social and Environmental Transformation

2.1 Early Planning in Chicago

One of the earliest examples of modern city planning for an existing U.S. city came after the 1893 World's Columbian Exposition in Chicago. The extravagant world's fair, dubbed the "White City," spanned 600 acres with enormous neo-classical buildings in electric lights, centered on a "court of honor" with grand boulevards, fountains and canals with gondolas (Rydell 2005b). While essentially a trade show, Chicago used the opportunity to establish itself as a world-class city highlighting its cultural gravitas and demonstrating its ability to reimagine and rebuild itself after the great fire.

After the fair, Daniel Burnham, who headed its construction, published the 1909 *Plan of Chicago* along with Edward Bennett and illustrator Jules Guérin. According to the Chicago Architectural Club, the plan "aimed to provide a comprehensive and coherent strategy to address the city's unregulated development creating conditions to improve commerce and reflecting on the way people live in a modern urban environment" (CAC 2020). Six improvements to the city's transportation, parks, and public spaces underpinned their vision of the future city, "so related as to give coherence and unity to the city" (Burnham and Bennett 1909):

1. Improvement of the lakefront
2. Creation of a system of highways outside the city
3. Improvement of railway terminals
4. An outer park system, and of parkway circuits
5. Systematization of the streets and avenues within the city
6. Development of centers of intellectual life and of civic administration.

The "Burnham Plan," as it is commonly called, garnered much attention as an example of the City Beautiful movement and raised awareness of the importance of a region's natural features, like the lakefront, and the creation of parks, rational transportation systems, and centers for public life in large cities.

By the 1930s, several parts of Burnham's Plan had been implemented, including enhancements to the lakefront (Grant Park and Buckingham Fountain), street widening (Roosevelt Road), street improvements (Wacker Drive and Michigan Avenue), including adding a double-decker bascule bridge (Michigan Avenue), and straightening a portion of the Chicago River (Smith 2005). Key cultural buildings, including the Field Museum, had also been added near Grant Park which was envisioned as the city center. While these improvements alleviated some of the urban problems of the time, including bringing some order to the city, the stock market crash of 1929 soon brought unemployment and poverty to the already crowded city. To bolster public morale, Chicago hosted a second world's fair in 1933–1934, the Century of Progress Exposition, in which nearly three times as many corporations participated in offering visions of the world of tomorrow. Collectively, they advanced "the message that cooperation between science, business, and government could pave the way to a better future" with "displays that insisted that Americans needed to spend money and modernize everything from their houses to their cars" (Rydell 2005a), promoting consumer spending while also creating jobs. Although these fairs became the examples for future fairs throughout the country, critics like Louis Sullivan and Frank Lloyd Wright condemned the architecture.

2.2 *Mid-Century Planning in the U.S.*

Caldwell began to develop the ideas behind his Plan of Chicago in the early 1940s, during which time more people were still living in U.S. cities than in rural areas and the economy was recovering (USCB 1999). The urban problems associated with the rise of industry and rapid urban growth, such as traffic, overcrowding, and environmental pollution, had paved the way for rapid suburbanization. In 1940, U.S. census reports showed for the first time that population growth in the suburbs was higher than in central cities (USCB 1999). A new planning problem emerged—urban sprawl.

Although research in agriculture was an active reflective practice at the time, urban planning itself was not often dealing with issues of food or of productive landscapes, let alone the relationship between land, food, and people. Instead, attention was focused on the "form and organization of economic and social activity in the community" (Lillibridge 1953), the interests of the urban elite and of commercial and civic power that were reflected in the world's fairs. The urban and rural were becoming two different worlds, reflective of both the distinct domains of practice for professional planners and the growing cultural distinction between urban and rural ways of life. This jurisdictional perspective tended to limit the urban planners' and researchers' attention to improvements that could be made within the city itself and, consequently, did not foster a more comprehensive understanding of the interdependent reality. Planning in the U.S. thus not only missed an opportunity to promote community and understanding among urban and rural dwellers, it may actually have inadvertently exacerbated civic problems by institutionalizing differences between the urban and rural. One of the lasting symptoms of this divide is that modern urban dwellers have forgotten where food comes from and how it is produced.

2.3 Toward an Ecological City

According to Hall and Tewdwr-Jones (2010), the European notion of town and country planning was more aligned with the comprehensive perspective embodied in the city and regional planning paradigm than the bifurcated urban–rural dichotomy. It may be significant to note, therefore, that Alfred Caldwell was introduced to planning from a European perspective, through Ludwig Hilberseimer who brought ideas from Germany to Chicago. Caldwell and Hilberseimer shared the view that the increasing specialization in city planning research and practice was problematic because they considered "the city as a whole, its zone of influence, its function in the region, and in the nation" (Hilberseimer 1944, 158; Caldwell 1948). In his City and Regional Planning program at IIT, Hilberseimer and his students approached planning by asking not only "What is wrong with our cities?" but further, "What is wrong with our world?" (Hilberseimer 1946); an example of transcending the level of thinking in which problems were created in order to solve them.

The notion of an "ecological city" might be the most appropriate term today to describe Caldwell's approach to urban development if an ecological city is understood as one that promotes "preserving and restoring the balance between the natural and the built environment through planning and design" (Platt 1994). The analogy is intended to capture the notion of organic order[3] in urban development advocated by Hilberseimer and Caldwell's appreciation of nature, of native plantings and natural ecology. The anachronism is also meant to signal that these ideas were radical forerunners to contemporary thought. Within the historical context, and even more so

[3] Hilberseimer contrasts organic with geometric urban form, of slow and planned development versus autocratic (Hilberseimer 1944, 22–23).

once urban renewal efforts were underway within cities, using the power of eminent domain for slum clearance, plans like Caldwell's Plan of Chicago were often misunderstood and dismissed either as fantasy or as unethical tabula-rasa design proposals, rather than being seen as a mode of critical practice highlighting systemic problems and proposing meaningful transformative change.

Today the plans are more apt to be rightly understood; as the chair of a recent international conference, *The Architect and the City*, noted, after the 2007–8 Global Financial Crisis and the period of intense critical reflection in architecture that followed, a more mature perspective on architecture and the city is now emerging:

> With an inclusive, transversal and revisionist nature, [it] incorporates and revisits concepts such as feminism, gender, childhood, shelter, migration, wealth, transversality, glocality, interculturality, multiculturality, and many more. (Cabrera and Fausto 2020).

While Caldwell's plans are a product of their time, they also seem to be more mature, in that they look back, incorporating enduring ideas on the relationship between land, food, and people, and they look forward, addressing social and environmental issues at the forefront of contemporary collective consciousness still today.

Some of the ideas[4] that influenced Caldwell's ecological vision of the city are experiencing renewed interest today. One was the influence of Patrick Geddes' more ethical proposal for urban renewal, the practice of "constructive peace" or "constructive surgery" as an alternative to wholesale redevelopment (Geddes 1930), and his "Valley Section" which depicted a way of life modeled after indigenous tradition[5] to promote the return to a harmonious relationship between people and nature (Thompson 2004). Such ideas provided an ethically and ecologically minded ideal that challenged modern ways of living and thinking in terms of social and environmental impacts. Many of the details about how best to incorporate nature into urban life were inspired by the concept of "permanent agriculture" as advocated by Wisconsin native, F.H. King, who studied forty centuries of farming in China, Korea, and Japan (King 2004). These case studies of permanent agriculture, or indefinitely sustainable agriculture, provided an alternative approach to the modern farming practices that were leading to desertification in the U.S. in the 1930s. Also of significant influence was the political and cultural ideal of an agrarian democracy, as initiated by Thomas Jefferson and echoed by Frank Lloyd Wright, which envisioned a democratic nation of farmers for whom land ownership promised individual security and welfare.

Motivated on one hand by environmental and cultural sustainability, and on the other by economic security and liberty, food and farming became an integral part

---

[4] The reader may refer to numerous references to Geddes, King and Wright in Caldwell's *The City in the Landscape: A preface for planning* and in Hilberseimer's *The New City* and *The New Regional Pattern.*

[5] The native Hawaiian concept of "ahupua'a" is an example of indigenous tradition akin Geddes' "Valley Section." It captures the island valley ecology, spanning from the mountain to the lower valley (plains) to the ocean including its natural resources like forests and rain, streams and coral reefs as well as its human settlement areas and farmlands, with work zones determined by the natural landscape, (e.g., hunting, farming, fishing).

of Caldwell's conception of "how to have security and a good life in America" (Caldwell 1987, 69). His work stands out as one of the few examples of mid-century planning research in which food production, sustainability, and the larger ecological landscape play a significant and integrated role. His plan is described in his master's thesis, *The City in the Landscape: A preface for planning,* and backing up his claim that in order to have livable cities we must first have "determination to live as a community" (Blaser 1984, 50), is his own farm in Wisconsin. A closer view of Caldwell's life and work experiences shows the more immediate context from which his plans developed.

3 The Planner: Alfred Caldwell

3.1 Biography

Alfred Caldwell had nearly 20 years of work experience before becoming involved with planning, including as a superintendent for landscape architect Jens Jensen (1926–1931), landscape architect in private practice (1931–1933), superintendent of parks and Eagle Park landscape/building architect in Dubuque, Iowa (1933–1936), Landscape Designer at the Chicago Park District (1936–1939), and Civil Engineer in the U.S. War Department, Army Corps of Engineers, (1940–1943) (Blaser 1984, 157). In the 1930s, he also had a few important visits with Frank Lloyd Wright (Caldwell 1987, 34–38).[6]

When Mies van der Rohe asked Caldwell to join the faculty of architecture at IIT in 1944, his drawing skills and talent for planning beautiful landscapes and structures already preceded him. He had completed the Lily Pool in Lincoln Park, Chicago, and Eagle Point Park in Dubuque Iowa, both award winning projects which illustrated the influences of Jens Jensen and Frank Lloyd Wright with such grace that they were often mistaken for the elder masters' (Caldwell 1987, 43). By then, he had also passed the state architectural board examination with high marks in engineering (Caldwell 1987, 66–68).

It was in assisting Hilberseimer with drawings for the book, *The New City: Principles of Planning* that Caldwell became acquainted with planning. Though he never did teach planning, and became known instead for his undergraduate courses in construction and architectural history, Caldwell was influenced by Hilberseimer's planning ideas and eventually considered himself an "architect—turned city-planner" (Blaser 1984, 48). In order to teach at IIT, Mies and Hilberseimer facilitated Caldwell's obtaining a bachelor's degree based on equivalency, counting his work experience and professional license in lieu of academic credentials (Caldwell 1987, 68–69).

[6] Caldwell's employment history is taken primarily from Blaser's monograph on Caldwell. How it unfolded, nuances of relationships, are primarily from Caldwell's words in the AIC Oral History and Blaser's monograph.

Later, with an eye to continuing working together at the university, Hilberseimer encouraged Caldwell to also complete a master's degree.

The war was part of the collective consciousness during that time, especially at the school being run by Mies, Hilberseimer, and Walter Peterhans, who were all former leaders of the German Bauhaus. In August of 1945, in retaliation for the 1941 attack on the U.S. at Pearl Harbor, the atomic bomb was dropped on Hiroshima and Nagasaki ending the war but leaving an indelible mark on humanity and the earth. On the day the news headline was released, Caldwell recalled his discussion with Hilberseimer, who said, "My God, the world is ruined. The world is ruined if man has that power" (Caldwell 1987, 76). In that moment, Caldwell said he was moved to write the paper, "Atomic Bombs and City Planning," which was published in the *AIA Journal* that year. In stating, "There's nothing more incongruous than atomic bombs and babies," he considered it to be the best and most prophetic paper he ever wrote (Caldwell 1987, 77). The paper aimed to raise awareness of the vulnerability of cities and was a call to action to begin decentralizing cities in order to bring fresh air and sunlight back into the city and make them safer places to live:

> [I]t would be very strange if a civilization, cunning enough to split the atom and release the primal energy of the universe, must somehow confess itself unable to provide the simplest need of mankind; a simple house in a safe place. (Caldwell 1945b)

Having lived through the economic hardships of the Great Depression and the atrocities of World War 2, Caldwell sought hope in the potential of planning to transform lives and set out to do something. He completed his thesis and secured his master's degree in 1948 while also starting his farm. He continued to teach in the architecture school at IIT until 1959, collaborating with Mies and Hilberseimer, when he resigned in loyalty and protest against the university hiring another architect to complete the IIT campus rather than Mies (Caldwell 1987, 87). He then worked as a City Planner for the Chicago Plan Commission (1960–1964), taught at Virginia Tech (1965) and the University of Southern California (1965–1973), before returning to private practice once again. In 1981, Caldwell was invited back to IIT where he taught until his death in 1998.

3.2 The Origins of the City in the Landscape

In an effort to advance scientific discourse in the relatively new realm of planning, Hilberseimer laid out theoretical principles of planning for the future of existing metropolises in his 1944 book, *The New City: Principles of Planning*. With this book, Hilberseimer aimed to make headway on dealing with questions that had been posed by Burnham and others at the beginning of the century and remained unanswered (Hilberseimer 1944, 141), such as:

> How are we living? Are we in reality prosperous? Is the city a convenient place for business? Is it a good labor market in the sense that labor is sufficiently comfortable to be efficient and content? Will the coming generation be able to stand the nervous strain of city life? When

> competence has been accumulated must we go elsewhere to enjoy the fruits of independence? If the city does not become better as it becomes bigger, shall not the defect be remedied? (Burnham and Bennett 1909)

By Caldwell's own admission, the foundations for *The City in the Landscape* were laid in *The New City* (Caldwell 1945a). In it, Hilberseimer developed his ideas for a rationally ordered system of the various elements of the city, an idea which he had introduced in *Großstadt Architektur* (Big city- or Vertical city Architecture) (Hilberseimer 1927). Using a mixture of historical precedent research and grounded-theory methodology, he argued the reasons and principles for planning existing cities to become decentralized. He analyzed relationships between urban functions (e.g., industrial, commercial, residential, and recreational), in terms of individual, social, and regional-environmental considerations. He also included his prior research on optimizing transportation, density, and solar insolation. In assisting Hilberseimer with plans and perspective drawings that explained these principles and how they apply to a variety of conditions existing in cities around the globe, Caldwell became familiar with the issues and interested in planning.

The phrase "the city in the landscape" which appeared for the first time in *The New City* was a reformulation of Chicago's motto, *Urbs in Horto*, (City in a Garden), (Hilberseimer 1944, 125), and was later used as titles of Caldwell's 1945 article in *Parks and Recreation* and his 1948 master's thesis. It was again referenced within Hilberseimer's 1949 book, *The New Regional Pattern: Industries and gardens, workshops, and farms* (Hilberseimer 1949, 140). The phrase served to express an ideal harmonious relationship between human settlement and nature, and to expand thinking about the relationship between nature and the city from the scale of the garden or park into a broader regional and environmental context.

The city in the landscape perspective encourages a reading of urban problems as "maladjustments" or symptoms of larger systemic issues. In the opening pages of his thesis, Caldwell reasoned that the cash economy had created vast world markets subject to collapse and that mass manufacturing had turned man into a commodity. He argued that impoverishment, unemployment, environmental pollution, respiratory disease, and continual noise could all be attributed to the effect of machines on humanity made possible by a cash economy run amok. He pointed out that in comparison with medieval cities and old agrarian villages, the modern city is relatively impermanent, in that it rises with military and economic advantage, and falls whenever political, social, or technical changes void those advantages (Caldwell 1948, 1). Caldwell likened the industrial city to a hungry beast:

> From the countryside to the city has come a steady tribute of food and materials, of boys and girls. Into the maw they have gone, goods and humans. This is the unhappy cost of cities. The substance of the earth has been taken to sustain the cities. The legendary streets are paved not with mere gold. (Caldwell 1948, 19).

Caldwell anticipated that in order to solve the problems, planners would need to shift their focus beyond increasing real estate value, which serves merely to "increase tempo and density" of cities (Caldwell 1948, 17). A sustainable solution would require a transformative approach involving structural and systemic change. With a nod to Hilberseimer, he claimed, "A true solution is concerned with the whole—with that relationship of parts to a whole we call order" (Caldwell 1987, 25).

3.3 The Meaning of Farming

Caldwell's planning ideas touch on issues of technology, the environment, and employment. One of the main undercurrents in his arguments on the causes of and solutions to urban problems concerns the relative connectedness or disconnectedness of the people and the land. Farms and farming are central planning issues for him because around them, all of these issues converge. His thoughts on the impact of industrialization on agriculture are sharply critical while remaining optimistic.

Industrialization of agricultural technology wound up displacing much manual labor first with steam-based machines and later with electric. Caldwell did not think that technological innovation would solve this problem, pointing out that just as railroads had superseded canals, ever newer and faster machines would outmode the previous (Caldwell 1948, 32–35). He recognized that with this mechanization, the ancient art of agriculture was being lost to industrial efficiency. Some of the disruptive technological innovations Caldwell cited included the mechanical reaper of 1833 which cut farm labor by 2/3, the centrifugal cream separators of 1910 and improved churns which shifted butter and cheese production from home to factory, and milking machines which increased the herd to farmer ratio, (one tractor could now plow 8–10 acres) (Caldwell 1948, 35–36). The scale of agriculture and profits may have been growing through the replacement of manpower by machines, but Caldwell pointed out that while someone was getting rich on these changes, it was not the individual farmer.

Farming had moved from a subsistence paradigm, where only surplus is sold, to a cash-based agribusiness. Ruled by competition and short-term efficiency, Caldwell explained that cash farming had created an economy in which farmers could lose money in both good times and bad (Caldwell 1948, 36). Burdened with heavy farm debt and competition, it became increasingly difficult for the individual farmer to hold on to his land. Machines had replaced jobs but neither mechanization nor increased efficiency had increased the individual wealth of the farmer, despite advertising to the contrary. Caldwell painted a vivid picture with the statistics. According to the USDA, in 1940, 40% of American farms were operated by tenant farmers and only one in 10 farmhouses had an indoor toilet (Caldwell 1948, 56–66).

In addition to the cost of industrialization paid by the individual farmer, Caldwell also highlighted the price to be paid by society and the environment, through the land. Caldwell claimed that since America had not followed a tradition of conservative agriculture since colonization, the environmental effects of industrialization were even more pronounced, intensified by a culture of wastefulness that was perhaps a remnant of pioneer culture (Caldwell 1948, 48). In precolonial times, there had been indigenous agricultural traditions in the American Midwest, including the practice known as the "three sisters" (maize, squash, beans) (Pauls 2017; Hurt 2021). Later, agricultural traditions from other regions were imported by U.S. plantations, for example, the southern rice culture from West Africa (Carney 2002; Ippolito 2021).[7] Still, the post-colonial and early-industrial practices of overgrazing and overcultivation overwhelmed 61% or 253 million acres of arable land leaving only 161 million acres to safely farm at mid-century. Caldwell, like F.H. King, believed that we should learn from conservative agricultural traditions, like China, where at the time 9 out of 10 people were still living off the land, or Egypt, who had enjoyed 4,000-years of fertile soil (Caldwell 1948, 42–43).

Caldwell's work was an act of resistance opposed to the "war against nature" fueled by cronyist industrial capitalism. He wanted his work to shine light on the profit motive behind what he called the "pseudo" agricultural sciences which had birthed a new agriculture—with more machines, chemical fertilizers, and one-crop farming, despite their known deleterious effects on the soil. Caldwell recognized the soil as perhaps our most precious resource. Instead of white picket fences and shiny cars, his work therefore presented for consideration, deserts, erosion, and deforestation, as evidence of the "spendthrift dissipation of the continent's capital resources" that we call progress (Caldwell 1948, 48–57).

In the 150 years prior to Caldwell publishing his plan, the urban population in the U.S. rose from 5 to 75% of the nation's population (Caldwell 1948, 56). Caldwell viewed this as an ominous sign. Although the causes of the peasant farmers' displacement from the land was different in the U.S. and in Europe, he recognized a parallel between the two in that the farmers' expropriation from the land and new status as part of a "free" labor market stemmed from the advance of capitalism and oligarchy (Caldwell 1948, 29). The consolidation of land ownership into the hands of a few created an imbalance of power and an unemployed group of farmers that led to their eventual oppression, in the U.S. and in Europe. The shift from subsistence to capitalistic farming practices also led to the depletion of the soil. The way farming was being industrialized meant that there were no stewards taking care of the land (Caldwell 1948, 57) and 11 million people living in the U.S. were unemployed and impoverished (Caldwell 1948, 6). Caldwell wrote, "The monopoly of land by the few, under whatever guise or circumstance, cannot but result in some form of slavery…Necessitous men cannot afford to be free. At the last ditch they will trade freedom for bread" (Caldwell 1948, 28; 1987, 56).

Caldwell saw these events, the widespread desertification of the Great Plains, the criminalization of and violence against displaced farmers who had become beggars in early-industrial Europe, as interrelated symptoms of a declining civilization akin to what had happened with the ancient latifundia in Carthage and Rome (Caldwell 1948, 22).[8]

---

[7] Authors explain that West African women were taken in great numbers particularly from the Senegambian region to farm plantations in the Southern U.S. because of their known expertise in rice culture.

[8] Based on citations in his thesis, Caldwell's perspective was influenced by Karl Marx's *Capital* (1932), W.E. Woodward's *A New American History* (1938), Charles and Mary Beard's, *A Basic History of the United Sates* (1944), and Theodor Mommsen's *The History of Rome* (1866). In

> Rome fell, not because of the barbarians, but because the crop of men failed... We have only to compare the prudence and wisdom of the early community of Rome, which could create the first of all free republics, with the popular immoderation and folly of the Ciceronian period, resulting finally in the slavish horror of the times of Caligula or an Elagabalus... Indubitably, there is an ecology of human societies; the endurance or decline of the culture of the past may have been determined, more than we suppose, by the maintenance of the fertility of the soil (Caldwell 1948, 41–42).

In Caldwell's eyes, our civilization was losing its humanity and its humanity is connected with the soil. By looking at these individual, social, and environmental problems as functions of systems, Caldwell was able to show how technology, the environment, and employment are connected, in such a way that land, food, and people become issues of central concern for urban planning.

What then, would Caldwell's city plan be like? What plan could solve these problems? The next section discusses Caldwell's Plan of Chicago in terms of what problems it aimed to solve and how, followed by his work on his own experimental farm, which developed in parallel.

4 The Plan: The Productive Living Landscape

4.1 Caldwell's City in the Landscape

"The City in the Landscape" is a proposal for replanning the region immediately west of Lake Michigan's shoreline, spanning roughly from Kenosha, Wisconsin, on the north, to Michigan City, Indiana, on the south (Blaser 1984, 48). Drawings of Caldwell's plan (Fig. 1) were first published in his 1948 IIT master's thesis, *The City in the Landscape: A Preface for Planning* (Caldwell 1948), and later in the monograph *Architecture and Nature: The Work of Alfred Caldwell* (Blaser 1984). Despite a similar aesthetic quality resulting from having been drawn by the same hand, Caldwell's plan was an original and unique application of Hilberseimer's planning principles for the Chicago metropolitan area, and was not the same as the proposal for Chicago published in Hilberseimer's book, *The New City* (Hilberseimer 1944, 142–47). Before mentioning the differences, it is necessary to understand Hilberseimer's planning principles which underpinned both plans.

Caldwell described the basis of Hilberseimer's ideas (Caldwell 1948, 83):

> Hilberseimer's conception is a city in the landscape composed of units, which we may think of as the size of small towns, each surrounded by parks and the fields and forests of the region, and all interconnected by highways and railroads.

The groups of rectangles shown in the drawings symbolize walkable neighborhoods. These "settlement units," which have been described amply elsewhere, (Hilberseimer

---

The New City, Hilberseimer had also quoted a related point made by Lewis Mumford about technology replacing natural means, like sunlight, or "turning our environment over to the machine" (Hilberseimer 1944, 55).

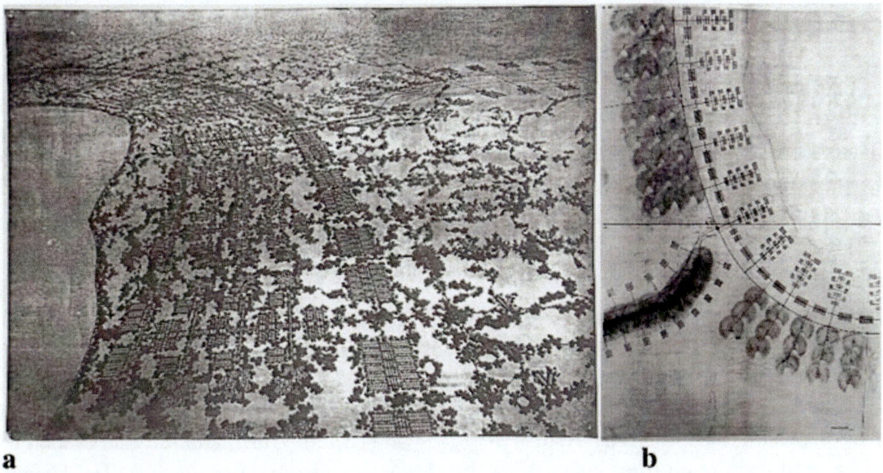

Fig. 1 **a** "The City in the Landscape" aerial perspective drawing, **b** "Plan of Chicago" plan diagram drawing, (north is up). (Reproduced from Caldwell 1948, 95–96).

1944, 126; 1949, 135; Swenson and Chang 1980, 103, 109; Jones and García-Requejo 2020) are a key principle of Hilberseimer's planning approach. Briefly here, the intention of Hilberseimer's settlement unit concept was to provide a flexible planning module that would contain all the functions needed to support daily community life, including residential, commercial, and light industrial areas, as well as schools. These units would be limited in size by walking distance, approximately ¾ mile long by ½ mile wide, but could vary in population density and exact configuration, depending on the community's focus of activity.

Another defining principle of the city in the landscape was the treatment of green space which was envisioned as the connective tissue of the city, situated between and around the houses and the settlement units, not simply isolated interventions within or belts around the city, and could be used for a variety of functions including gardens, parks, farms, and forested areas.

An innovative conception of the transportation system was another principle. No throughfares were provided in the settlement units so that green spaces between and behind the houses would always remain free of car traffic and children could walk to school not along the streets, but within the green parks and gardens, without crossing any streets at all. Beyond the scale of the walkable settlement unit, the commercial and industrial work areas, natural landforms and features, and transportation within and beyond the city were part of a larger regional conception.

The plan diagram (Fig. 1b) describes the relationship between the settlement units, industry, and agriculture with natural and other significant features in the region. It is a clear application of Hilberseimer's principles (Hilberseimer 1944, 144):

> [W]e must distinguish between two main parts of our city plan: The commercial area and the industrial area. The industrial area may be subdivided into two parts; heavy and light manufacturing industries. These parts, together with their respective residential districts,

> form the city. The suburbs, and also the areas containing country residences, are parts of this whole. A replanning of the city must begin with the reorganization of these areas. It must establish a proper relationship between them and provide for their connection with one another by adequate means of transportation.

Therefore, walkable settlement units would be linked by a main transportation artery for car and rail traffic along which the commercial work areas would be located. Locations of heavy industrial plants were also carefully considered in relation to other areas and to natural topography and wind currents to mitigate the impact of smoke and pollution on residents (Hilberseimer 1944, 42–43). A linear approach to reorganizing the city was the backbone of Hilberseimer's planned decentralization. Notably, instead of a center for civic and cultural life, as in Burnham's plan, the only central feature provided in Hilberseimer's plan was a centrally located airport.

Hilberseimer acknowledged that his linear city concept was a development of the ribbon system first noted by Arturo Soria y Mata (Spain 1882) and later advanced by N. A. Milyutin (Stalingrad 1930) and the M.A.R.S. group (London 1942).[9] Hilberseimer decided that the linear form was the ideal development pattern because it permitted the city to be free to grow laterally, avoiding the constrictive problems of centralized city forms, like traffic congestion, overcrowding, lack of sunlight. He reasoned that compared with centralized city plans, which evolved in cities that began as defensible fortresses, linear city plans evolved as a result of transportation and communication between towns. Therefore, Hilberseimer considered the linear form to be more consistent with the technology and circumstances of the current epoch.[10] For one, it offered better defensibility against current military technology (i.e., nuclear warfare) and, for another, it seemed a natural form for a city in the age of electricity, with technology like high-speed rail and power distribution systems increasing practical distances more and more. (Hilberseimer 1944, 66–68).

As mentioned, Caldwell's Plan of Chicago (1948) was an evolution of the original plan published by Hilberseimer in *The New City* (Caldwell 1948, 89). It endeavored to solve "the five major maladjustments of this city—smoke, dangerous streets, traffic congestion and distance to work, slums, lack of parks" (Caldwell 1948, 92). Hilberseimer's original 1944 plan offered perhaps a more immediately accessible solution, by leaving heavy industry near its existing location directly on the lake and concentrating it slightly toward the south, whereas Caldwell's plan offered a more idealized, nature-forward solution that transformed the entire lakefront into a park. Caldwell identified four functional criteria in his plan: (1) smokeless industries, (2) commercial area, (3) smoke-producing industries, and (4) heavy industry. Principal factors in organizing these functions were (1) Lake Michigan, (2) the Des Plaines River, and (3) a new main transportation artery for car and rail traffic running north–south around the lake, roughly 15–20 miles off shore.

---

[9] The references can be found in Hilberseimer 1944, pp. 68, 70 and 152.

[10] In *The New City*, Hilberseimer compares the centric and linear plans in detail, referencing multiple examples of centric planning from Howard's "Garden Cities" to Raymond Unwin's "Satellite Towns" to Le Corbusier's "Une Ville Contemporaine" and traffic patterns by Ludwig Sierks and Peter Friedrich.

The area immediately adjacent to the lake, east of the main artery, was planned for smokeless industry to preserve and take advantage of its natural beauty. Ten clusters of 36 settlement units, or "settlement aggregates," were shown branching off east of the main artery, with schools and continuous parks between. Where each branch intersects with the main artery a minor rail station was planned. Smaller settlement units, good for single family homes with gardens, were organized closest to and along the main branch street, with larger settlement units located around the perimeter of the cluster, good for houses with small farms or a mixture of single and multi-family dwellings. A corridor of settlement units was planned on both sides of a local highway, envisioned to be a commercial area, running parallel to the main transportation artery. Ample open space on both sides of the corridor provided for plenty of sunlight and air and access to nature.

The major share of industrial activity in Caldwell's plan was located west of the main artery. Moderate smoke-producing industry was shown branching off of the main artery westward, at intervals determined by the anticipated zone of pollution. Fan-shaped settlement units were planned at the south side of each zone, in the area expected to be least affected by smoke according to prevailing winds. Heavy industry was relocated along the Des Plaines River, which was considered a more ideal location. The new location provided connections to both the U.S. inland river system and the Great Lakes, making it ideal for shipping, and it moved the pollution off the lakefront. Since the heavy industries are ideally in close proximity to each other, which did not allow space for the fan-shaped settlements, Caldwell planned the associated residential communities to simply be connected by longer streets so that residences would be located beyond the anticipated zones of pollution. It was at the juncture between the main artery and the heavy industry zone that the main railroad station and airport were located.

Like Hilberseimer, Caldwell envisioned that the city could be reformed gradually over time and that planning ought not to be based on economic convenience, but on what is best for humans.

> To build such a city nothing of worth need be destroyed, nothing prematurely torn down. The useful of today could be used. The new city would simply be built, according to reasonable plan, by the gradual process that even now customarily replaces obsolescence… No new billions of dollars are necessary. All that is necessary is the vision to see it and the heart to make it. (Caldwell 1948, 88).

The year after Caldwell's master's thesis was completed, Hilberseimer published both his original planning solution and the one from Caldwell's thesis in *The New Regional Pattern* (Hilberseimer 1949, 159–170). The ideas were presented using new diagrams and descriptions alongside new variations. Of the concept presented in Caldwell's Plan of Chicago, Hilberseimer wrote:

> Here the population density is very low… The open space between the houses is increased considerably and consequently the settlement structure becomes freer, more open, and merges to a much higher degree with the landscape. (Hilberseimer 1949, 164).[11]

[11] In his next book on planning, *The Nature of Cities*, Hilberseimer acknowledged Caldwell's "generous help" with the work (Hilberseimer 1955).

4.2 The Productive Living Landscape

Turning next to Caldwell's ideas on the productive aspect of the city in the landscape, we look beyond the two drawings that were included in his thesis. His aerial perspective drawing was republished in the monograph, *Architecture and Nature: The Work of Alfred Caldwell* (1984),[12] along with five related drawings and a number of scale model photographs. The continuous landscape was conceived of as a productive living landscape where timber forests could coexist with farms of all sizes and productive parks that might contain recreational facilities as well as they might contain community vegetable gardens. Caldwell explained (Blaser 1984, 54):

> [I]t would be a nation of ten or twenty million small farms, and of productive parks and vegetable gardens in parks, of sun-filled houses and sun-filled apartments—and all as one continuous part of the living landscape—the new city: the city of the future which could be now.

The five other drawings provide an increasingly closer view of the productive landscape. One drawing is an illustration of a regional park (Fig. 2), showing how agriculture and recreation might be integrated within a large green space. Forests of Oak, Ash, and Black Walnut for timber, firewood, and recreation are shown organically intermingled with clearings of varying sizes that could be used for recreational purposes, gardens, or farms. The farms were shown utilizing the practice of contour plowing to minimize sheet erosion, or loss of soil due to excessive runoff, a sustainable farming practice that had been advocated by the USDA Soil Erosion Service (USDA n.d.). Also integrated into the living landscape was a zoological garden, at the top of the drawing, and settlement units, shown at the bottom of the drawing. Another drawing, (Fig. 3), provides a closer aerial perspective view of the houses and school within the landscape (Hilberseimer 1944, 127; Blaser 1984, 53).

Caldwell saw in the productive living landscape a more resilient way of life—one with greater physical, economic, and environmental security implemented through planned decentralization, diversified independent or family farms, and regional integration of industry and agriculture. He claimed that "small farms could solve mass poverty, mass unemployment and inflation—our great national disasters" (Blaser 1984, 54). He recalled a conversation with Wright (Caldwell 1987, 122):

> A modern architect cannot live in this world anymore. Alfred, I have my uncle's lands, there's about two hundred acres, and we will all be here together. We will stick together. You bring your wife and we will all stick together and farm the land. We will not give an inch. We will not destroy architecture to get jobs. I give it up, I give it up. It's hopeless. There's no work.

Although Thomas Jefferson, Ebenezer Howard and Patrick Geddes held similar beliefs about the connection between people and the land, it was unique in the middle of the twentieth century to be revisiting kitchen gardens and small farms with any seriousness considering that high-rises and high-speed rail presented

[12] In 1984, the landscape appears to be more filled in, in the upper right corner of the drawing, compared to the drawing in the 1944 thesis.

Fig. 2 Regional Park, 1957 (Canadian Centre for Architecture, Gift of Alfred Caldwell)

novel and formidable challenges, and that America was experiencing tremendous economic prosperity after the war.

However, it was complementary to Caldwell's understanding that land ownership was the bedrock of a free contemporary society (Caldwell 1948, 58–66). He considered agriculture and industry, growing and making, as natural and complementary means to provide for the physical needs of life. To temper the impact of the growing capitalist economy on individuals, he, like Hilberseimer, advocated a more balanced and secure economy through part-time industrial work made available to the farmers, and part-time agricultural work to the industrial worker. Not everyone would need to be a farmer, but the ability to provide food for oneself and family in desperate times might make it possible for work to get closer to becoming a genuinely free enterprise (Blaser 1984, 54). In his own impassioned words:

> I believe that unemployment and poverty—which create modern wars—can be eliminated by part time small farms and gardens for industrial and clerical workers. I believe that poverty can be transformed into what used to be called the good life. I believe that unemployment can be replaced by self-reliance. (Blaser 1984, 60).

Fig. 3 Landscape perspective for the Settlement Unit, 1944 (Canadian Centre for Architecture, Gift of Alfred Caldwell)

4.3 The Experiment: Caldwell's Farm

In parallel with working on drawings and writing about the productive living landscape, Caldwell also put his ideas into action with his own experimental farm. In about 1943, Caldwell purchased half of an 80-acre tract of land in Bristol, Wisconsin for \$83 per acre (Caldwell 1987, 124). His reason for purchasing 40 acres was based on the amount of arable land available in the U.S. and what he calculated to be his fair share (Caldwell 1987, 123). Based on census figures at the time, the calculation probably looked something like this:

144 million people × 5% (population of farmers) = 7.2 million farmers.

161 million acres of arable land / 7.2 million farmers = 22 acres per farmer.

22 acres per farmer × 2 farmers = approximately 40 acres.

When he started to look at different ways this land could be used, he knew from F.H. King that it was possible for a family of six to twelve people to support themselves on 2 ½ acres (Caldwell 1948, 43). He considered a variety of options (Fig. 4), from subdividing the tract into smaller farms of 2 ½ acres (1 hectare), 10 acres (4 hectare), and 40 acres, to using the whole 80-acre tract[13] for a farmstead with space

[13] Photographs of a scale model of the 80-acre lot are shown in *Architecture and Nature*.

for animals and a large vegetable garden (Blaser 1984, 54–59). Each farm drawing shows a house, vegetable garden, barn, and yard for animals, with an orchard or field for crops. Some are terraced, some have ponds for fish, each one is individualized, but all are surrounded with trees and greenery. Caldwell showed his own 40-acre farm in an incomplete state (Fig. 4c), which he said was a symbol of the creative force which never ends (Blaser 1984, 61).

In 1948, Caldwell began to build a house on that land, "with his own hands" (Caldwell 1987, 58). It is a house built of native stone,[14] unearthed from the site as much as possible, supporting a laminated timber roof. The rustic natural materials today have a deep patina which, contrasted with a polished terrazzo floor and large glass openings, provide a rich aesthetic experience that recall Wright's material palette on one hand and Mies's clear space on the other.[15] The drawings and scale model that Caldwell made of the house and site show his vision of a large farm villa with stone walls that define different parts of the plan and integrate the whole, (Fig. 5). He built the east wing of the house which contains the bedrooms and fireplace. Other parts of the plan, including the barn, were not built; however, much was accomplished including a beautiful working studio with a fireplace and a pond in front, a grape arbor, and garage. Complete with ditches and culverts, he also constructed a meandering gravel road to get to the house that feels like it's always been there (Fig. 6).

The majority of trees on the wooded property were planted by Caldwell, his family, friends, and students. A topographical survey which he had prepared shows the mature trees that existed originally on the property, including White Oak, Burr Oak, and Black Walnut and the varieties of fruit trees that were planted in the orchard, such as Bartlett Pear, Early Richmond Cherry, and numerous apples including Delicious, Jonathan, McIntosh, Cortland, and Wealthy Apple. The farm started with the orchard in 1944, and later included experimenting with grapes and new techniques for high density growing using support wires and leader management which are still on the property. Among the improvements to the richly contoured landscape are also a clearing in the woods, in the spirit of Jensen, and a council ring which is reached by crossing a footbridge.[16]

Over time, Caldwell built a place which has a magical and poetic feeling, perhaps not unlike his own experience of Taliesin in the 1930s—"The feeling of architecture that floats in space—I shall have that all my life, that dream" (Caldwell 1987, 38). Perhaps this sense of place is what comes of a place that is built deliberately and as a reflection of ideas, hopes, and dreams. A man of action, when Caldwell was asked whether he still thought it was a viable solution, he said:

---

[14] "This stone is granite and various igneous rocks, left on the land by the receding glaciers. The interior cores of the walls are poured with concrete as the work proceeds, thus making solid stone walls about 22 inches thick." (Blaser 1984, 116).

[15] The influence of Caldwell's mentors is described in more detail in Jones and García-Requejo, 2020.

[16] According to Richard Polansky, friend and property caretaker, "Alfred's last physical effort on the place lies in the leftover council ring stones he set on the earth down the middle of the path, marking its entrance." The council ring, he said, was "placed so as to beckon you into the forest." (Personal correspondence, November 3, 2020.).

Fig. 4 80-acre tract divided into: **a** 32 small farms (2 ½ acres or 1 hectare each), **b** 8 small farms (10 acres or 4 hectares each), **c** two small farms (40 acres each); (north is up). (Canadian Centre for Architecture, Gift of Alfred Caldwell)

a

b

c

I wanted to see what it would actually come to. When I made my plan for the city of Chicago, I put my farm on this plan… Thousands and thousands of city plans have been made in America and they just gather dust on the shelves and are entirely forgotten. Nobody ever initiates any of these plans. Only my city is under construction (Caldwell 1987, 71).

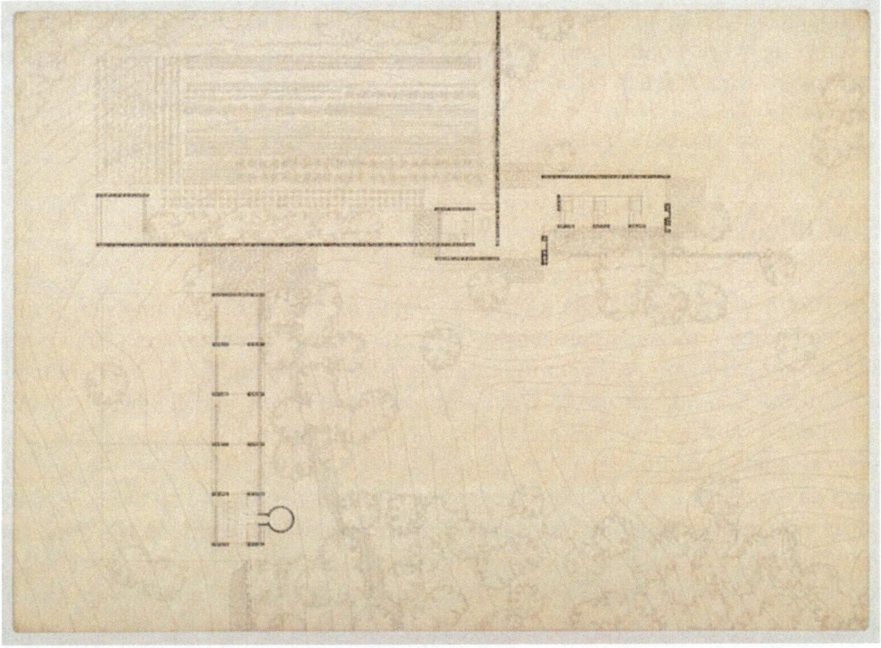

Fig. 5 Farmhouse at Bristol, Wisconsin, 1948, (see Fig. 4c for site orientation) (Canadian Centre for Architecture, Gift of Alfred Caldwell)

Fig. 6 Caldwell's farm, looking from the studio toward the house, November 2020 (Photo by author)

Although he no longer intended to get the cows and tend the farm himself, his hope was that his work would stand as a means of education, a statement on humanity, an example of what one can do with a dream and in one lifetime (Caldwell 1987, 106, 125, 140).

5 Conclusions

The City in the Landscape is a prime example of a modern planning perspective that embodied a more comprehensive view of the city encompassing social needs and ecology. How do we begin to reimagine a more resilient productive landscape now, in the first quarter of the twenty-first century? Plans like Caldwell's, which incorporate food and agriculture within the urban-regional planning milieu, may suggest a path forward.

Today we will need to take on a regional-global perspective. If we assess the current situation with a focus on the food landscape, according to the UN Food and Agriculture Organization, 9% of the world population—around 697 million people—are severely food insecure and one-in-four people globally—1.9 billion—are moderately or severely food insecure (UN 2020). The world population is increasing and as of 2018, 82% of the U.S. population (up from Caldwell's cited 75%) and 55% of the world population lives in urban areas (UN 2018). These statistics imply that more people than ever are disconnected from the land.

International goals for 2030 include making cities and human settlements inclusive, safe, resilient, and sustainable (UNESCO 2015). Architectural education in the U.S. is attempting to answer the call by promoting ecological knowledge and responsibility, more diverse history and theory offerings, and social equity and inclusion (NAAB 2020). However, within the discipline, the study of food security and urban community gardens and farms remain novelty practices generally addressed from site-specific rather than systemic perspectives. Planners are encouraged to work across different scales, (e.g., site, district, and regional). They are expected to develop an integrated understanding of land use planning, geography, and social sciences. This puts planners in a unique position to make an impact. Current accreditation criteria include issues of equity, diversity, social justice, governance structures and systems, elements of sustainable communities (e.g., environmental, economic, and social/political), factors in urban and regional growth and development (e.g., economic, infrastructure, social, and cultural), and implications for planning on health and the built environment (for the individual and community) (PAB 2017). A resilient productive landscape will need to factor in all of these issues in a modern way.

The Global Pandemic of 2020–2021 has provided us an opportunity to reflect on land use in an unplanned and unexpected way. With compulsory urban lock downs, more people have been able to spend time away from the city and develop innovative means to remain employed and connected. Because the pandemic occurred at a time when, for many, communications technology and collaboration software were available to handle the pivot to virtual, many of these technologies and approaches to working are likely to be continued into the future. After this experience, a decentralized workplace does not seem out of the question. One might easily envision now living on a small farm within driving distance of a central rail or air transportation hub, communicating remotely for "urban" work while also tending a farm part-time.

As we continue to reflect and move toward a better situation for all, perhaps Caldwell's work can provide a bit of guidance. It is a peaceful and transformative example that looked to a productive landscape to provide individual, social, and environmental health and resilience. It also stands as a reminder about the impact of capitalism on humanity, and a call to invest in earth's soil. Last century, a statement on a comprehensive urban planning theory without specific mention of agriculture or food might have seemed reasonable. However, as the field of planning matures, so does the awareness of the complexity of planning issues, and we are looking once again to restore a healthy relationship between land, food, and people. These are timeless relationships which are waiting to be resolved in today's technology and reflecting our communities' determination.

References

APA, American Planning Association (2021) "Planning History Timeline." American Planning Association. 2021. https://www.planning.org/timeline/.

Blaser W (ed) (1984) Architecture and Nature: The Work of Alfred Caldwell. Birkhäuser Verlag, Basel; Boston

Burnham, Daniel Hudson, Edward H Bennett (1909) Plan of Chicago. Edited by Charles Moore. Chicago: The Commercial Club of Chicago

Cabrera Fausto, Ivan. 2020. "Introduction : The Architect and the City." In The Architect and The City. EAAE-ARCC International Conference & 2nd VIBRArch. Editorial Universitat Politècnica de València

CAC, Chicago Architectural Club (2020). "2020 Burnham Prize." Chicago Architectural Club. 2020. http://chicagoarchitecturalclub.org/2020-Burnham-Prize.

Caldwell, A (1945a) "The City in the Landscape." *Parks and Recreation* 28 (March-April 1945): 59–64

Caldwell A (1945) Atomic Bombs and City Planning. Journal of the American Institute of Architects 4(December):298–299

Caldwell A (1948) The City in the Landscape: A Preface for Planning. Illinois Institute of Technology, Chicago

Caldwell, Alfred.. 1987. Oral history of Alfred Caldwell: Chicago Architects Oral History Project, Department of Architecture, the Art Institute of Chicago Interview by Betty J. Blum. https://digital-libraries.artic.edu/digital/collection/caohp/id/1758/.

Carney, Judith A. 2002. Black Rice: The African Origins of Rice Cultivation in the Americas.

Ford, Henry. 2013. My Life & Work - An Autobiography of Henry Ford. CreateSpace Independent Publishing Platform.

Geddes, Patrick. 1930. "Ways of Transition—Towards Constructive Peace." *The Sociological Review* a22 (2): 136–41. https://doi.org/10.1111/j.1467-954X.1930.tb01744.x.

Harvard, GSD. n.d. "The Department of Urban Planning and Design." Harvard Graduate School of Design. Accessed June 6, 2021. https://www.gsd.harvard.edu/urban-planning-design/.

Hilberseimer, Ludwig. 1927. *Großstadt Architektur*. Hoffmann.

Hilberseimer L (1944) The New City: Principles of Planning. P. Theobald, Chicago
Hilberseimer L (1946) Review of Review of New Architecture and City Planning–A Symposium, by Paul Zucker. College Art Journal 6(2):165–166. https://doi.org/10.2307/773305
Hilberseimer L (1949) The New Regional Pattern: Industries and Gardens, Workshops and Farms. P. Theobald, Chicago
Hilberseimer, Ludwig 1955. *The Nature of Cities: Origin, Growth, and Decline | Pattern and Form | Planning Problems.* Chicago: P. Theobald. http://archive.org/details/natureciti00hilb.
Hurt, R. Douglas. 2021. "Agriculture, American Indian." In *Dictionary of American History.* Encyclopedia.com. https://www.encyclopedia.com.
Ippolito W (2021) Horticulture, History, and Immigrants: Systems Altering This Land.
Jones, Kristin, and Zaida García-Requejo. 2020. "The City in the Landscape : Alfred Caldwell's Broader Perspective on Urban Design." In *The Architect and the City*, 1:42–51. Editorial Universitat Politècnica de València.
King FH (2004) Farmers of Forty Centuries: Organic Farming in China, Korea, and Japan. Dover Publications, New York
Lillibridge RM (1953) City Planning Research in the United States. The Town Planning Review 23(4):296–307
NAAB. 2020. "NAAB Conditions for Accreditation." *National Architectural Accrediting Board* 2020 Edition: 15.
PAB. 2017. "2017 Accreditation Standards." Planning Accreditation Board. https://www.planningaccreditationboard.org/.
Pauls, Elizabeth P., ed. 2017. "Native American - Prehistoric Farmers." In *Encyclopedia Britannica.* https://www.britannica.com.
Platt, Rutherford H., ed. 1994. *The Ecological City: Preserving and Restoring Urban Biodiversity.*
Rydell, RW (2005a) "Century of Progress Exposition." The Electronic Encyclopedia of Chicago. 2005. http://www.encyclopedia.chicagohistory.org/pages/225.html
Rydell RW (2005b) "World's Columbian Exposition." The Electronic Encyclopedia of Chicago. 2005. http://www.encyclopedia.chicagohistory.org/pages/1386.html
Smith C (2005) "The Plan of Chicago." The Electronic Encyclopedia of Chicago. 2005. http://www.encyclopedia.chicagohistory.org/pages/10537.html
Swenson A, Chang P-C (1980) Architectural Education at IIT, 1938–1978 /. Illinois Institute of Technology, Chicago, IL
Thompson CW (2004) Geddes, Zoos and the Valley Section. Landsc Rev 10(1 & 2):115–119
UN (2018) "World Urbanization Prospects: The 2018 Revision, Online Edition." UN Department of Economic and Social Affairs, Population Division. File 1: Population of Urban and Rural Areas at Mid-Year (thousands) and Percentage Urban, 2018. https://population.un.org/wup/Download/
UN (2020) "Suite of Food Security Indicators." Food and Agricultural Organization of the United Nations (FAO). https://ourworldindata.org/hunger-and-undernourishment#moderate-food-insecurity
UNESCO Qian Tang (2015) Rethinking Education: Towards a Global Common Good? UNESCO Publishing, Paris
USCB, ed. (1999) "20th Century Statistics." In Statistical Abstract of the United States, 867–89. U.S. Census Bureau
USDA (n.d.) "Conservation Choices: Contour Farming." Natural Resources Conservation Service. Accessed June 20, 2021. https://www.nrcs.usda.gov/wps/portal/nrcs/detail/null/?cid=nrcseprd414214

From Rural Lands to Agribusiness Precincts: Agriculture in Metropolitan Sydney 1948–2018

Joshua Zeunert and Robert Freestone

Abstract Sydney is both the birthplace of the nation of Australia as well as its western agricultural tradition. Since its founding in 1788, Sydney has been a continually growing settlement whose significant metropolitan expansion has come at the expense of its immediate agricultural hinterland. While this phenomenon is not unique per se, Sydney's natural context presents a constrained and finite area of immediate and proximate land available for local production of fresh food. Official metropolitan planning commenced in Sydney from 1945 and the first scheme in 1948 provided genuine but only rudimentary strategic and statutory consideration for the preservation and continued use of its agricultural lands. Sydney's subsequent eight metropolitan plans have given agriculture lip service at best and complete disregard at worst, largely treating it as land awaiting 'higher' economic development. Maximizing short-term financial windfalls through urban land development continues to undermine Sydney's local food supply security, while gearing food supply to an energy-intensive 'just-in-time' system sourcing from further afield. The latest (2018) metropolitan plan has evolved to high-tech indoor food production for lucrative overseas markets via a new 'aerotropolis' precinct. Sydney's metropolitan agriculture has thus been remade into a late fossil fuel era expression of corporate and neoliberal industrial agribusiness.

Keywords Planning · Metropolitan · Urban · Agriculture · Sydney

1 Introduction

The interdependence between city growth and securing adequate food supplies goes back to ancient times. At the genesis of modern metropolitan planning in the late nineteenth century, the symbiosis was keenly appreciated with the influence of the

J. Zeunert (✉) · R. Freestone
University of New South Wales (UNSW), Sydney, Australia
e-mail: j.zeunert@unsw.edu.au

R. Freestone
e-mail: r.freestone@unsw.edu.au

C. Brisotto and F. Lemes de Oliveira (eds.), *Re-Imagining Resilient Productive Landscapes*, Cities and Nature, https://doi.org/10.1007/978-3-030-90445-6_12

garden city movement. Ebenezer Howard's treatise was as much about the agricultural estate as the planned town it encircled, stressing its beauty, productivity, and economic viability with a captive urban population near at hand (Howard 1898). The idea of the green belt to articulate a crisp and healthy divide between city and region while honoring the historically close relationship between them became a staple of planning thought alongside other totemic planning innovations like land use zoning and satellite towns from the 1910s. The British town and country planning tradition encompassing these themes was globally influential into the early 1960s until overtaken by other paradigms better geared to the rampant sprawl denoting what Self (1961) termed 'cities in flood'. When that happened, managing urban space was emphasized and the agricultural dimension of broader metropolitan strategies was diminished such that Budge (2013) concluded that numerous post-war plans paid little explicit attention to sourcing food for the metropolis. While that may be overstated, this awareness has returned in some cities with the intersection of modern tropes of sustainable development, food security, resilience, localism, and food urbanism (Parham 2015).

Against that backdrop, this chapter considers the intertwined narrative of agricultural production and metropolitan planning in Sydney, Australia post-World War Two (WWII). The ever-expanding footprint of the metropolis, along with the seeming unwillingness of metropolitan and local government plans to truly protect agricultural land from development, is the recurring story of Sydney's agriculture. We focus on the 1948 County Cumberland Council (CCC) Planning Scheme—the first metropolitan scale plan prepared for Sydney—and specifically appraise its approach to the nexus of agriculture and the built environment. We then link this discussion to the eight subsequent metropolitan plans for Sydney since 1948—giving prominence to the most recent in 2018—specifically focusing on the dynamic of a growth-addicted city.

We emphasize the diminished metropolitan production area available in a topographically constrained setting and the lack of planning vision post-1948 to attempt to address this issue, hence increasing reliance on long-chain food supplies.

The initial metropolitan vision for Sydney embedded in the town and country planning tradition of a proximate productive countryside was steadily eroded through urban expansion and lack of policy foresight. Explicit concern with peri-urban planning drifted in and out of official thinking. The inevitable outcomes were, on the one hand, an irreversible loss of agricultural land and, on the other, evolution towards a new understanding of agricultural activity within urban planning. While vestiges of foundational values of ready access to fresh food supplies remain, the contemporary policy framework acknowledges multiple competing demands beyond the city fringe apart from agriculture and a seemingly inevitable prospect of transition to large-scale corporate agribusiness geared to the massive infrastructure investments of the ever-expanding metropolis.

We first give some background concentrating on the physical environment of Sydney, the early history of agriculture in the region, and the lead-up to the development of the landmark 1948 metropolitan planning scheme. The provisions and impact of that scheme are considered, and specifically its approach to agricultural land planning. We then move on to briefly chart the shifting treatment of agricultural issues across subsequent metropolitan plans from the late 1960s to the mid-2000s, before turning to a more detailed consideration of the latest (2018) strategy to emphasize the changing relationship between the ever-expanding contemporary city and its progressively shrinking rural background.

2 Sydney's Historical and Geomorphological Context

Sydney is Australia's largest city, presently containing around 5 million residents, and it was Australia's founding European settlement. The continent now known as Australia was colonized by the British from 1788, branded as a 'land of no one' (*terra nullius*) to establish sovereignty—in part owing to the lack of a visible Aboriginal agriculture to British eyes (Ma Rhea 2017). Like subsequent settlements in Australia, a dramatic subjugation of its landscape ensued (Benson and Howell 1990; Martyn 2018) along with decimation of clans of Aboriginal (Koori) cultures who had successfully managed these landscapes for many millennia. Transformation was significantly catalyzed through imported agricultural pursuits of mixed farming and pastoralism, which inevitably involved frontline conflict with the prior custodians.

For agricultural production in Sydney, as well as its wider supply lines for food, it is crucial to note that Sydney's topographic and geological features form an encircling basin that challenges cross-country movement (Fig. 1). The geographical perimeter of the Sydney region—effectively the County of Cumberland as named by colonial administrators—is defined by sandstone barriers, deep gullies, and aquatic river valleys. The Hawkesbury River lies to the north, the edges of the Blue Mountains of the Great Dividing Range to the west, the rugged country of several National Parks to the south, and the South Pacific Ocean eastward.

Sydney's topographic basin (roughly 5,000 km^2) limits the immediate available area for western agricultural pursuits in direct proximity to the metropolis. Sourcing food outside of this area requires traversing a limited number of extra-regional routes, the majority crossing challenging terrain, to reach agricultural regions starting 100 km away. A proportion of Sydney's substratum is sandstone and unsuited for productive agriculture. Figure 2 shows sandstone geology encircling 'Wianamatta' (shale), sedimentary and alluvium, indicating that apart from the sandstone areas a significant proportion of the now metropolitan area was suitable for agriculture.

3 Background History of Western Agriculture in Sydney

Sydney's agriculture presents a chequered history. Its challenges began with near starvation in the first five 'hungry years' (1788–1793) stemming from difficulty in imposing British and European agricultural techniques and food preferences in a

Fig. 1 Three-dimensional topographic model of Sydney showing its enclosure by hydrology and landform (the darker the tone, the higher the elevation). *Source* J. Gowers, J. Zeunert & S. Stankiewicz

frequently unreceptive ancient and weathered landscape. This was compounded by a lack of place-specific agricultural skills, knowledge, resources, and infrastructure (such as manure from few livestock, a lack of mills, and knowledge in operating them).

Events such as large and regular early floods on the fertile Hawkesbury River flats were devastating. A lack of hygienic practice causing water pollution, rat infestations from urban abattoirs, typhoid outbreaks, and foul odors saw agriculture being included in the 1848 list of Noxious Trades (and targeted in an 1894 Act) pushing it out of and away from the expanding city whose population was about 50,000 at mid-century. Pests and diseases would also prove troublesome to the stone fruit industry from 'fruit fly' in the late nineteenth century onward and to the previously thriving viticultural industry from phylloxera by the early twentieth century.

The dominant Anglo-Celtic culture was unwilling to substantively take up and persist with smallholder agriculture, although this was not the case with numerous culturally marginalized and hard-working migrant groups from China, southern Europe, and South-east Asia (Christie 1988; James 2016). These groups faced other challenges, for example, the exclusion of many well established and industrious

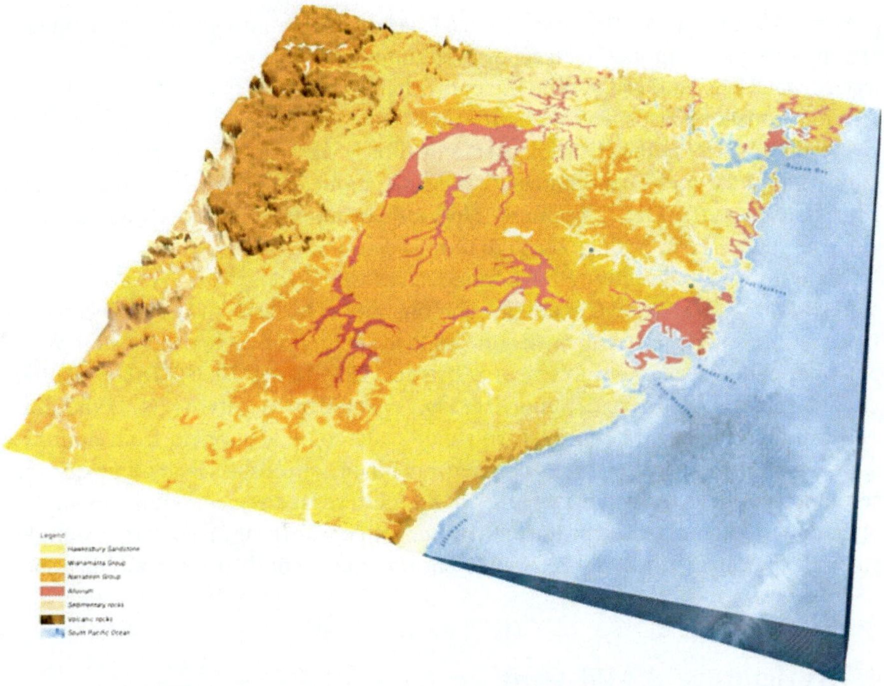

Fig. 2 Three-dimensional geological map of Sydney. This shows that while the original settlement site (near Sydney's CBD) and the metropolitan area are encircled by sandstone geology (Hawkesbury and Narrabeen categories), a large proportion of its metropolitan spatial extent is land suited to Western agriculture (Wianamatta, Alluvium, and Sedimentary categories). *Source* J. Gowers, J. Zeunert & S. Stankiewicz

Chinese market gardeners through the 1901 'White Australia Policy' as well as racist campaigns against long hours worked by Italian market gardeners.

Although historical accounts tend to focus on dramas and failures in agriculture (McIntyre 2009), there were agricultural success stories (Zeunert 2022). Sydney is the birthplace of Australian agriculture. Emancipated convict farmer James Ruse and others identified receptive lands for farming and extracted yields despite very limited infrastructure. Green Hills (now Windsor) and nearby Pitt Town Bottoms at the outer rim of the County of Cumberland support the longest continuing farms in Australia. A pastoral industry was long the foundation of Australia's economy and a key catalyst in the nation's expansion beyond its more notorious role as a dumping ground for felons by the British (Hoorn 2007). It was established in the Sydney region by colonial pioneers such as Samuel Marsden, John and Elizabeth MacArthur, and William Charles Wentworth (Deniehy 1853).

Sydney was integral in establishing Australia's viticultural industry and, despite phylloxera, continued to produce a notable proportion of the state of New South Wales' (NSW) total grape production into the mid-twentieth century (with several small wineries still existing). There were over 500 metropolitan dairies across all regions of Sydney in 1901 (Lewis 1980). Livestock were initially a very visible part of urban Sydney and its streetscapes including abattoirs, which now form a much less conspicuous land use even at the peri-urban fringe. The patterned geometries of thriving stone and citrus fruit orchards once draped across extensive areas of rolling hills—particularly in the Hills District in the north-west—persisted as a commercial industry into the mid-twentieth century (McClymont 2013). Again, a few remnants still exist (Zeunert 2022). Cropping—especially on alluvial river flats—also achieved ongoing successes in Sydney through to mid-century.

Many forms of agriculture have found success in Sydney in a *given* location for a *given* time. Better-known examples demonstrate polar ends of the socioeconomic spectrum: extensive pastoralism in contrast to the comparatively postage-stamp-sized plots of vegetable market gardeners (Christie 1988; James 2016). The agricultural middle-ground is more challenging to grasp. As noted by McIntyre (2009), the stories of Australian history tend to exclude small-scale colonial farmers from notions of nation-building in preference to swagmen, bushmen, pastoralists, and miners. Sydney certainly acted as a testbed for an array of agricultures to be successfully applied and realized (often at increased scales in regions with cheaper land); an early example is Australia's first agricultural irrigation district in now suburban Liverpool in 1856.

4 Agriculture at Mid-Century

The inexorable growth of the metropolis between the wars erased much early agricultural activity. Before WWI, suburbs on the railway line leading south to the Illawarra region boasted many small farms and vegetable gardens that disappeared with residential expansion. Sydney's built-up area 'expanded enormously, engulfing many erstwhile agricultural areas'; the urban–rural fringe was on the move and 'during the same period.... new farms were being opened up farther out', in places such as Prospect and Castle Hill (Wills 1945: 35).

WWII lifted Sydney out of the lingering effects of the Great Depression and stimulated economic activity and growth (Christie 1988). Agriculture was significantly activated, with the armed forces becoming the priority channel for locally processed fresh food production to supply troops with necessary nutrition across significant distances (Christie 1988). McGill writes that 'local farmers felt they were the backbone of the nation during the wartime emergency' (McGill 2017). Rationing for civilians was in place, although some self-interested citizens found ways to bypass and even profit from wartime food arrangements (Christie 1988).

Two geographic studies in the early mid 1940s capture the vitality of activity but also clouds on the horizon. Harold Maze's (1943) study of the Kurrajong-Windsor district showed farm production oriented not only to the central Sydney market but also Parramatta and locally. Arable farming on the Hawkesbury River's alluvial flats, extensive citrus orchards, ample pasture land, and upland poultry farms were all thriving. He concluded that 'it would seem likely that the area is approaching its maximum agricultural development' (Maze 1943: 168). Neville Wills (1945) highlighted the significance of near city farming with over 40% of most vegetables consumed by Sydneysiders coming from farms within 40 km of the Central Business District (CBD). Two major trends were evident. One was an expansion of intensive farming as irrigation and fertilizers were more liberally applied to compensate for lagging natural soil fertility. The other was the further disappearance of market gardens in the middle suburbs. Botany-Mascot was declared 'a spectacular example of a declining rural–urban area. Until recently, it was a flourishing center of small vegetable gardens, but is now being rapidly absorbed into the South Sydney industrial belt' (Wills 1945: 35).

Planning studies undertaken by the CCC between 1945 and 1948 provide the most authoritative account of Sydney agriculture to that time (summarized in Fig. 3 and Table 1). The bulk of production occurred on the (then) peri-urban fringes of the city, encircling the core built-up areas (albeit missing some remnants in hatched areas such as Botany near the coast). The accompanying table quantified the total *horticultural* area for Sydney as 38,581 acres (15,613 ha). Cropping (including for fodder) was the most expansive (horticultural) land use at 17,275 acres (6,991 ha), followed by market garden vegetables at 8,815 acres (3,567 ha), then citrus fruits at 6,718 (2,719 ha), non-citrus fruit 3,434 acres (1,390) (although passionfruit strangely received its own category at 459 acres/186 ha); grapes at 1,174 acres (475 ha); nurseries 670 (271 ha); and 'other' totaling 36 acres (15 ha). The Baulkham Hills Local Government Area (LGA) had the largest area of cultivated horticultural land at

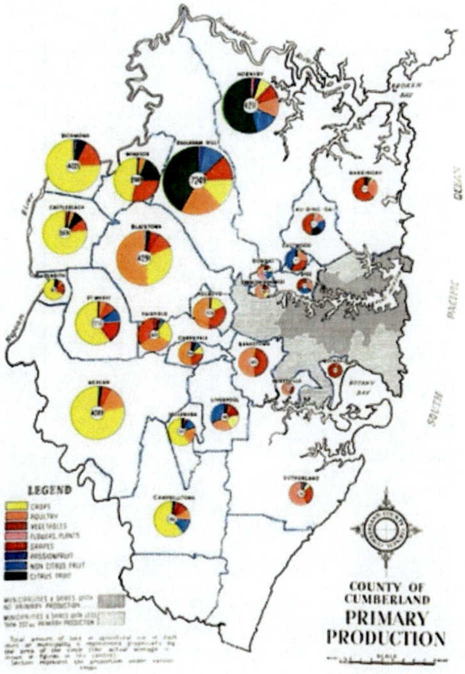

Fig. 3 A map of agricultural production across eight categories including poultry (but no other livestock/grazing) in Sydney in the mid-late 1940s. *Source* Taken from CCPS (1948)

Table 1 Details of horticultural production for each municipality and shire in Sydney in the mid-late 1940s.

Municipality or Shire	Citrus Fruit*	Non–Citrus Fruit*	Market Garden Vegetables	Crops and Fodder	Passion fruit*	Grapes	Nurseries	Other	Total area cultivated
Baulkham Hills*	3,165	720	801	1,060	247	7	82	–	6,082
Richmond	143	56	706	3,108	–	4	–	5	4,022
Hornsby	2,352	598	432	321	156	–	159	–	4,018
Nepean	7	52	232	3,150	1	14	1	–	3,457
Castlereagh	205	55	399	2,193	–	23	–	–	2,875
Windsor	563	19	873	1,293	3	–	–	–	2,756
St. Mary's	4	244	354	1,470	–	398	–	–	2,470
Blacktown	83	96	426	1,187	1	307	13	1	2,114
Campbelltown	8	225	272	1,182	28	95	–	30	1,840
Fairfield	6	116	632	278	–	61	17	–	1,110
Ingleburn	2	158	37	702	11	38	–	–	948
Liverpool	7	384	97	247	–	165	4	–	904
Holroyd	2	15	480	236	–	–	21	–	754
Warringah	21	7	526	39	7	–	45	–	645
Penrith	5	28	164	395	–	23	–	–	620
Eastwood	6	283	206	35	–	–	50	–	580
Bankstown	–	9	543	25	–	–	–	–	577
Ku-ring-gai	104	93	293	11	4	–	56	–	561
Cabramatta/Canley Vale	–	58	158	101	–	30	3	–	350
Sutherland	13	30	214	76	1	–	15	–	349
Ryde	4	115	117	16	–	1	11	–	264
Rockdale	–	–	225	–	–	–	18	–	243
Dundas	12	52	42	32	–	3	78	–	219
Randwick	–	–	128	–	–	–	10	–	138
Botany	–	–	122	–	–	–	10	–	132
Mascot	–	–	92	5	–	–	12	–	109
Canterbury	–	–	91	6	–	–	–	–	97
Ermington/Rydalmere	3	8	15	41	–	–	29	–	%
Hurstville	–	–	47	11	–	–	32	–	90
Granville	–	2	33	24	–	–	–	–	59
Kogarah	–	–	36	–	–	–	–	–	36
Parramatta	2	11	–	16	–	–	–	–	29
Drummoyne	–	–	10	–	–	–	2	–	12

(continued)

Table 1 (continued)

Municipality or Shire	Citrus Fruit*	Non–Citrus Fruit*	Market Garden Vegetables	Crops and Fodder	Passion fruit*	Grapes	Nurseries	Other	Total area cultivated
Homebush	–	–	8	–	–	–	–	–	8
Lidcombe	–	–	–	7	–	–	–	–	7
Willoughby	–	–	2	–	–	–	1	–	3
Auburn	–	–	–	3	–	–	–	–	3
Bexley	–	–	1	–	–	–	1	–	2
Alexandria	–	–	1	–	–	–	–	–	–
Hunters Hill	1	–	–	–	–	–	–	–	1
Total	6,718	3,434	8,815	17,275	459	1,174	670	36	38,581

Source Taken from CCPS (1948)

over 6,000 acres, followed by Richmond, Hornsby, and Nepean all exceeding 3,000 acres (1,214 ha); with Castlereagh, Windsor, St Marys, and Blacktown all greater than 2,000 acres (809 ha) of cultivated land.

Two disjunctions are evident between the map and the table. Firstly, 122 acres of market gardens are listed against Botany in the table but not shown on the map. Secondly, poultry appears to be the second-largest spatial activity in the map, but does not appear in the table. On closer inspection, the data table only included horticultural activities. The lands used for grazing of livestock (principally cattle, dairying and other animals such as sheep and goats) were thus excluded from the study. Lands used for growing fodder crops for animal consumption, however, are included. The significance of this statistical reporting is that Sydney's actual agricultural acreage—if livestock activities were included—would be considerably larger than 15,613 ha at the time, especially given that over 70% of NSW's poultry farms and 18% of milk production for Sydney were located within the County (CCPS 1948: 125) and that of all agricultural activities, grazing livestock commands the largest area. This is evident in a 1957 map showing that pastoral uses were indeed the largest in the County (Winston 1957: 10).

What this analysis did convey was the value of agriculture to the metropolis even as late as the mid-twentieth century. The significance of the 10,300 County of Cumberland agricultural workers was recognized as 'immense …. beyond their numbers' (CCPS 1948: 125). The employment survey further revealed that over 34,000 factory workers were tied to 'food' production, the third-largest employment category in manufacturing (CCPS 1948: 51).

5 Trends in Urban Agriculture at Mid-Century

Despite some inconsistencies, this detailed snapshot of Sydney's agriculture provides an overview which coupled with further research (Zeunert 2022) reveals some important trends. Overall, due to an urban footprint undergoing accelerated expansion, existing agricultural areas were increasingly threatened and disappearing. In the immediate post-war period, the composition and dynamics of each key agricultural element in metropolitan Sydney primarily comprised:

- significant loss and fragmentation of prior districts of market gardens;
- declining (irrigated and non-irrigated) areas of cropping;
- steady ongoing decline in citrus and stone fruit orchards and fruit production;
- ongoing decline of urban dairies and their metropolitan depots;
- increasing poultry farming at increased stocking rates, transitioning from free-range to large indoor (shed) operations (later controlled through planning-stipulated agglomeration sizes and buffer zones);
- declining commercial scales of cattle and sheep grazing with decreasing pastoral property-holdings from subdivisions allied to a declining role in local fattening of livestock and holding before slaughter;
- declining viticultural production;
- ongoing government support for agricultural education, research facilities and institutions;
- a steady decline in local processing (e.g., canning) of fresh foods and food manufacturing;
- increasing ability to import 'fresh' foods from elsewhere through ongoing improvements in refrigeration, freezing, plant breeding (for durability and storage), post-harvest technologies and transportation; and
- the impending emergence of supermarket retailing which, over time, would favor larger rather than smaller producers.

6 The Cumberland County Council Planning Scheme 1948

For the first half of the twentieth century, metropolitan Sydney had no guiding planning machinery as it grew from a population at the turn of the century of around 480,000 to 1.7 million persons. Town planning advocates had promoted over many decades the efficiency, economies, and social dividends that would come from greater controls and coordination over market-driven development, but this was only realized with new legislation enacted by the state government in 1945. As well as enabling local authorities to prepare zoning schemes across NSW, for the Sydney region a new body called the Cumberland County Council (CCC) was established to prepare and oversee the implementation of a legally binding metropolitan plan.

In March 1948 after three years' preparation, the CCC exhibited *The Planning Scheme for the County of Cumberland New South Wales* (CCPS). The scheme contained metropolitan scale land use zoning recommendations, a major roads scheme, proposals for new suburban and rural district centers and inner city renewal, aspirations for considerable open space acquisitions, and a sharp town-country boundary intended to prevent urban sprawl and conserve the agricultural and undeveloped lands beyond. The innermost portion of the outer region of 'rural lands' was dubbed the 'green belt' (Fig. 4).

Here was a plan very much of its time, and squarely in the British town and country planning tradition. Outlined in a substantial report sprinkled with empowering garden city references, its graphical form immediately channeled Patrick Abercrombie's *Greater London Plan* (1944) with its green belt and prospect of new satellite towns on a rural backdrop to absorb long term urban growth. The notion of constraining growth, recentralizing activity, and utilizing regional green space to articulate a spatial framework was close to the conventional wisdom in Australian planning circles by this time (Freestone 1989). The standout local precedent was A.J. Brown's vision for Sydney surrounded by a 'green girdle' published just before the war:

> As social and economic forces respect no arbitrary bounds, the planner must also ignore them and, acknowledging the interdependence of town and surrounding country, provide for ample play of those forces along well-ordered lines. The bustling city cannot be considered as separate from its rural background, which is essential to its existence, and the converse is also true (Brown 1937).

The CCPS (1948, 123) quoted F.J. Osborn's remarks to a UK Town and Country Planning Association conference in March 1941 to underline this point: planning for cities and countryside had to be 'sanely balanced'. In the immediate lead-up to work commencing on the Sydney plan, the NSW Department of Main Roads had produced a future road plan which marked an 'assumed future limit beyond which continuous urban development will not extend' (DMR 1945). The influence of this statement was sealed when a senior roads engineer S.L. Luker was appointed first chief county planner, although according to Harrison (1983: 230) 'almost all' of the County plan's 'intellectual and professional content' originated with his deputy, R.D.L. Fraser. Fraser assumed the leadership role upon Luker's premature death and remained a strong advocate for valuing the 'primary producing value of land in the County' until the CCC's abolition in 1963 (Fraser 1957).

The Minister for Local Government speaking in State Parliament in 1951 when the CCC's plan was legally gazetted clearly captured how these ideals were to be translated into action:

> The rural areas of the country will be preserved not only for vital food production and as a countryside to serve the metropolis, but also for soil conservation, irrigation, afforestation and the development of towns and villages with services, amenities and opportunities for local employment. A "green belt" will be established around the urban development of the metropolitan area, the principal function of which will be to preserve a distinct, permanent line between town and country and to retain some portion of the county at a higher standard of rural development within reasonable reach of all. It will generally take the form of farms, forests and rural institutions (Cahill 1951: 9).

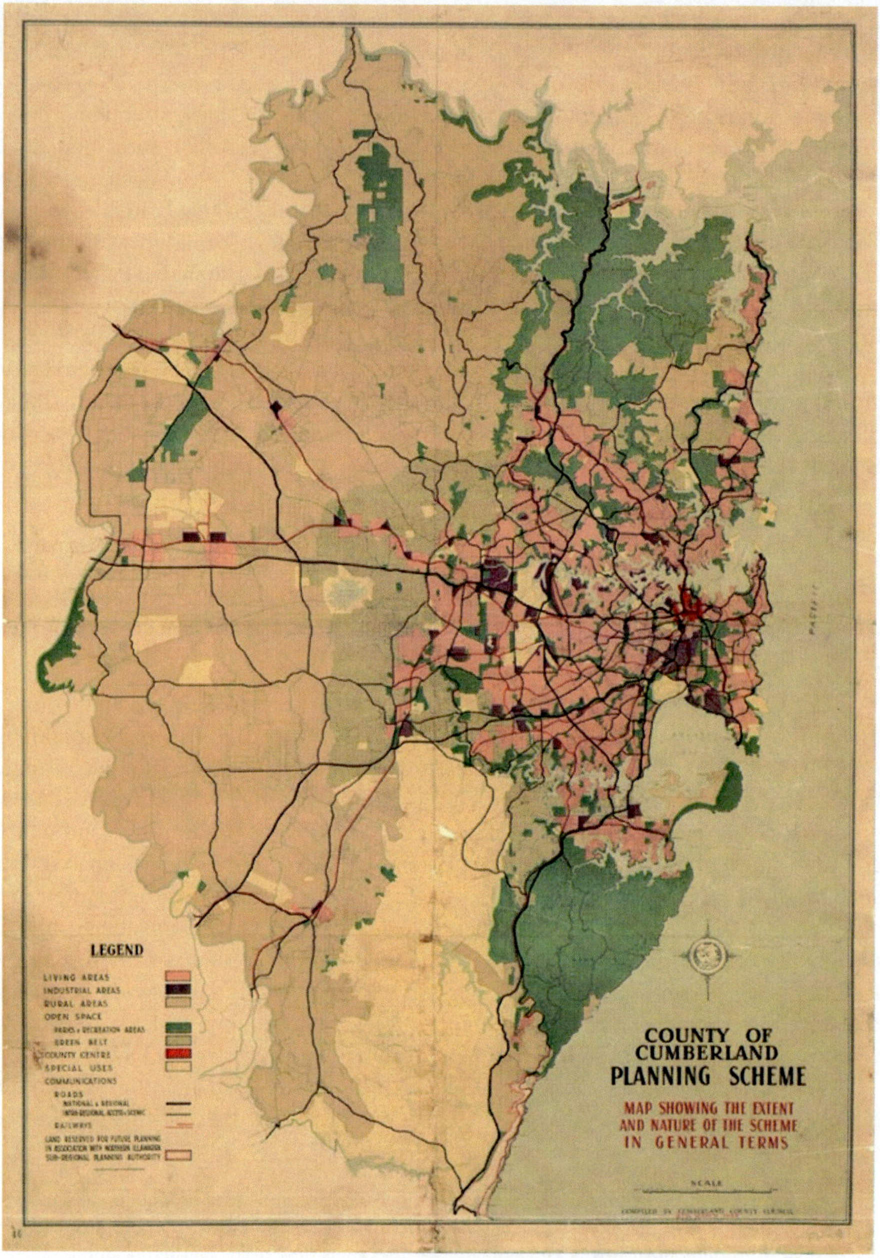

Fig. 4 The functional plan of the County of Cumberland 1948. *Source* CCPS (1948)

The planning scheme report devoted considerable space to the importance of conserving rural lands, indeed giving this issue more attention than any subsequent metropolitan plan, as we shall see. Apart from the land use and production statistical surveys drawn from earlier, a generous photographic record illustrated key areas of Sydney's agriculture. Some agricultural regions were specifically identified as threatened by an expanding city (e.g., Cabramatta and Lurnea in the south-west) or their value made clear (e.g., Eastwood in the north). Many of these areas would be associated with particular cultural groups especially with the influx of refugees and migrants immediately after WWII, although planning policy did not recognize the significance and strength of the social ecosystem (James 2016).

Denis Winston's account of the County of Cumberland scheme in *Sydney's Great Experiment* (1957) repeatedly reiterates the importance of supporting metropolitan agriculture through regional planning, chiefly for supplying urban citizens with nutritious, fresh food. A key takeaway from his book is that our current food system, with its 'just-in-time' logistics processes and subsequent food-mile laden products (Gaballa and Abraham 2008), was not yet feasible. To furnish the city with fresh vegetables and to a lesser extent, fruit, local production was necessary, as cold storage, post-harvest technologies, corporate chain food retailing, along with networks of cheap fossil fuel powered aviation, trucking and shipping were not nearly as developed as now.

The rural areas would be 'primarily devoted to the production of the County's fresh food' and declared to be 'as essential to the County as its industry and its port' (CCPS 1948: 65). The green belt was the more accessible inner rural zone of high quality and the front line defence against 'promiscuous urbanization' (CCPS 1948: 129). The CCPS (1948: 129) recognized the need for 'stabilization of rural land use and tenure', particularly for improving land management practices. It proposed a 'rural district system' which would be 'defined mainly by farming activity' and focused on re-planned townships as employment hubs and to define resilient communities of interest (CCPS 1948: 130). The primary development control was a proposed minimum subdivision size of 2½ acres, considered 'the smallest area from which a livelihood could be obtained by full agricultural development' and to 'discourage ribbon development along arterial roads' (CCPS 1948: 129).

7 The Legacy

While the CCC's aspirations for Sydney's agricultural industries were progressive and welcome, the regulations set in place were flawed and ultimately ineffectual. The subdivision control was early recognized as manifestly inadequate. The minimum rural landholding area was doubled when the scheme became law in 1951 on the recommendation of senior state government planner Norman Weekes, 'reportedly' based on the standards used by the Water Conservation and Irrigation Commission in the Murrumbidgee Irrigation Area in southern NSW (Ashton 1983: 150). While the CCC recognized metropolitan agriculture overall as well as various existing areas under threat of development, it made no nuanced provisions—outside of green belt and rural area zonings—to zone or protect high-value agricultural lands from loss to urban development. There were no specific initiatives targeting rural tenure beyond zoning which worked to some degree in the rural lands (Golledge 1960). The rural district system organized around renaissance rural centers to counter commuting to the central city was stillborn. The green belt, the signature planning proposal of the scheme, was progressively eroded by Ministerially approved land releases under pressure from both public and private developers seeking sizeable broad acre land for new low-density housing.

Broader cultural influences were also at work. Urban containment to conserve farmland was less a driving policy force in Sydney than in Great Britain. This can partly be attributed to the significantly larger population in the UK and the fact that food rationing, pressures on food supply lines, dependence on food imports, constraints on spatial availability of agricultural land, and wider British participation in Victory Gardening initiatives were all greater there than in Australia. Despite having a small proportion of arable landmass, the Australian mindset was (and arguably still is) to perceive available land for agriculture in extensive supply. British planning ideals of containment and countryside protection were less compelling in the antipodes, and agricultural land conservation in Sydney became a fringe and sporadic ideal at best. Further, the widely pervasive notion of 'countryside' being a civilizing respite largely preserved for the aristocracy in Britain was never replicated in Sydney. Instead, the deeply embedded historical perspective of the Australian landscape—led by those who have exercised power over it—aligns with an extractive paradigm for speculative enterprise to leverage financial gain; 'a camping ground for profit' (Lines 1991: 54), routinely in self-interest over that of a common good.

From a planning perspective, a sprawling metropolis represented much more than a threat to agricultural land. The CCC's proposed green belt had good intentions to preserve peri-urban agriculture at the time but it was nonetheless representative of a problematic planning ideology for metropolitan agriculture, whereby 'the aesthetic argument was to prevail over the agricultural' (Freestone 1992: 72). That is, planners at the time (and still evident today) routinely regarded metropolitan agriculture in relation to its amenity traits (e.g., countryside character) over its foundational role in nourishing the captive urban population. The green belt also disregarded existing and established urban agriculture situated in Sydney's interstitial spaces, such as in

Botany and Brookvale, which were being rezoned for industrial use. Specifically, its shortcomings represent the problematic 'heroic' planning gesture, over-scaled and for a single use. The green belt was a simplistic arcing element which didn't functionally address the inherent land and soil capacity for agriculture and its dendritic and patchwork nature following the alluvial creek and river valleys and pockets of superior soils.

Not surprisingly, then, the Cumberland scheme proved a more satisfactory urban than rural plan. The CCC Chairman in the late 1950s declared that 'development of the Rural Area Zone should be as much a matter for positive planning as the development of living and industrial areas' (Scutts 1958). Yet his survey of conditions in Outer Sydney identified various problems from large grazing properties 'somewhat neglected and not used to anything like their potential' and many smaller lots 'only partly occupied with little or no rural use and with roads and services in a rudimentary condition'. The response was to progressively increase the minimum property holding in rural areas to 40 ha by the 1970s, but by then the damage had been done. Rural land was fragmented by speculative pressures, the green belt was no more, and indeed the CCC itself was long gone.

8 Sydney's Metropolitan Planning Strategies from Late 1960s to 2010s

Despite the shortcomings of the 1948 CCPS's approach to agriculture, it offered a serious consideration of metropolitan agricultural needs when compared to subsequent metropolitan planning schemes. Later plans would pay lip service at best, and complete disregard at worst. Metropolitan planning acting as a facilitator of urban development increased, subsequently becoming the norm in Sydney.

Table 2 lists all seven of Sydney's official metropolitan strategies developed after the Cumberland plan and before the latest strategy released in 2018. The key elements relating to metropolitan agriculture and rural lands in the published plans are identified.

When it came time for a serious review of the County Plan in the late 1960s ahead of a new metropolitan strategy, the contribution to preserving 'much of the rural character' of the Sydney region was acknowledged (SPA 1967: 75). At the same time, the review was troubled by the lack of strong and diverse employment in Sydney's outer region and stressed the need for future planning to bear in mind 'the necessity for encroachments to meet the needs of urban expansion.' Underlain by growing awareness of intrinsic environmental values and constraints on development, these twin concerns became major themes in all strategic planning exercises over the next four decades—although ironically, the first post-CCC plan was virtually silent on agricultural lands.

These schemes do recognize Sydney's overall constrained physical setting in relation to the cultivation and transport of agricultural produce in theory. Although

Table 2 Urban Agriculture in Sydney metropolitan strategies 1968–2014.

Metro strategy	Year	Core strategy	Rural territory	Agricultural and rural lands elements
Sydney Region Outline Plan	1968	Linear growth corridor and green wedge strategy to replace green belt and satellite towns	'Non-urban' zone	Agricultural industry marginalized as of low economic importance to the regional economy
Sydney into its Third Century	1988	Growth corridors, urban infill, and environmental quality	Non-urban areas	'High quality agricultural lands to be kept in productive use for as long as possible for the economic benefit of the Region' but pressure from greenfield land releases on the urban fringe
Cities for the 21st Century	1995	Integrated urban management	Non-urban areas	As for 1988. However large tracts of land in western Sydney were mapped as 'potential future urban'
Shaping Our Cities	1998	A 'peak plan' combining government directions and directives	Recognizes 'protected agriculture and other rural production'	A companion 'Shaping Western Sydney' strategy supported implementation of the 'Strategic Plan' for Sustainable Agriculture—Sydney Region (1998) from NSW Agriculture
City of Cities	2005	A polycentric global and growing city for 1.1 m more people: '5 cities—3 corridors—1 global city'	Rural and resource land	Policy foci to maintain rural activities and resource lands, exclude incompatible and inappropriate uses, and encourage investment. Conflicts identified with vegetable production and greenhouses in proposed north-west and south-west growth centers

(continued)

Table 2 (continued)

Metro strategy	Year	Core strategy	Rural territory	Agricultural and rural lands elements
Metropolitan Plan for Sydney 2036	2010	A 'sustainable, affordable, liveable, equitable and networked' vision for metropolitan Sydney. Updating 2005 strategy and integrated with the Metropolitan Transport Plan	Rural/resource lands	Major strategic direction was 'F: Balancing Land Uses on the City Fringe' with four main objectives: contain Sydney's urban footprint, maintain and protect agricultural activities and resource lands, encourage investment in agriculture and resource lands, and maintain Sydney's soil health. Accentuation of conflicts with greenfield development in the suburban growth centers with loss of agricultural land and residential complaints about agricultural activities
A Plan for Growing Sydney	2014	A strong global city/A great place to live	Metropolitan rural area	A 'rural and bushland backdrop' to be managed to balance local growth needs, environmental protection, resource management, agriculture, tourism and culture, research activity, military uses, and community safety. Economic value acknowledged along with need for a new 'strategic framework'

Source Extracted and paraphrased from the listed plans

the engagement of strategies with the agricultural sector was highly variable, as others have noted (Wilkinson 2011; Wynne et al. 2020), all have failed to truly realize the preservation of agricultural lands in practice. This is evidenced by the continuing disappearance of agricultural land throughout the late twentieth and early twenty-first centuries, despite government planning rhetoric espousing its value to the city

(Zeunert 2022). The one strategy not included in Table 2 is the most recent, to which we now turn to convey the flavor and substance of the contemporary planning agenda. This affords a comparison with the 1948 strategy from some 70 years earlier.

9 Greater Sydney Commission's 'A Metropolis of Three Cities' (2018)

The 2014 *Plan for Growing Sydney* (Table 2) foreshadowed a major overhaul of metropolitan planning bureaucracy and in late 2015 the Government duly established the Greater Sydney Commission (GSC) as a new statutory body dedicated to leading metropolitan planning, a departure from decades of departmental-led governance (Bunker et al. 2018). A year later the GSC released a draft 'amendment' to the 2014 plan that reconceptualized metropolitan Sydney into three urban realms—western, central, and eastern—a vision duly confirmed in the comprehensive plan published in March 2018 (GSC 2018a). This scheme for a metropolis of three major cities envisages a productive, livable, and sustainable poly-centric '30 min' conurbation of 8 million people by 2056.

Western Sydney is reimagined as the 'Western Parkland City' (WPC). The WPC upsizes the two growth centers announced in the 2005 Metropolitan Strategy into what could amount to an 80 km north–south by 35 km east–west (c3,000 km^2) ovular conurbation, with the supplementary western district plan showing an even larger area (GSC 2018b). The WPC remains the least developed realm in relation to built form but the most significant in relation to pre-existing agriculture, with only small agricultural remnants remaining in the other two more urbanized agglomerations comprising the 'central' and 'eastern' cities. Adopting the nomenclature of the 2014 strategy, a 'Metropolitan Rural Area' is identified as largely framing the WPC and including the floodplains of the Hawkesbury-Nepean River, more elevated broken territory, and somewhat anomalously, the major urban corridor westward across the Blue Mountains (Fig. 5). This part of the Greater Sydney region has been said to account for five per cent of NSW total agricultural output by value (GSC 2018a: 140), although this is arguably inflated as noted below. Much of the WPC area is presently rural, semi-rural lands or hobby-scale allotments, unavoidably demonstrating peri-urban characteristics (Padgett Kjaersgaard 2020); specifically, it routinely exhibits traits of lands awaiting 'higher economic development' (Mason and Knowd 2010: 63–64).

The plan itself is a functional document descendant from the CCC tradition with a major focus on the integration and interconnection of critical structural elements in spatial terms. The aspirations for agriculture are concentrated under two main imperatives: sustainability (Objectives 27, 28 and 29) and productivity (24) (Table 3).

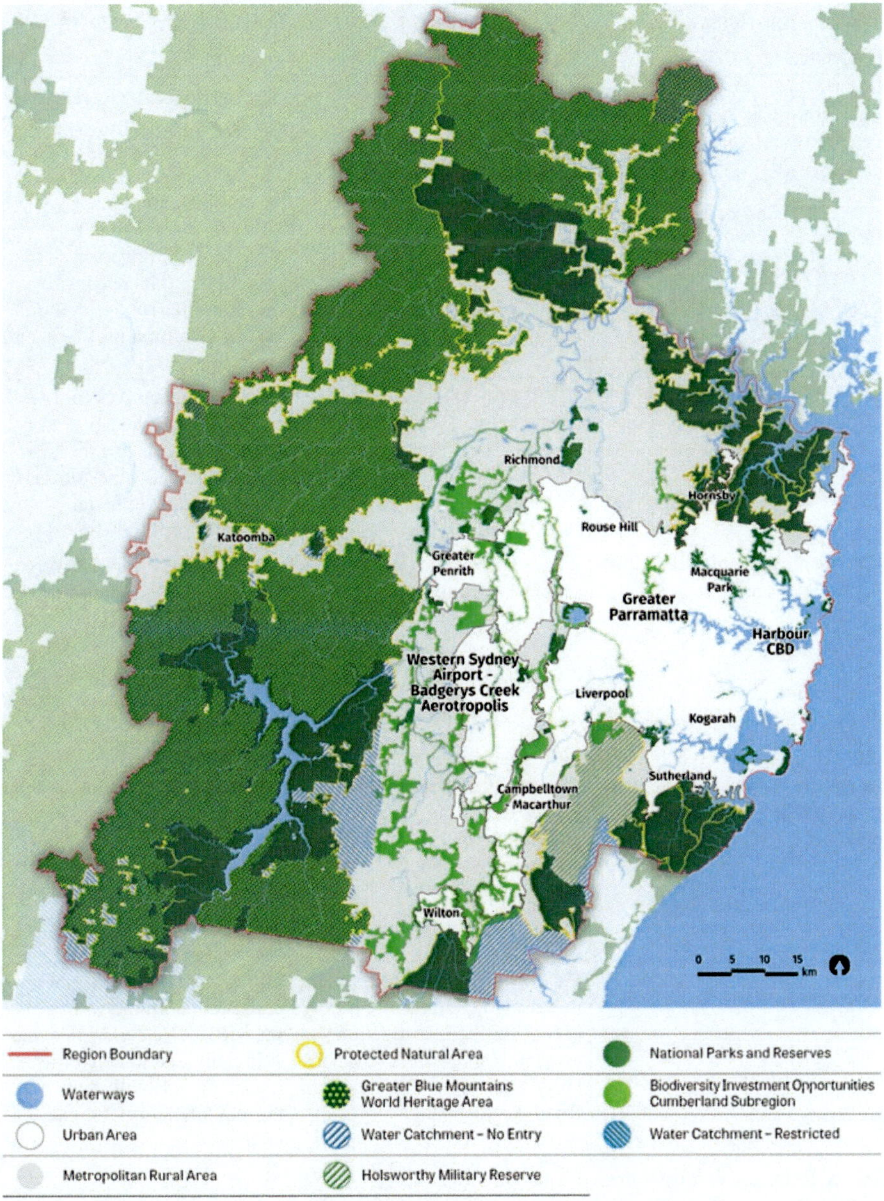

Fig. 5 The Metropolitan Rural Area (light grey) in the Metropolis of Three Cities (2018). *Source* GSC (2018a)

Table 3 Agricultural-related content in Sydney's 'Metropolis of Three Cities' plan (2018)

Objectives	Strategies	Intentions
24: Economic sectors are targeted for success	24.3: Protect and support agricultural production and mineral resources … by preventing inappropriately dispersed urban activities in rural areas	The Metropolitan Rural Area (MRA) is of specific importance for poultry, eggs, vegetables including mushrooms, nurseries, cultivated turf, and cut flowers. Importance of retaining and increasing opportunities for agricultural and horticultural uses to keep fresh foods available locally
27: Biodiversity is protected, urban bushland and remnant vegetation is enhanced	271.1: Protect and enhance biodiversity	Providing incentives for landowners in the MRA to protect and enhance the environmental values of their land. Can deliver better outcomes for biodiversity
28: Scenic and cultural landscapes are protected	28.1: Identify and protect scenic and cultural landscapes	Rising demand for biodiversity offsets and continuing support for traditional forms of agriculture within the MRA will realize more opportunities to protect and enhance natural landscapes
29: Environmental, social, and economic values in rural areas are protected and enhanced	29.1: Maintain or enhance the values of the Metropolitan Rural Area using place-based planning to deliver targeted environmental, social, and economic outcomes	Place-based planning can be used to identify, maintain and enhance the environmental, social, and economic values of rural lands. Rural-residential development is not an economic value of the MRA and is not supported
	29.2: Limit urban development to within the Urban Area, except for the investigation areas at Horsley Park, Orchard Hills, and east of The Northern Road, Luddenham	Urban development is not consistent with the values of the MRA. Greater Sydney has sufficient land to deliver its housing needs within the current boundary of the urban area

Source Extracted and paraphrased from GSC (2018a)

10 Western Sydney Agribusiness Precinct

The economic and sustainability aspirations of the GSC plan intersect decisively with the major turn to agribusiness concepts as a future land use planning strategy for western Sydney. This thinking is most directly connected to the development of a new airport conceived as the major economic driver of the WPC. Influenced by the gravitation of employment clusters including 'AgTech' to airports (Wastnage 2019) and the Netherlands approach to intensive horticulture and especially the Schiphol Airport region (Viviano 2017), this approach reflects an enthusiastic adaptation of the economic growth-oriented aerotropolis thinking of American aviation management expert J.D. Kasarda (Kasarda and Lindsay 2012). Representing a significant departure from the socio-economically and socio-culturally marginalized subsistence legacy of Sydney's urban horticulture (James 2016), it is proposed that protected and highly technological automated food production—namely robotically run greenhouses and warehouses—will produce crops and outputs largely decoupled from the ground (and thus the intrinsic land capability and natural climatic limitations). These will be value-added and exported to lucrative 'premium' markets (targeting China), in provenance-traceable long-chain food systems. In other words, Sydney's contemporary planning vision for its metropolitan agricultural future is to feed the wealthiest non-Sydney-based consumers in an aviation-enabled market. Land use plans have been produced to guide future investment decisions (Fig. 6), although it is unclear the extent to which the built form shown is devoted to actual food production and capable of actually supporting large-scale agribusiness concerns (NSW Government 2020).

11 Trajectory

Given the alignment of the seven preceding metropolitan plans for Sydney with facilitation of the growth and commercial development agendas of state governments, the GSC's plan unsurprisingly propels that into a supercharged neoliberal philosophy. The planning form of the WPC region reflects an evolution from a town and country paradigm, through satellite cities to urban corridors and an increasingly ubiquitous greenfield suburbia, albeit filtered through new urbanist thinking at higher densities (although continuing private vehicle dependence seems assured).

The WPC presents numerous challenges for urban development. Its drainage characteristics are prone to flood, which also makes provision of wastewater infrastructure challenging, its landlocked basin geography has a propensity for housing poor air quality, and a hot summer climate is exacerbated by the urban heat island effect. While 'progress' through technology and development will likely address these challenges and create flagship projects and job creation in the process—raising dam walls; major earthworks; diversions, pipelines, and pumping; air-conditioned housing—such directions are at odds with over half a century of prevalent sustainability discourse and recent liveability, healthy cities and biophilic design consensus in planning, design, and development (Wheeler 2013; Zeunert 2017).

The WPC houses nearly all the remnant vestiges of agriculture in Sydney's immediate basin, along with its most fertile lands in the alluvial fans and at times volatile floodplains of the Nepean and especially Hawkesbury Rivers (such as Windsor and

Fig. 6 Vision of a new Agribusiness precinct for western Sydney. *Source* NSW Government (2020)

Pitt Town Bottoms—which most recently flooded in March 2021). A notable proportion of the WPC known as the Cumberland Plain supports thousands of farm dams and market gardens, however, many of these are disappearing under the new Western Sydney aerotropolis and other greenfield developments.

Statistics on more recent levels of Sydney's food production and value come with a range of caveats, and in short, are indicative at best, or misleading at worst. An appraisal by James et al. (2010) was that official Australian Bureau of Statistics (ABS) data from at least the early 1990s until 2005–06 underreported and failed to adequately capture the value of metropolitan and peri-urban agriculture in Sydney. Specifically:

> Peri-urban farming...has been systematically under-recognised by an ABS data collection methodology focused on broad scale Agriculture...Despite the need for clear and coherent data, however, previous studies of Sydney's agricultural land have consistently failed to create a coherent picture of Sydney's agricultural industry and its changing nature and extent (James et al. 2010: 6).

A review of the latest ABS census data (2016) for Sydney's agriculture reveals a continuation of these issues; this dataset is also inaccurate due to 'mis-reported locations of farming propert[ies]' (Spencer 2021). This calls into question the prior statistic of the GSC that Greater Sydney accounts for five per cent of NSW total agricultural output by value (GSC 2018a), as this is based on ABS data.

A 2015–16 project titled *Sydney Food Futures* in quantifying the output of Sydney's agriculture, stated that 'in 2011, the Sydney Basin produced half a million tonnes of food—enough food to feed 20% of its population, including 40% of its demand for eggs, 10% of vegetables, 38% dairy, 55% meat and 2% of fruit' (SFF, n.d.). This project, however, used the 'Sydney Peri-Urban Network' (SPUN) area combined with the metropolitan area in its calculations (Wynne 2021). This means a total production area of approximately 19,710 km^2 (SPUN is approximately 16,800 km^2) was used in its calculations, equating to an area about four times larger than usually conceived as Sydney and its topographic basin as shown in Fig. 1.

Despite the GSC's numerous nods to agriculture summarized in Table 3, these will be tokenistic if greenfield developments eviscerate productive agricultural land, as well as if residential land use is allowed to encroach upon pre-existing agricultural areas, which inevitably leads to conflicting land uses. Peri-urban areas containing both existing agriculture and proximate greenfield developments include Camden (e.g., Spring Hill), (North) Richmond, Windsor, North Kellyville, Austral, Rossmore, Leppington (Emerald Hills, Denham Court, Gledswood Hills, Catherine Field), Oran Park, Spring Farm, Horsley Park, Luddenham, Badgery's Creek, Mulgoa (north), Marsden Park, Schofields, Rouse Hill, Riverstone, Box Hill, Pitt Town, North Richmond, The Oaks, and Orchard Hills. Numerous other agriculturally active regions are also potentially threatened by future growth in the WPC (Zeunert 2022).

While a notable proportion of the WPC is deemed Metropolitan Rural Area, similar failings to the 1948 CCC scheme are risked with a lack of nuancing to existing activities and to land and soil capability classifications. The concern is that almost all other land uses (all forms of urban development and primarily housing, employment precincts, sporting facilities, and nature conservation lands) are ultimately valued above agriculture.

12 Conclusion

Since the city's founding in 1788, Sydney's agricultural lands have demonstrated fluidity in relation to the expanding urban footprint, whereby the latter has consistently dictated the former. Into the early twentieth century, as the city maintained a radial urban form around a single major center on the eastern seaboard with total population only breaking the 'million' milestone in the 1920s, rural land uses dominated in the Sydney region. The CCC's mid-century audit captured a still vital and productive land use that faced strengthening challenges. Its position was nonetheless clear: 'The rural background of the County must be preserved and strengthened to the utmost practicable degree, having in mind the necessity for encroachments to meet the needs of urban expansion' (CCC 1948: 64).

But the pressures would only intensify with an ongoing ballooning population and rising land prices driving new development and thus conflicts between pre-existing agricultural activities (especially intensive animal agriculture) and new activities, notably housing. The strong demand for rural lifestyle living and rising domestic and global competition are more recent disruptors (Clarke 2017). Although these trends are mirrored widely elsewhere, they have been exacerbated over a long period in Sydney's case given the physical constraints presented by its topographic basin setting.

Despite the shortcomings of the CCC's 1948 Metropolitan Planning Scheme,it did attempt to address rural issues more substantively than all of its successors. Admittedly, this was an era prior to the developed cold-supply chain logistics systems for the emerging agribusiness sector, the consequences of planning decisions in a post cheap fossil fuel era had not yet surfaced, and it contemplated a modest rate of population growth extrapolating pre-war trends. All these premises would soon prove unrealistic. The CCC fought a losing battle to preserve the green belt and by the early 1960s the predominant rural flavor of the outer city was receding, at least along main transport corridors.

All Sydney's subsequent metropolitan plans enabled the reduction of land available for agricultural production while concurrently facilitating swelling of the population of the city. They placed their faith in the continuity of an energy-intensive food supply system geared to 'just-in-time' supply from further afield. The latest GSC plan—while recognizing the value-add possibilities of modern agricultural practices—seals the displacement of family concerns in working the land and short-chain local food systems to instead reimagine food production in Sydney as corporate industrial agribusiness servicing lucrative export markets: the new neoliberal metropolitan techno production landscape.

Agriculture results in a range of issues relevant to metropolitan planning that require a strategic and political balancing. Many of these issues relate to specific eras but there is an overall narrative evident in the Sydney experience. An historical perspective reveals planning concerned primarily with the facilitation of tangible urban development and economic growth rather than intergenerational sustainability.

References

Ashton N (1983) New South Wales: Notes on the Evolution of Planning from 1945 to 1974 with particular reference to Governmental Involvement. Unpublished. Sydney: NSW Department of Environment and Planning

Benson DH, Howell J (1990) Taken for Granted: The bushland of Sydney and its suburbs. Kangaroo Press, Kenthurst

Brown AJ (1937) City of the Future. Sydney Morning Herald 4:13

Budge T (2013) Is Food a Missing Ingredient in Australia's Metropolitan Planning Strategies? In: Farmar-Bowers Q et al (eds) Food security in Australia: challenges and prospects. Springer, New York, pp 367–379

Bunker R, Freestone R, Randolph B (2018) Sydney: growth, globalization and governance. In: Hamnett S, Freestone R (eds) Planning Metropolitan Australia. Routledge, London, pp 76–100

Cahill JJ (1951) 2nd reading, Speech delivered by the Hon JJ Cahill MLA in the Legislative Assembly on the County of Cumberland Planning Scheme. Local Government (Amendment) Bill. From Parliamentary debates, 29 May 1951. NSW Parliamentary Debates

Christie M (1988) The Sydney Markets 1788–1988, Flemington Markets, NSW: Sydney Market Authority

Clarke M (2017) Values of the Metropolitan Rural Area of the Greater Sydney Region, AgEconPlus Consulting for the NSW Department of Planning and Environment and Greater Sydney Commission, Accessed 10 December 2018. https://gsc-public-1.s3.amazonaws.com/s3fspublic/Values_of_the_Metropolitan_Rural_Area_of_the_Greater_Sydney_Region_(Ag_Econ_Plus).pdf

CCPS (1948) County of Cumberland Planning Scheme. Cumberland County Council, Sydney

Deniehy D (1853) The Constitution of New South Wales. Speech documented in The Sydney Morning Herald, New South Wales, Australia, 16 August 1853. p 4. Accessed 2 February 2021 https://trove.nla.gov.au/newspaper/article/12948089

DMR (1945) Main Road Development Plan for Sydney Metropolis and County of Cumberland. NSW Department of Main Roads, Sydney

Fraser RDL (1957) Minto. Chief County Planner's Report No 1546. 6 June 1957. Cumberland County Council. NSW Department of Environment and Planning records, AK415. NSW State Archives

Freestone R (1992) Sydney's Green Belt 1945–1960 Australian Planner. 30, 70

Freestone R (1989) Model Communities: The garden city movement in Australia. Melbourne: Thomas Nelson

Gaballa S, Abraham A (2008) Food Miles in Australia: A Preliminary Study of Melbourne, Victoria, Melbourne, Australia: Centre for Education and Research in Environmental Strategies

Golledge RG (1960) Sydney's metropolitan fringe: a study in urban-rural relations. Aust Geogr 7:243–255

GSC (2018a) Greater Sydney Region Plan: A Metropolis of Three Cities, Greater Sydney Commission: NSW Government. https://www.greater.sydney/publications

GSC (2018b) Western City District Plan, Greater Sydney Commission: NSW Government. https://www.greater.sydney/publications

Harrison P (1983) Profile of a purposeful planner: Roderick David Lovat Fraser, 1911–1983. Australian Planner 21:29–33

Hoorn J (2007) Australian Pastoral: The making of a white landscape. Fremantle Press, Fremantle

Howard E (1898) To-Morrow: A Peaceful Path to Real Reform. Swann Sonnenschein, London
James S (2016) Farming on the fringe: Peri-urban agriculture, cultural diversity and sustainability in Sydney. Springer International Publishing, Cham
Kasarda J, Lindsay G (2012) Aerotropolis: The Way We'll Live Next. Farrar, Straus and Giroux, New York
Lewis M (1980) Milk, Mothers, and Infant Welfare. In: Roe J (ed) Twentieth Century Sydney: Studies in Urban & Social History. University of Sydney Press, Sydney, pp 193–207
Lines W (1999) Taming the great south land: a history of the conquest of nature in Australia. University of Georgia Press, Athens, GA
McClymont J (2013) Pictorial History: The Hills District. Kingsclear Books, Sydney
McGill J (2017) Pictorial history Campbelltown & District. Kingsclear Books, Sydney
McIntyre J (2009) Not Rich and Not British: Philip Schaeffer, "Failed" Colonial Farmer. Journal of Australian Colonial History 11:1–20
Ma Rhea Z (2017) Frontiers of Taste. Springer, Singapore
Martyn J (2018) Rocks and trees: a photographic journey through the rich and varied geology, scenery and flora of the Sydney region, STEP: Turramurra, NSW
Mason D, Knowd I (2010) The emergence of urban agriculture: Sydney, Australia. Int J Agric Sustain 8(1–2):62–71
Maze WH (1943) Land utilization surveys in the Kurrajong-Windsor District. New South Wales, Australian Geographer 4:155–174
NSW Government (2020) Overview of the Agribusiness Precinct Plan, Western Sydney Partnership. Sydney, November 2020.
Padgett Kjaersgaard S (2020) Learning from the Edge: An Inquiry into the Peri-urban Territory of Perth'. PhD Thesis, University of Western Australia
Parham S (2015) Food and Urbanism: The Convivial City and a Sustainable Future. Bloomsbury Academic, London
Scutts L (1958) Subdivisions in the Rural Areas of the County of Cumberland. NSW Department of Environment and Planning records,AK415, NSW State Archives
Self P (1961) Cities in Flood: The problems of urban growth, 2nd edn. Faber & Faber, London
Spencer J (2021) Personal communication, 21 April 2021
SFF, (n.d.) Interactive food maps, Sydney Food Futures, Institute for Sustainable Futures, University of Technology Sydney, accessed 13 April 2021 http://www.sydneyfoodfutures.net/interactive-maps/
SPA (1967) Prelude to a Plan. State Planning Authority, Sydney
Viviano F (2017) How the Netherlands Feeds the World, National Geographic, https://www.nationalgeographic.com/magazine/2017/09/holland-agriculture-sustainable-farming/
Wastnage J (2019) Bringing Industry to Airports: How the US Experience can help Western Sydney Take Off. United States Studies Centre, University of Sydney, Sydney
Wheeler S (2013) Planning for Sustainability: Creating Livable. Routledge, Equitable and Ecological Communities, New York
Wilkinson J (2011) Agriculture in the Sydney region: historical and current perspectives, e-brief 2/2011, NSW Parliamentary Library Research Service, Accessed 12 May 2018. https://www.parliament.nsw.gov.au/researchpapers/Documents/agriculture-in-the-sydney-region-historical-and-/Agriculture%20in%20the%20Sydney%20region%202.pdf
Wills NR (1945) The rural-urban fringe: Some agricultural characteristics. Aust Geogr 5:29–35
Winston D (1957) Sydney's Great Experiment: The progress of the Cumberland County Plan. Angus and Robertson, Sydney
Wynne L (2021) Personal communication, 29 April 2021
Wynne L, Ruoso L-E, Cordell D, Jacobs B (2020) 'Locationally disadvantaged': planning governmentalities and peri-urban agricultural futures. Aust Geogr 51:377–397
Zeunert J (2017) Landscape Architecture and Environmental Sustainability: Creating Positive Change through Design. Bloomsbury, London
Zeunert J (2022) Sydney's Food Landscapes: Past, Present, Future, forthcoming

Brøndby Allotment Gardens: Building Social Cohesion in Danish Peri-Urban Spaces

Kathryn R. Terzano

Abstract Allotment gardens in Denmark are cluster garden developments where year-round residency is prohibited. Similar gardens exist in numerous countries and are especially prominent in areas with a dense urban fabric where private green space is scarce. Allotment gardens tap into Ebenezer Howard's concept of the magnetic pull of both the town/city and the country; in some places, allotment gardens are even given the name of garden city. Although these allotment gardens are not true Garden Cities that function as stand-alone settlements with year-round residents, allotment gardens such as Brøndby Haveby, located in the greater Copenhagen metropolitan area, are private outdoor spaces that hint at the appeal of Howard's third magnet. In Denmark, allotment gardens were originally designed to provide access to nature for city dwellers, but the purpose has changed over time to provide both recreation and important food subsidies, especially in times of crisis, such as during wars when resources are scarce and expensive. This chapter discusses Erik Mygind's design of Brøndby Haveby as a model for gardening allotments in peri-urban spaces serving as places to cultivate land for food, to provide economic resilience for the middle classes, and to build social cohesion through communal space and shared governance.

Keywords Allotment gardens · Denmark · Peri-urban · Finger Plan · Garden associations

1 Placing Danish Allotment Gardens in Context

1.1 History of Allotment Gardens in Denmark and, Briefly, Elsewhere in the World

Small gardens in the outskirts of Copenhagen have existed since at least the end of the sixteenth century. The Danish Allotment Garden Federation has chronicled the history of these early gardens and defines them as productive gardens located

K. R. Terzano (✉)
Arizona State University, Arizona, USA
e-mail: kterzano@asu.edu

C. Brisotto and F. Lemes de Oliveira (eds.), *Re-Imagining Resilient Productive Landscapes*, Cities and Nature, https://doi.org/10.1007/978-3-030-90445-6_13

outside of city walls, supplying food to those living within the walls (Damin & Palmer 2003; Kolonihaveforbundit n.d.). In the late eighteenth and early nineteenth centuries, the Danish government formalized peri-urban gardening spaces as "haugekolonier," or mound gardens, meant as productive spaces to supplement the diet of less well-off Danes who lived in dark, cramped apartments (Petersen et al. 2011). The growth of allotment gardens was also associated with the interrelated phenomena of industrialization, urbanization, and the rise of the labor movement, which meant that in Denmark, as well as in other countries, people were increasingly moving away from the countryside—where productive spaces were conveniently nearby—and moving into cities to work in factories (Galschiøt 1893; Hohenberg and Lees 1995). The allotment gardens themselves were a form of compensation for poor housing and working conditions in the city and a way for members of the laboring class to achieve some self-sufficiency in growing their own food and to have the rights to a plot of land with some fresh air. These gardens began to be referred to as pauper gardens, a term that shows the close relationship between poverty and gardening in Denmark at the time, as the gardens were intended to provide relief specifically to poor families (Damin and Palmer 2003). Wealthier families had better access to fresh produce and did not require governmental intervention to provide gardening space.

The secondary purpose of these gardens was to provide leisure activities through gardening, which was seen as a good and moral way to spend one's free time (Tolstrup 1980). This is similar to German allotment gardens and "their focus on food production and the creation of village-like communities" (Bigell 2015, p. 103). This also mirrors the rise of industrialization as, indeed, the work ethic needed for a successful garden is compatible with the work ethic expected in an industrialized society (Lawson 2004). The concern with how people spent their leisure time—especially with how the working classes spent their time—was much like what occurred with the simultaneous City Beautiful movement in the United States. The City Beautiful movement was a movement where architects, planners, and upper-class lay people believed that through beautifying a city, its residents would become more civil, as defined by upper-class values. The mere look of these beautiful spaces as well as the time spent beautifying them were believed to have a civilizing effect on people. Similar to how the City Beautiful movement was characterized in part by a concern for the morals and activities of the poor, the Danish provision of allotment gardens was seen as a way of encouraging the poor, especially fathers, to spend their time in their garden instead of going to the pub. Near the end of the nineteenth century, Copenhagen created its first garden association called Arbejdernes Værn, soon followed by the garden association Vennelyst, which is now the oldest garden association in Copenhagen that is still in existence. From the beginning, these garden associations provided the management structure, and they governed allotment gardens for working families to use for outdoor recreation and for the cultivation of food. Many more garden allotments were built in the early part of the twentieth century, although as land became more scarce they were often on undesirable pieces of land such as former landfills, where gardening was, at best, difficult and, at worst, hazardous (Kolonihaveforbundet n.d.).

Europe in general suffered from food scarcity throughout the First World War, and produce was imported from the United States during the war and for several years after it ended. Victory gardens, also called war gardens and liberty gardens, were vegetable, fruit, and herb gardens planted at private residences and in public spaces such as former parks, golf courses, or on vacant land, in Australia, Canada, Germany, the United Kingdom, the United States, and elsewhere during the First and Second World Wars. In wartime, governments encouraged people to plant victory gardens not only to supplement their rations but also to boost morale by contributing to the war effort through the use of volunteer labor (especially the labor of women and sometimes children) to produce food, and to reduce the burden on transportation networks by consuming locally produced food whenever possible. In the United States during the Second World War, victory gardens produced more than 40 percent of the vegetable supply in the country (Lawson 2004). In the United Kingdom during the same time, under the Dig for Victory campaign, there was a great expansion in both the number of allotment gardens (from less than one million before the war to 1.7 million at the start of 1943) and the number of private gardens producing vegetables (Ginn 2012).

In Denmark during the First World War, the lack of produce for sale meant that people increasingly came to rely on growing vegetables in their own gardens. However, as Copenhagen grew in both population and its need for more residences, areas of land that had been used to cultivate food, such as gardens, were taken by the government to be redeveloped for housing. The former landowners were usually given land on the outskirts of the city in exchange. During the Second World War, and especially during the German Occupation of Denmark, the Danish government allocated land specifically for food-productive gardens and citizens who wished to have such a garden were given one for free. Over time, this led people becoming accustomed to having an allotment garden as a place to grow food and, to a limited extent, to spend their leisure time. Allotment gardening was at its height during this era with over 100,000 gardens in Denmark (Jensen 1996). These gardens were no longer the pauper gardens of earlier eras, but rather were in high demand among the middle classes as well, mirroring a trend elsewhere, such as in Germany, where allotment gardens became the domain of the middle classes and served a social, communal purpose (Bigell 2015). Additionally, compared to the earliest allotment gardens, many allotment gardens during the Second World War were purposefully designed as they had become the domain of landscape architects, whose role had broadened professionally to also include the design of playgrounds, sports facilities, parks, and residential areas (Yang 2013). Prior to landscape architects' involvement in allotment gardens, designs were haphazard and created by lay people.

After the Second World War, the Danish standard of living improved and Danes enjoyed their increasing wealth as well as leisure time—thanks to new laws that introduced additional holidays and shortened work hours. Danes were able to use their allotment garden homes for the length of their children's summer breaks, which was approximately seven weeks, and sometimes even as full-time, summer residences (Bech et al. 1980). In other areas of Denmark aside from the Copenhagen region, allotment gardens were still often only used during the daytime because they were closer to the owners' residences and thus a day trip was feasible, but in the Copenhagen metropolitan area, the gardens became something of a holiday destination. Over time, some of the allotment gardens stopped being used to produce food but rather were used as holiday homes for Danes who did not own homes by the sea, the preferred location for a summer home. However, with the 1970s came oil crises that made it unaffordable to travel a great distance to a summer house; allotment gardens located only a moderate distance away became a compromise (Jensen 1996). This resurgence of interest in allotment gardens, especially those located nearer to Copenhagen where developable land was increasingly scarce, meant that allotment gardens were in competition with urban development, and this in turn meant that some allotment gardens were redeveloped for the built environment. By the end of the 1970s, the Danish Parliament recognized the importance of allocating space for allotment gardens, both because of their potential for food production and also for the social demand for them, and passed legislation to protect allotment gardens; moreover, in 2007 the Allotment Gardens act was passed to further establish the allotment gardens as permanent (Brøndby Kommune 2016). A large call for allotment gardens still exists, especially in the Copenhagen area, turnover is rare, and waitlists can be long (Jensen 1996).

1.2 Copenhagen's System of Green Spaces

Since around the Second World War, Danish leaders have recognized the need to plan appropriately for Copenhagen's growth—both in terms of its human population and its expanding city limits. The private firm, Urban Planning Lab, with planners Steen Eiler Rasmussen and Christian Erhardt "Peter" Bredsdorff created Copenhagen's original Finger Plan in 1947 to set objectives for directing growth and constructing a system of green space in the region (Yang 2013). The Urban Planning Lab would later become the Regional Planning Office, co-financed by the local municipalities, the regional government, and several ministries, allowing for greater authority for implementing and adapting the Finger Plan over the years. Since its inception, the Finger Plan (see Fig. 1) has guided urban development to grow out like fingers along rail and radial road networks from Copenhagen, rather than allowing for unrestricted growth in all directions, with green wedges between the fingers that would be largely undeveloped (Kommune 2016). The Finger Plan consisted of both green space that is internal to the city, such as urban parks, as well as green space in the wider region. This plan, including its hierarchy of plots, lines, lumps, wedges, and rings, which each corresponding to specific kinds of green spaces, has guided urban development and provided protection for green space in the region since its creation, although the 1947 version of the Finger Plan has been updated and adapted throughout the decades (Danish Nature Agency 2015).

One of the key elements of the Finger Plan is that all residences in the Five Finger area should be within 300 m of green space. Allotment gardens in the region are integrated into the Finger Plan and are consistent with municipal planning, greater

Fig. 1 Original Finger Plan from 1947 (open access)

regional planning, and with the government's concern with the provision of green space. With notable allotment gardens such as Brøndby Haveby being located in the green wedges between the "fingers," Copenhagen residents have easy access not only to green space but to productive and recreational space.

It is only because of the allocation of space to the green wedges that allotment gardens in the greater Copenhagen area have been able to survive and thrive. Had the region developed in a more traditional, sprawling manner, without reserving space explicitly for the green wedges, allotment gardens would likely have been destroyed to make space for the built environment. Indeed, this is what had begun to happen in the early years before the implementation of the Finger Plan.

1.3 Inspiration from Ebenezer Howard's Garden City Concept

Ebenezer Howard's work predates the Finger Plan of Copenhagen by decades. Howard also lived in a different country (the United Kingdom, with a brief stay in the United States). However, Howard's work serves as an example of early recognition of the importance of access to nature, and this importance is what is stressed in the Finger Plan and undergirded the creation of it.

Howard witnessed the overcrowding of rapidly urbanizing London in the second half of the nineteenth century and into the twentieth century, in stark contrast to the peaceful if dull countryside (LeGates and Stout 2015). London was not alone in its experience of rapid growth; cities throughout Europe, including Copenhagen, were also feeling the effects of urbanization and industrialization. Howard famously concluded that an alternative form of settlement was needed. In Howard's alternative, the town-country, a greenbelt would allow easy access to nature, the need for a commute (e.g., to London) would disappear, and the population size would be capped to prevent overcrowding. Howard's Garden City concept spread throughout other countries but in most places, these garden cities were built as loose interpretations of the concept, rather than recreations of the bona fide Garden Cities of Letchworth and Welwyn. Places such as the greenbelt towns of the United States have been well studied as deriving inspiration from Howard's Garden City concept (Parson 1998; Sturzaker and Mell 2016).

Howard explicitly intended his Garden Cities to serve as fully developed settlements where the population would reside year-round, whereas Denmark's allotment garden developments were never intended to become a full-time alternative to living in Copenhagen or other established cities. Instead, Danish allotment gardens have drawn inspiration from Howard's "third magnet" by offering an alternative to having to choose between being an urbanite and having one's own plot of land in the picturesque countryside. Indeed, allotment gardens in the outskirts of Copenhagen fulfill Howard's goal—albeit only part time—of having a location that offers the best of both the town and country; they are near enough to the city to function as summer homes with space for gardening and they function as social spaces outside of the city. Although no proper garden city was ever established in Denmark, the advent of the new Danish suburbs and the growth in allotment gardening represent the same kind of recognition of the draw of both the "country magnet" and the "town magnet." Danes residing in cities like Copenhagen may not have the ability, or inclination, to move to the countryside since city life generally offers higher wages, greater activity, and access to cultural amenities. However, through their allotment gardens, these Danes are able to reap the benefits of country life as well.

Although Howard's concept of garden cities with full-time residents was realized in the United Kingdom and, to an extent, in the United States, the concept never fully caught on in Denmark; nevertheless the phrase "garden city" and many of Howard's ideals, including communal ownership, live on (Sverrild 2016). Indeed, one of Brøndby Haveby's several nicknames is Brøndby Garden City. The origins of this nickname are lost to time, but may have come from a bird's eye view of the allotment gardens. Viewed from above, Brøndby has the same circular design as Howard's garden city schematic, with communal space at the middle and agricultural space further out. Although there are many differences between Howard's design and Brøndby—for example, the scale is much different, with Howard's concept being one of 30,000 residents—the nickname of Brøndby Garden City pays homage to Howard's ideals.

2 Erik Mygind, Landscape Architect

In contrast to the rest of Scandinavia and certain other parts of Europe, Denmark has no untouched natural area left in its landscape. The entire country was deforested by the end of the Middle Ages, making room instead for the built environment and agricultural land uses. This lack of wild, undisturbed nature influenced the evolution of landscape architecture in Denmark, including on the larger scale of the Finger Plan. Danes came to appreciate the rural countryside as synonymous with nature, even when that rural countryside was cultivated agricultural lands. Perhaps it is because of the perceived equivalence of agricultural land with nature that allowed Danish landscape architects to easily turn their focus toward small, productive gardens. While other Scandinavian countries cultivated open gardens set within woodlands, Danish gardens have historically been enclosed spaces, most often enclosed by hedges (Woudstra 1995). The majority of these enclosed designs are private spaces, favoring the individual garden over a community design. Among Danish landscape architects who have worked on allotment gardens, Erik Mygind stands out for his work on Brøndby Allotment Gardens and its joining of private gardening with communal space.

2.1 Biographical Background

An entry in a book about Danish gardens, *Danmarks Havekunst: 1945–2002 [Danish Garden Art: 1945–2002]*, gives the most complete account of Erik Mygind (Lund 2002), about whom little else has been written. According to Lund (2002), Erik Mygind (1916–78) was a Danish landscape architect who worked in the firm of renowned landscape architect Carl Theodor Sørensen from 1943 to 1946. C.T.

Sørensen is among the most significant landscape architects in twentieth century Denmark, with the diamond-shaped design of Hans Christian Andersen garden in Odense among his more famous works (Andersson and Høyer 2001). Mygind, Sørensen, and other landscape architects of the time were undoubtedly influenced by G.N. Brandt's garden construction, which heavily utilized geometric patterns. Other landscape architects, such as Morten Klint, Sen Hansen, Ingwer Ingwersen, and George Boye also created designs that incorporated geometric shapes, with the height of this geometric design period being 1930 to 1950. Furthermore, Brandt and Sørensen were influential instructors at the Royal Danish Academy of Arts, School of Architecture, from which many practicing Danish landscape architects graduated, including Mygind (Yang 2013).

Mygind ran a design studio with his wife, Agnete Muusfeldt, to whom he was married from 1948 to 1961 (Lund 2002). During their time as design partners, their design focus was on children's spaces within gardens, including an exhibition for the Royal Danish Garden Society (Royal Danske Haveselskab) that they called the children's gardens ("børneneshaver"). Even within these early designs, which were created years before more famous circular designs such as those at Brøndby Haveby, circular patterns emerge; in the children's gardens there were two circles that they created based on gender norms of the time: flowers and a playhouse in one circle, meant for girls, and a sand pit, tree stumps, and a rustic hut in the other circle, meant for boys (King and Newstead 2017). In 1956 in Åbrinken, Virum, which was designed as a terraced housing development, they expanded this theme, continuing to emphasize interaction and engagement with outdoor spaces, designing a food-producing garden, playground, and playhouse within three willow-fenced rotundas. Lund (2002) speculates that Mygind may have been inspired by historian Lewis Mumford's book, *The Social Basis for Construction in the Post-War Period,* noting that Mygrind wrote an article about the book in the early 1950s. Undoubtedly, Mygind also continued to be influenced by Sørensen, who served as the first chair of the Danish Playground Association (1959) and the first President of the International Play Association (1961), promoting public spaces in which children could be active and play (Yang 2013). Evidence of this influence is seen in Mygind's championing of spaces for older children to be active outside—in playgrounds, in green spaces between buildings, and in formal garden designs. In a 1956 article from *Havekunst* (Garden Art) magazine, Mygind argues that existing spaces like playgrounds were adequately provided for young children but not for older children, who needed opportunities for development and engagement. Lund (2002) reports that Mygind was indignant over a particular park design where children were given only 1,500 square meters of space (mostly in the form of sand playgrounds) whereas the adults had over 9,000 square meters of grassy area to enjoy. Mygind believed that when children behaved badly in public, it was because they lacked opportunity for activity and engagement. Mygind was a colorful character who would attend on-site construction meetings in knee-high leather boots and with a horn to blow reveille, and that he was an optimist in the post-war years who engaged passionately with all areas of his profession. Lund (2002, pp. 151–152) quotes Mygind as saying, "Videre fremad kommer vi kun ved at skabe børnesamfund, spejlbilleder af vort eget, hvor børn opdrages og skoles til at lege by, gartneri, landbrug, håndværkere eller sportsfolk" [We will only move forward by creating children's communities, mirror images of our own, where children are brought up and trained to play town, horticulture, agriculture, artisans, or athletes].

2.2 *Productive Landscapes in Peri-Urban Areas*

Like many architects in Scandinavia in the post-war era, Mygind could be characterized as an idealist. He criticized the design of open spaces and Danish park policies for providing inadequate experience opportunities for both children and adults (King and Newstead 2017). He saw these spaces as inactive, a sterile theater landscape. Children's engagement with gardening and farming was further realized in a design in Brede in the late 1950s. In a small community, Mygind engaged around one hundred children to help to build 35 small huts to raise chickens, pigeons, rabbits, and goats as livestock. Children assisted in the slaughtering of livestock and the preparation of meat, in accordance with local farming traditions. Mygind and Muusfeldt's idea that land should be productive (e.g., garden spaces should include fruit and nut trees) can also be seen in 1958 in the design of the open spaces at Aalborghus Gymnasium. The spaces were designed for aesthetics as well as productivity with a large green yard (about 25 by 60 m) containing square hedges and walnut trees and two other green gardens with apple trees as well as chestnut and ash trees to frame the surrounding buildings. Mygind believed deeply in creating lush environments for people of all ages but especially children and young adults (Lund 2002).

After the dissolution of Mygind and Muusfeldt's design studio, Mykind designed three allotment gardens, including Brøndby Haveby, as well as open spaces for the many new residential areas built by non-profit housing associations in the Copenhagen area (Rødovre, Hvidovre, and especially Herlev), open spaces around schools, and park spaces in Aalborg. In Mygind's allotment garden designs, he was seeking to create active, food-productive spaces in peri-urban spaces that were also social and community-oriented. His approach was partly aesthetic, with firm ideas about the geometric design of the allotment gardens, as he rose within his profession at the height of these geometric designs in landscape, but he also revealed his beliefs in creating social spaces. His designs set strict standards for fruits, vegetables, and nuts to be grown in each of the gardens.

3 Brøndby Haveby

3.1 Overview and History

Mygind designed Brøndby Haveby (also occasionally referred to as Brøndby Garden City, the Colony Gardens, and Green Area) in 1964. Located in a peri-urban area to the southwest of Copenhagen, between the "thumb" and the "index finger" in the Finger Plan, these allotment gardens in the original part of the development are organized in clusters rather than in a traditional design of rectangular gardens laid out in rows. There had been previous allotment gardens in the area but Mygind was brought in by the municipality to design a new garden association with an initial design on the former land of the Hersemose farm. The new garden association required memberships to be bought for DKK 150 each (equivalent to about 20 euros) and grass seed was bought by the board for DKK 3 per kilogram. The association took out a bond loan of DKK 721,000 to cover startup costs; in turn, to rent a garden for the first year cost a member DKK 44.50 per month, of which DKK 27.00 went toward the repayment of the loan and DKK 17.50 for operating costs. In the early years of Brøndby Haveby, it was possible to obtain a home loan through the Garden Colony Association for either DKK 1,000 or DKK 2,000, and 18 people received such loans, which were repaid at a rate of DKK 25 or DKK 50 per month, depending on which size of loan was taken (Brøndby Haveby III 2004).

In more than 50 years since the founding of the Brøndby Haveby gardens, the association has financed various improvements for the good of the community, including building and repairing roads, purchasing equipment (such as tractors, a green harvester, and rain pumps), building a playground (which has since been demolished because it was improperly permitted), erecting a flagpole, installing sewers, building a communal greenhouse, and purchasing horse manure as fertilizer for the gardens; it has continued to earn money through collecting association fees, land rent, and waitlist fees, as well as through the operation of a flea market, and has offered numerous free classes (e.g., winter pruning of apple trees) (Brøndby Haveby III 2004).

In the 1980s, the Brøndby Haveby allotment gardens area was expanded; in total, two of the six different garden areas consist of the characteristic circular allotment design as originally created by Mygind and an additional four other allotment areas are consistent with the more traditional character of rectangular gardens (Kommune 2016). These latter gardens were not designed by Mygind and likely do not have the same communal feel as the circular designs with their social spaces at the center.

3.2 Specific Design

The Sørensen School's concepts and its Modernist influence of geometric shapes became the artistic basis for Mygind's projects, just as a strong social commitment became a distinct feature. Although other geometric patterns besides circles and ovals were used by other landscape architects in their designs, Sørensen's modernist landscape design of "runde haver" (round gardens) in Nærum (1948–52) was likely the chief influence in Mygind's design of Brøndby Haveby (Dee 2012). Whereas the Nærum gardens are each an individual oval enclosed by a hedge, and thus are still individually private gardens, Mygind's design for Brøndby was communal. Each circular unit (or rotunda) is subdivided so that the individual gardens are the wedges that comprise the circle and at the center of the circle lies a common area meant to provide community space and activities as well as parking. Each garden was designed to allow a small (no larger than 40 square meters) house on it with additional space allowances for small sheds and greenhouses. The houses are not permitted to be occupied during the winter months as they are intended to be used during the growing season. Furthermore, the area is not part of the collective heat supply of the region; the buildings are unheated and therefore unsuitable for year-round tenancy for this reason as well (Kommune 2016).

The design of Brøndby can be compared to a pie or cake. An early newspaper article refers to the design as a "layered cake"; although there are no actual layers to the gardens, unless one considers the grass and soil to be layers, a layered cake refers to a cake that is cut into wedge slices (Sidste skrig i kolonihaver: Lagskager 1964). Each circular unit is roughly 400 square meters in size with some variation in the sizes of the circular units with some being slightly smaller than 400 square meters and others being slightly larger. The circles themselves are bounded by hawthorn hedges that were planted as part of the original design. Each circular unit has between 16 and 30 wedges, and there are 24 circular units in total, with 196 gardens in one section and 284 gardens in the other section (see Fig. 2).

The allotment gardens were built on both sides of a major road (Søndre Ringvej) where the terrain is completely flat, and the individual circular units were landscaped to slope down slightly toward their centers, where the communal space and a parking lot is situated within each rotunda (Lund 2002). (See Fig. 3 for a sketch of Brøndby's plan.) Brøndby's common area has roots in the traditional socio-spatial patterns of Danish villages, using the center as a place for social exchange. Mygind wanted to avoid large parking spaces and at the same time place the cars nears the gardens for convenience as well as to use the height of the hedges to mask the cars from view from outside of the gardens (Sidste skrig i kolonihaver: Lagskager 1964).

3.2.1 Current Status of the Brøndby Circular Gardens

Physically, the gardens have been maintained to stay true to their original design. The distinctive circular-shaped allotment gardens have retained their tight geometric shape over the years. The majority of the allotment garden houses orient with the gable facing the center of the rotunda. The current local planning document, the Local Plan 405, dictates that the gardens must be each surrounded by a hedge, all of which are cut equally high and in a way where the height decreases toward the center of the rotunda, which conforms to the original design by Mygind. The side hedges

Fig. 2 Aerial view of Brøndby, courtesy of Henry Do, photographer @www.henrydo.com

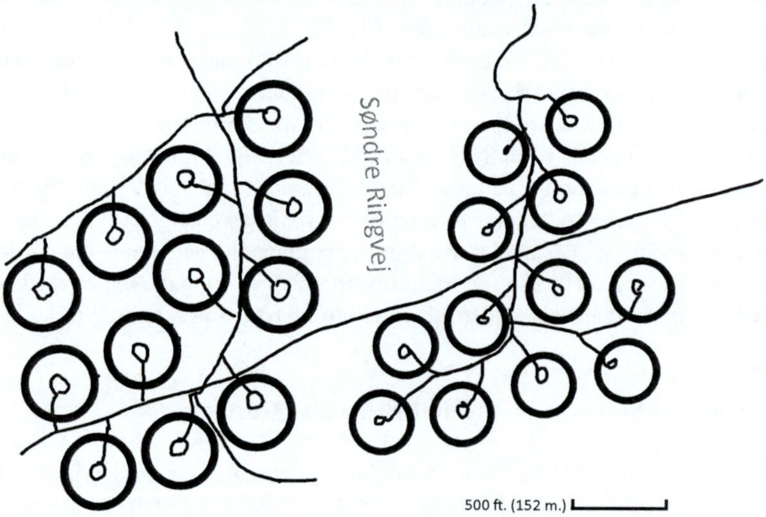

Fig. 3 Brøndby plan sketch with individual gardens represented by circles (by author)

that run the length of each wedge must be maintained at 0.8 m at the center of the rotunda, with the height gradually increasing to 1.8 m at the rear hedge (Kommune 2016). This causes the view as well as the perspective to be enhanced for the viewer when standing in the middle of the rotunda.

According to the original design of Brøndby Haveby by Mygind, each wedge garden was required to plant four fruit trees, four currant bushes, and eight black-currant bushes as well as one plum tree. These gardens were intended to be used as productive space, hence the requirement for fruit-bearing trees and bushes. This requirement is no longer present in the current local plan for the area (Kommune 2016). Some of the gardens are no longer used to grow food but instead are used purely as recreational summer homes. However, many of the gardens continue to be used productively and food continues to be grown, including different fruits, vegetables, and herbs than originally planned, such as cucumbers, tomatoes, and gooseberries. Pesticides are not permitted to be used on the property (Brøndby Haveby III 2004).

Restrictions still exist on the number and height of buildings, including each wedge's main building as well as any playhouses, greenhouses, toolsheds, or other smaller buildings within a given wedge, with additional restrictions on the siding and roofing materials, which helps to maintain the character of the original design, serves to prevent the spread of any fires between structures, and also ensures that structures are not visible from outside each circular unit. Parking is restricted to designated areas, and recreational vehicles, boats, and commercial vehicles may not be parked without permission from the garden association (Kommune 2016).

In addition to the physical aspects of the gardens, the social aspects of the allotment gardens and the communal feeling, with its origins as a planned morale booster and a fosterer of good work ethics, have also been preserved. In a newsletter reflecting on the first fifty years of the Brøndby garden association, members shared their stories of having found and supported a sense of community with others in their rotundas, gratitude for friendships found within their communities, and memories of communal activities. In one part of the newsletter, owners of a garden relayed their story of having bought their garden in 1982 and went on to describe their relationships with neighbors who had become friends over the years, remarking on the feeling of "hyggeligt"—an arguably untranslatable Danish word that means something close to cozy togetherness. Others wrote about parties of various kinds (summer celebrations, birthday parties for children and adults alike, theme parties like Hawaiian Night, and wedding anniversaries), the pioneering spirit of the earliest members of the Brøndby garden association, and the garden as a sanctuary; there was even a mention of members who have winter birthdays enjoying a party thrown for them on the 1st of August instead, since the gardens are largely vacant over the winter (Brøndby Haveby III 2004).

4 Conclusions

Within the Danish landscape, nature is equivalent to the countryside, which is dominated by agricultural lands. For Danes, the opposite of nature is the built, urban environment. Within the urban environment, there is little space allocated to green space or productive space. Allotment gardens such as the Brøndby gardens provide a means of bridging this urban and rural divide by offering mostly sized, food-productive spaces in peri-urban areas. Brøndby Haveby enhances this contribution by showing a model where these peri-urban productive spaces can be used to boost social cohesion and create a place where not just food but also a sense of community can grow and flourish. Danish researcher Peter Dragsbo examined the importance of the growth of suburban areas of Copenhagen, noting that social housing in Danish suburbs was associated with garden cities, and he saw an explicit connection between single-family house gardens and allotment gardens, both serving a need for people to connect with growing food, but only the latter serving a communal purpose (Sverrild 2016).

The formal recognition that allotment gardens serve a vital function in the Danish landscape came about in the late 1970s. A debate in the Danish Parliament at that time concluded with a decision that local governments had a responsibility to ensure existing allotment gardens would be protected from redevelopment efforts and that additional allotment gardens should be built in areas where there was demand for them. Urban development in the greater Copenhagen region occurs in areas adjacent to and surrounding the allotment garden areas, as they are protected areas of green space, consistent with the green wedge design from the Finger Plan. Without formal protection through the Finger Plan and by the regional government, these green wedges and the allotment gardens within them would undoubtedly succumb to development pressures.

The demand side of the supply-and-demand equation for allotment gardens comes from the middle classes of Copenhagen, who continue to want a place where they can grow their own food. This is especially true in more recent years with a resurgence in interest in local, organic produce. This trend is reminiscent of the wartime gardens, where the provision of food was more important than the use of gardens for idle relaxation. At the same time, these allotment gardens and their formal associations, which sponsor social events and issue newsletters, also provide a feeling of community for members. The supply side of the equation is land in peri-urban areas that are administered by local governments, who serve both their own local citizens but also the larger region. To mediate this relationship, the Municipality of Agriculture oversees these efforts through regional planning.

For planners, landscape architects, and policy-makers, allotment gardens in general provide an option where green space can be created in peri-urban areas, especially where urban areas lack sufficient space for gardens. The wartime surge in allotment gardening offers guidance for other times of crisis when resources are in short supply. Allotment gardening can help to provide food security for urbanites and to save them money in the process, especially if autumn harvests produce a surplus that can be tinned or jarred for later use, and the ongoing challenge has become to find sufficient space for allotment gardens; in Denmark, Austria, Canada, Germany, the UK, and elsewhere, waiting lists can be long for space in an allotment garden (Göttl and Penker 2020).

The Brøndby gardens, like many Danish allotment gardens, do not allow the use of pesticides, and this approach to food is in line with the growing interest in healthy food and small-scale agriculture that is organic and relies on human hands rather than mechanization. In addition to being inspired by Howard's "third magnet," allotment gardens—especially when designed with community in mind—can also focus on serving as a "third place," places that are neither work nor home but fulfill a separate function and provide socialization (Oldenburg and Brissett 1982). Brøndby Haveby specifically offers a model for creating food-productive spaces that also promote social cohesion.

References

Andersson SI, Høyer S (2001) C. Th. Sørensen landscape modernist. Danish architecture pressure, Copenhagen

Bech SC, Kjersgaard E, Danielsen J (eds) (1980) Københavns historie. City of Copenhagen, Copenhagen

Bigell W (2015) Allotment and community gardens: Commons in German cities. In Luccarelli M, Bergmann S (eds) Spaces in-between: Cultural and political perspectives on environmental discourse, Koninklijke Brill, Leiden, p 102–130

Brøndby Haveby III (2004) From the collections of Forstadsmuseet, Hvidovre, Denmark. Accessed 29 September 2020

Brøndby Kommune (2016) Lokalplan 405: Lokalplan for kolonihaverne i Den Grønne Kile. https://broendbyhavebyafd4.dk/wp-content/uploads/2019/06/PDF-Med-lokalplan-405.pdf

Damin A, Palmer JF (2003) Rural life in the city: the chalet garden in Denmark. In Schuster R (ed) Proceedings of the 2002 Northeastern Recreation Research Symposium, vol 302, p 59–64

Danish Nature Agency (2015) The Finger Plan: A strategy for the development of the greater Copenhagen area. Ministry of the Environment, Denmark. https://danishbusinessauthority.dk/sites/default/files/fp-eng_31_13052015.pdf

Dee C (2012) To design landscape: Art, nature and utility. Routledge, New York

Galschiøt ML (1893) Danmark i Skildringer og Billeder af danske Forfattere og Kunstnere. PG Philipsens, Copenhagen

Ginn F (2012) Dig for Victory! New histories of wartime gardening in Britain. J Hist Geogr 38(3):294–305

Google (n.d.) Map of Brøndby, Accessed 24 June 2021. https://www.google.com/maps/place/2605+Brondby,+Denmark/@55.6364844,12.4037422,121m/data=!3m1!1e3!4m5!3m4!1s0x465256d9392ef63b:0x4b282aa8663e4051!8m2!3d55.6473055!4d12.414886

Gött, I, Penker M (2020) Institutions for collective gardening: A comparative analysis of 51 urban community gardens in Anglophone and German-Speaking countries. International Journal of the Commons 14(1)

Hohenberg PM, Lees LH (1995) The making of urban Europe, 1000–1994. Harvard, Cambridge

Jensen N (1996) Kolonihaver Guide: København og Omegn. Copenhagen

King P, Newstead S (2017) Researching play from a playwork perspective. Taylor & Francis, New York

Kolonihaveforbundet (nd). *Kolonihavebevægelsens Historie.* https://www.kolonihaveforbundet.dk/kolonihaveforbundet/om-os/kolonihavebevaegelsens-historie/ Accessed 29 September 2020.
Lawson L (2004) The planner in the garden: A historical view into the relationship between planning and community gardens. J Plan Hist 3(2):151–176
LeGates RT, Stout F (eds) (2015) The City Reader. Routledge.
Lund H (2002) Danmarks Havekunst: 1945–2002. Denmark: Arkitektens Forlag
Mygind E (1963) *Fritidsbyer.Havekunst: nordisk tidsskrift for planlægning af have og landskab* [Garden art: Nordic journal for planning gardens and landscapes], 44: 124–129
Oldenburg R, Brissett D (1982) The Third Place. Qualitative Sociology 5(4):265–284
Parsons KC (1998) Clarence Stein's variations on the garden city theme by Ebenezer Howard. J Am Plann Assoc 64(2):129–130
Petersen JH, Petersen K, Christensen NF (eds) (2011) Dansk velfærdshistorie, vol II. Syddansk Universitetsforlag, Copenhagen
Sidste skrig i kolonihaver: Lagskager. (Last word in allotment gardens: Layer cakes.) (1964, November 7). *B.T.*, Copenhagen
Sturzaker J, Mell I (2016) Green belts: Past; present; future? Taylor & Francis, New York
Sverrild P (2016) Periurban Phase and Sphere. Aalborg University Press, Copenhagen
Tolstrup I (1980) *Kolonihaven*, PhD thesis, Ethnology, University of Copenhagen, Copenhagen
Woudstra J (1995) Danish landscape design in the Modern Era (1920–1970). Gard Hist 23(2):222–241
Yang B (2013) Urban Green Spaces for Quality Life: Case Study: The Landscape Architecture for People in Copenhagen, Frederiksberg: Copenhagen Business School (CBS), Copenhagen

In a Crisis, Re-Start from… Patch! Possible Learnings from P.M.'s Social Ecological Utopia *Bolo*

Silvio Cristiano

Abstract The ongoing social, economic, ecological, and now health crises are not always present in urban and regional planning. Emergencies suggest cities are more fragile than we believe, exhibiting poor sustainability and resilience. Urban dwellers live far away from where livelihoods are produced, have scarce control on their food, depend on waged jobs that can just be lost, and rely on averagely weak community links. In such a context, this chapter digs out the urban–rural implications of a utopia, "bolo", described by Hans Widmer (also known as P.M.) in the aftermath of the 1970s economic crisis, when the first ecological warnings were also emerging. A bolo is a spatial organisation concept that can be applied to either the city or the countryside, suggesting forms and scales that may ease social cohesion and local food production. It is not a proper plan for a given area; it is instead a flexible model, addressing the cultural, social, economic, and geographical aspects of human settlements that supposedly pursue sustainability and resilience. In systems thinking, we can say that a bolo targets our mental models. Its claimed leveraging potential is here explored towards possibly groundbreaking planning and hopefully deeper debates on cities and food production.

Keywords Urban and regional planning in the twenty-first century · Resilience · Systems thinking · Food geographies · Innovative societal forms

---

S. Cristiano (✉)
Department of Environmental Sciences, Informatics and Statistics, Università Ca' Foscari Venezia, Venice, Italy
e-mail: silvio.cristiano@unive.it

Department of Civil, Environmental and Architectural Engineering, Università degli Studi di Padova, Padua, Italy

Interuniversity Department of Regional and Urban Studies and Planning, Politecnico di Torino, Turin, Italy

C. Brisotto and F. Lemes de Oliveira (eds.), *Re-Imagining Resilient Productive Landscapes*, Cities and Nature, https://doi.org/10.1007/978-3-030-90445-6_14

1 Why Going Utopian to Imagine Resilience?

The COVID-19 health emergency adds up to interconnected social, economic, and ecological crises that were already ongoing yet not always effectively present in debates on urban and regional planning (Xue 2021; Cristiano 2020a). At the outbreak of the pandemic, rushes at grocery stores in many cities revealed a fear that food might not be granted even for the more privileged. Such instinct can suggest that urban systems are more fragile than we tend to believe: cities lie far away from where livelihoods are produced, over which urban dwellers have scarce control (Cristiano et al. 2020); instead, urban areas offer pollution able to worsen the effects of old and new diseases, a living usually made out of waged jobs that can just be lost, easy crowding, and averagely few strong community links. Matched with the escalation of climate change effects, such pandemic has persuaded many that the Global Northern approach to the world needs to radically change if we really want to be sustainable and resilient.

A previous mix of crises can be traced back in the 1970s, when a major economic depression (Issawi 1978; Bini et al. 2016) coexisted with the first major contemporary ecological warnings. Among the most prestigious voices, the report "The Limits to Growth" (Meadows et al. 1972) may be cited: by means of systems- and geo-biophysically based simulations, some scenarios were depicted, clearly admonishing humanity that growing economic rhythms would unavoidably lead to resource depletion and increasing pollution in the twenty-first century (and—we might now add—climate change), in turn causing economic collapse, food scarcity, and abruptly decreasing human population. Besides such ecological concerns, economic growth was also associated with some social limits (Hirsch 1977; Georgescu-Roegen 1977): increasing material wealth would accentuate inequalities and competition. Nonetheless, the first (and quite slow) responses by international politics and economics suggested to paradoxically pursue sustainability through further economic growth. This was the paradigm of sustainable development, as formulated by The World Commission on Environment and Development (1987) and adopted on the occasion of the United Nations Conference on Environment and Development, held in Rio de Janeiro in 1992. In the light of the above-mentioned ecological remarks, Georgescu-Roegen defined sustainable development as a toxic recipe (Bonaiuti 2001), Latouche (2004) as an oxymoron, and Springett and Redclift (2019) as a threat to sustainability. Yet today sustainable development still defines the global agendas. If we observe the global trends in material and energy consumption and in greenhouse gases release (see graphs by Krausmann et al. 2016; and by Wiedmann et al. 2020), no real major discontinuity seems to be found from 1992 on: nor in the light of the sustainable development paradigm, nor following other major international actions supposedly inspired by sustainability.

Through a parallel with graphs that are somewhat familiar in pandemic times, it looks like we cannot notice any drop in the infection rate following a lockdown period or control measures on virus spread. 1972's simulations were later repeated, still predicting very similar scenarios (Meadows et al. 2004). In the stream of unlikely growth-oriented sustainability, the Kyoto protocol failed (Rosen 2015), and the Paris agreement was forecast to change nothing and thus fail as well (Spash 2016). More recently, the North American proposal for a Green New Deal was also blamed for prioritising "special interests and money over the safety and wellbeing of the Earth, plants, animals, and ultimately humans" (Miller 2020). Similarly, the new European Green Deal was read as a pretension to keep dominating nature instead of re-imagining regional and global communities to synchronise with the ecosystems of which we are part (Huber 2020). So, we do not seem to be on the right path at all. Nonetheless, this is what lies at the basis of the current international agendas for oxymoronic growth-inspired sustainability and resilience (with the latter borrowed by ecosystems principles; Holling 1973). Criticism for such an oxymoron is frequently addressed by claiming that it would be possible to decouple economic growth and ecological damage, but—after the first ecological studies of the 1970s—this thesis has been recently demolished by Parrique et al. (2019): no empirical evidence is found to support such a decoupling, which is anyway argued as unlikely to arrive even in the future.

Something as vital for us humans as food and agriculture is either directly or indirectly important in all the Sustainable Development Goals (SDGs) for 2030 (United Nations 2015; FAO 2020a); however, facts report we were "off track" to meet them (FAO 2020b) even before the pandemic, with the latter possibly able to worsen the situation (*ibid.*). SDGs also involve cities and communities (goal #11), yet their intrinsic search for economic growth might undermine the desired results. In particular, urban sustainability and resilience were proposed to be biased by an underlying and overarching profit-maximising goal (Cristiano and Gonella 2020). Apparently, after undergoing a series of interconnected crises in the 1970s, the world (at least its dominating portions) embraced the wrong turn, or simply did not leave a dead-end street. Half a century later, in maybe even more uncertain decades, new interconnected crises might be the occasion for exploring another road more wisely. Choices and facts have led us far away from sustainability and resilience. This is the rationale for resorting to an old utopia (P.M., 1983), imagined at a previous crossroads nearly 40 years ago, to now hopefully inspire new choices and ideas to face the rest of this difficult twenty-first century out of the current fragility of our human systems. In recalling and analysing the main concept of such a utopia—i.e. the *bolo*, a sort of loose matrix to seek socio-ecological sustainability in a given human settlement in a given geographical context, a focus is dedicated to the foundations of human life: food, addressed from social ecological, spatial, and political economic geographical views. To better undertake such a transdisciplinary venture, systems thinking will be used to explore the roots of the problem, with a clear focus on food, on urban and rural landscapes, and on the nexus between the patch where agricultural food is grown and its final recipients.

2 Pills of Systems Thinking

Systems thinking is a transdisciplinary epistemological approach. It was originally to contribute to von Bertalanffy (1969) and Forrester (1973), and later refined by several authors, developing different approaches aimed at meeting different scopes, in different fields ranging from social sciences and management to ecological and environmental accounting (Cristiano and Gonella 2020). Systems thinking was used to study the societal organisation of human societies based on the resource upon which they can rely (Odum 1971; 2007) and later applied—through Odum's same approach—to cities and urban planning (Cristiano 2018a; Viglia et al. 2018; Cristiano et al. 2020). In these cases, a system is described by means of the energy systems language (Odum, 1983), adopting stock-and-flow symbols (Sterman 2012). From an epistemological perspective, we can say systems thinking shifts the attention from a linear description of the factual features of a system, addressed as chains of cause-effect processes, to the appraisal of the structure of configuration patterns, linked—through regulating feedback mechanisms—to the stability, resilience, and adaptability of a system (Sterman 1994; Meadows 2008; Cristiano and Gonella 2020). The short systems thinking remarks that are made in the present chapter resort to Odum (1994, 2007) and Meadows (2008). Through the metaphor of an iceberg, the latter (*ibid.*) assigns increasing levels of systemic complexity from the visible peak down to the submerged basis: events (i.e. what is happening), patterns of behaviour (i.e. what trends are present over time), systems structure (i.e. the relations among the parts of a system, or what influences the patterns of behaviour), and—finally—mental models (i.e. the values, assumptions, and beliefs that shape the system). Together with the systemic complexity, the leverage potential increases too, i.e. acting at that level in a system is likely to yield greater consequences with a smaller effort (Meadows 1997).

3 P.M.—A Quite Mysterious Author

The author of the utopia addressed in this chapter can be hardly described as a conventional figure, and surely not as a professional planner. The above-mentioned pamphlet in which the utopia is described—"*bolo'bolo*"—was signed by the pseudonym P.M. (sometimes written in lowercase: p.m.). Now we know this pseudonym belongs to Zürich-based German-speaking author Hans Widmer, to whom more works can be attributed: essays, pamphlets, radio dramas, and theatre performances. Prior events leading to the writing of his utopia had P.M. as part of some squatters' movements in Europe: in a moment of collective despair after suffering from strong repression, the author aimed at providing his comrades with an optimistic vision of a possible socially and ecologically desirable future (Widmer and Schneider 2019). Since its release, the pamphlet has inspired several hands-on neighbourhood projects (*ibid.*). Hans Widmer was himself one of the initiators of some applied projects, at least partly inspired on his utopia *bolo*, mostly through an association called *Neustart Schweiz* (Schneider 2018): a housing venture of these, *Kraftwerk 1*, hosts nearly 250 permanent residents, public areas with workshops, small stores, and meeting places, and—compatibly with local laws—is somehow linked to agricultural lots producing organic vegetables through self-standing cooperatives (*ibid.*).

4 Origin and Essentials of P.M.'s Utopia

The basic elements of P.M.'s utopia are introduced here, as presented in his 1983's pamphlet and in the preface to the second and third editions (P.M. 1993, 2011). *bolo'bolo* depicts a utopia, which may even be applied; as a matter of fact, the author presents possible milestones to reach it. The roots of the interconnected crises of the previous decade are there found in human lives being dominated by the industrial economy and its related societal organisation schemes, to which P.M. (1983) refers to altogether as the Planetary Work and War Machine. Originally written in the three-world era, this utopia stands as an alternative to both socialist and capitalist values, models, and institutions, and to their social and spatial implications in human groups and settlements. In *bolo'bolo*, the relations among humans, goods, and labour are redefined while trying not to repair but to rather break free from such a Machine. Envisaging a future collapse for both the "Western" and the "Socialist" models, P.M. proposes "not to build a better industrial society and to realise the affluent universal socialist consumer family, but to tie direct relations of material exchange between farmers and city-dwellers", not mediated by the big industry and the state (*ibid.*). [Most people might find it odd and bizarre to imagine the end of the industrial era as we know it. Nevertheless, this might be associated in scientific terms with a later invitation to get actually prepared for a global collapse (Odum and Odum 2008) as per the worst forecasts for the present century if no major change is made (Meadows et al. 1972, 2004)—as it is apparently the case. Such a collapse would be related to both the depletion of resources and the negative effects of an endless pressure onto the planet, and can be now explained in nature and systems ecology by resorting to the pulsing paradigm (Odum et al. 1995).]

The criticised economy is defined as "[i]mpersonal, indirect exchange of crystallised life-time" (P.M. 1983), and meant as the "expansion of control by the Machine over its parts, making the parts more and more dependent on the Machine itself" (*ibid.*). Such a dependency might be associated with poor resilience (Cristiano and Gonella 2020). In the preface, justifying some change, the social and ecological dimensions are relevant. On the one hand, among others, concerns are expressed for the inability to feed all humans, for "people being put up against each other", and for accepting to live often displeasing and dissatisfying lives (P.M. 1983); on the other hand, the consequences of "200 years of accelerated industrial progress" are denounced as being able to ultimately swallow humanity: through a metaphor, a hotel terrifying its guests, who are also hosts at the same time (*ibid.*). However, the utopia at issue is not even an invitation to go back to the village, especially whenever this means to live in misery. Nevertheless, if a yet unpleasant collapse is awaiting and imposing radical changes in the world as we know it, then P.M. notes that—when minimum life conditions are met—the least industrialised parts of the world might be more prepared for "a new way of life based on self-sufficiency" and to solve shelter, energy, and food problems locally instead of in difficult large metropolitan areas.

P.M.'s utopia is presented as "a modest proposal"—that "might not be best and the most detailed or certainly a definitive" one—for "a new arrangement of human settlements, economies, and lives after the Machine's disappearance" (*ibid.*). This would be supposedly able to guarantee "a soft landing" into the next era, where the author expects that "[n]obody will starve, freeze, or die earlier than today" (*ibid.*). It is curious that the image of a soft landing is also present in Odum and Odum's call for a "prosperous way down" (2008), opposed to the alternative of a crash (Ulgiati 2004). P.M. invites not to dream of a new world to be created from scratch, but to rather fill in "empty" zones with new elements: "free interstices, abandoned areas, conquered bases", etcetera, which he imagines as places to be converted. The coexistence of "construction" and "deconstruction" practices is offered as part of the intangible prerequisites leading to the utopia at hand.

In a few words, such a utopia presents a non-coercive proposal for the free flourishing of culturally independent forms of socially and spatially organised rural/urban subsistence (*bolo*), framed in inclusive and democratic structures, thought to "function beyond the law of value", whose constituency is imagined to "guarantee systemic stability" (P.M. 2011).

The third edition of the pamphlet (*ibid.*) includes some comments by the author. On the one side, he observes a larger common sense to start imagining a world necessarily or voluntarily beyond economic growth, experimenting with transition towns, cooperatives, climate justice, and producer–consumer cooperation (including Community-Supported Agriculture). On the other side, he recalls more recent proposals—e.g. by Vandana Shiva and Helena Norberg-Hodge (Campbell 2008)—to at least partly re-ruralise the world, and their implications in terms of (in)compatibility with productivist and/or profit-seeking economies and of urban revitalisation.

4.1 The Bolo as a Spontaneous Basic Collective Agreement and as a Context for Life and Production

Introduced above as a utopian form of socially and spatially organised rural/urban subsistence, the *bolo* is meant as an autonomous community corresponding to the anthropological unit of a tribe (300—500 individuals), proposed as the unit for a socially just and ecologically sustainable society: a spontaneous basic agreement among individuals and a "direct, personal context for living, producing, dying" (*ibid.*). Therefore, a *bolo* is not presented as a mere neighbourhood, a self-help network, or a tribe; yet a *bolo*'s size is reported as a recurring one in the history of humanity since back in the societies of gatherers and hunters (Leakey and Lewin 1979, as cited in P.M. 1983). Through speculations and reconstructions, it is right in the history of humans and food that the author retraces some basic evolution of societal structures: after the *homines* in the Old Stone Age, living on game and spontaneous vegetables and not really having a word for "work", the arrival of agriculture is suggested as a first step towards specialised and often dominating social organisms that still guide and condition our contemporary societies; yet the second edition of the pamphlet acknowledges recent researches and reports "no inevitable logics of authoritarian development in agriculture" (P.M. 1993). Back to the basic unit of nearly 500 individuals, this seems to let at the same time a fair diversity to be met, large enough age groups, and every individual to be familiar with one another (P.M. 1983):

> The number of 500 persons seems to be a kind of upper level limit for "spontaneously" functioning larger social organisms. It corresponds to the inhabitants of typical older urban neighbourhoods in a lot of countries, to an infantry batallion, to the capacity of a larger hall, to the size of a medium enterprise, to a medium-sized school, etc. The reasons are not purely genetic or traditional. The number of 500 persons permits a minimal diversity of age, sex, interests, a basic division of work. At the same time, self-organisation is still possible without special organisms, anonymity is not a necessary consequence (you can still know personally all members of the community, but without necessarily being close friends).

Some figures like this are suggested in *bolo*, inspired by previous utopias and actual experiences, yet a remark is made. P.M. observes that utopian theorists—including Thomas More (1971 [1516])—tend to imagine communities in prescriptive ways and from administrative or technical/ecological viewpoints only. The author deplores the fact that most modern utopias would be "totalitarian, mono-cultural models organised around work and education" and that, ironically, "utopian elements have been used for the conception of prisons, hospitals, and in totalitarian regimes" (P.M., 1983). Conversely, the utopian *bolo* rejects compulsory prescriptions in its basic dimensions (e.g. education, work schedule, clothing, sexuality, etc.) and predefined forms of internal organisation, even if these are inspired by or oriented to something theoretically shareable (*ibid.*):

> […] in a *bolo* culturally defined people live together and their motivations are not determined by a compulsory set of moral laws. Each *bolo* is different. Not even a perfectly democratic structure can guarantee the expression and realisation of the desires of the participating persons. This is also a basic flaw of many proposals for self-administration (block councils, neighbourhood-defense committees, soviets, grassroots democracy, etc.), especially if such grassroot organisations are initiated and controlled by state or party organisms. Only cultural identity and diversity can guarantee a certain degree of independence and "democracy". This is not a question of politics.

Starting from its title, P.M.'s utopia introduces a new vocabulary, which is here reported in italics. A *bolo* relies on the concept of subsistence, which is here explored as a source of inspiration for possible new societal metabolisms and their consequences in rural and/or urban food landscapes. Subsistence is aimed at reaching some degrees of self-sufficiency, to which the next sections dedicate a prominent role, and whose two main elements are here anticipated: the buildings and equipment for housing and crafts (*sibi*)—also containing the "supplies of fuel, electricity, and water", and "the production of tools and machineries"—and a piece of land (i.e. the "patch" appearing in the title of the present chapter) for the production of most of its food (*kodu*). The structural and the agricultural dimensions of a *bolo* are connected, with the latter feeding the former, but not necessarily adjacent (Fig. 1). The varying agricultural bases and geographical conditions for the supply of basic food would differentiate the economic practices of a *bolo*: "pastures, mountains, fishing and hunting grounds, palm tree groves, algae cultures, gathering areas, etc." (*ibid.*). In the author's view, a *bolo*'s self-sufficiency can be largely met when talking about the daily supply of basic food and about repairing and maintaining tools and buildings. Extra livelihoods are also considered in order to accommodate and feed additional 30–50 travellers or guests out of a *bolo*'s own resources (*ibid.*), i.e. nearly 10% of its permanent population.

It might now be clear that a *bolo* is not meant as a system, but rather as a patchwork of micro-systems, and that it does not require proper planning from scratch, but rather allows re-using and converting existing structures and environments, according to its own plural cultural features. To the cultural aspects of a *bolo* an entire section of the pamphlet is dedicated. It was already explained that P.M.'s utopia is not only about neighbourhood planning and practical arrangements. Similarly, it is not primarily aimed at defining a way to prevent and/or survive a collapse. Indeed, behind external, technical aspects, the motivation for many individuals to gather and live together is

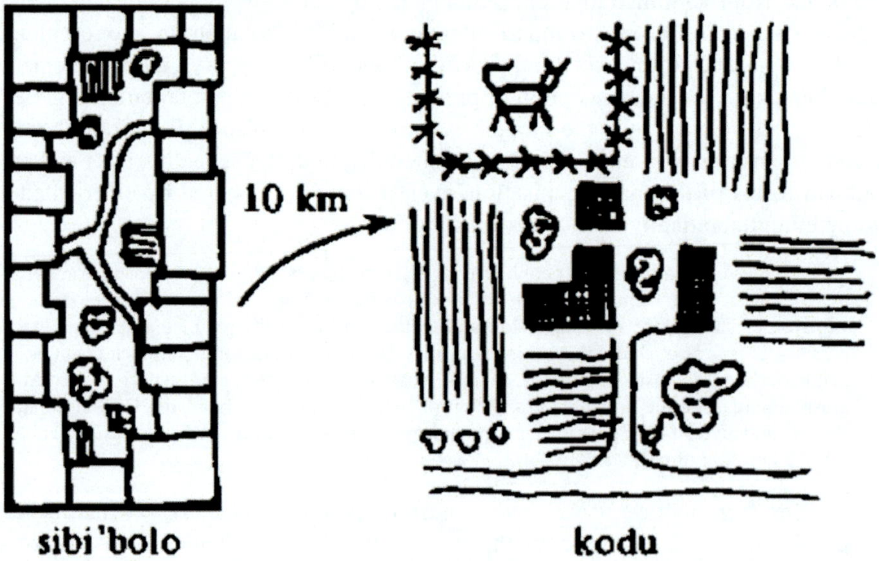

Fig. 1 The two dimensions of a *bolo*: the built environment with equipment (*sibi*) and the food production areas (*kodu*) (P.M. 1983, anti-copyright)

mostly intangible: i.e. the community relies on some common cultural background (*nima*), anyway not meaning a homogeneous identity (*ibid.*):

> The *bolos* are a framework for the living-up of all kinds of life styles, philosophies, traditions and passions. *bolo'bolo* is not a life style in itself, but only a flexible system of limits (biological, technical, energical, etc.). As for the knowledge of such limits, ecological and alternativist materials can be quite helpful, but they should never serve to determine the content of the different life styles.

Individuals all have their own beliefs and visions of life, but some aspects make them like-minded (*ibid.*):

> The nima contains habits, lifestyle, philosophy, values, interests, clothing styles, cuisine, manners, sexual behaviour, education, religion, architecture, crafts, arts, colours, rituals, music, dance, mythology, body-painting […]. The nima defines life […] in its practical everyday form.

Together, a *nima* can be lived, transformed, and completed, but absolutely not defined a priori by the utopia at hand (*ibid.*). An ever-changing culture looks consistent with its systemic description (Cristiano 2018b). P.M. (1983) imagines that individuals whose cultures exclude social forms (e.g. hermits, misanthropists, individual anarchists, etc.) would still find their way to live by themselves in the free spaces among *bolos*: after all, not everyone necessarily wants to live in a society, and living in a utopian community would surely not be compulsory (*ibid.*).

It is in this context that the feeding potentials and the food-oriented social and spatial features of a *bolo* were originally sketched by P.M., and that are presented, re-elaborated, and revised in the next section of the present chapter, based on the guiding thread of the whole book of which this is part.

4.2 Realistic Food and Utopian Foodscapes

As anticipated, one of the two elements upon which the self-sufficiency of a *bolo* lies is represented by some land adequately large to meet most of the community's food requirements. The pamphlet addresses these aspects through the concept of *kodu*. Relocalising food production closer to the final recipients can be matched with the notion of resilience (Cristiano 2020a; Cristiano and Gonella 2020): an increased control over one's livelihoods would strengthen a system and let it resist and react more easily to external shocks. *bolo'bolo*'s utopian approach to food has both cultural aspects and spatial implications. The type and amount of land required by a *bolo* depends on the features of its surrounding environment. This brings P.M.'s utopia closer to discourses regarding possible social and ecological transitions or transformations, warning that new desired settlements cannot prescind from what already exists and from its peculiarities—in other words, that geography matters (Kraehmer 2019).

When it comes to quantifying the required size of land to averagely feed one person, P.M. resorts to the then most recent data by United Nation's Food and Agriculture Organisation, as elaborated by utopian architect and planner Yona Friedman

(1982), and additional data by Seymour (1976) and Lappé et al. (1977): these figures vary between hundreds and thousands of square metres per capita. This is now close to the global average arable land available per capita, i.e. 2,100 m^2 (FAO 2020c).

The concept of *kodu* also is seen as "the agricultural basis of the *bolo*'s self-sufficiency and interdependence" and the cropping types and methods vary according to the cultural background of each *bolo* (P.M. 1983):

> Agriculture is part of a *bolo*'s general culture. It defines its way of dealing with nature and food. Its organisation cannot then be described on a general level.

However, new skills might be required to improve current agricultural practices (*ibid.*):

> New knowledge in the field of biodynamic methods and the intensive combination of different factors (crops + animals, animals + methane production, alternation of crops, etc.) is indispensable for a new start.

P.M. (*ibid.*) suggests that a *bolo* reaches high fractions of self-sufficiency in its food requirement supply. However—he adds—self-sufficiency does not necessarily imply self-restraint or isolation: agreements can be made to exchange food and other products and services among different *bolos*, within a de-centralised spirit complying with the involved cultural features of the parties (*ibid.*). Independently of a *bolo*'s own culture, every individual is imagined as getting a minimum of know-how of cropping skills, since this might improve the *bolo*'s overall independence (and, in the light of the remarks above, we might add resilience). This is admitted as a minimum level of autarchy needed "to enter into a network of exchange without being exploited" (*ibid.*).

Such an approach to agriculture is part of *kodu*'s aim to abolish the separation between producers and consumers "in the most important domain of life: the production of food"; indeed, each inhabitant of a *bolo* is imagined to have access to at least 2,000 kcal a day.

Without a certain level of food independence, P.M. admonishes that a *bolo* would be "exposed to blackmailing", for "guarantees", "agreements", and "rights" alone do not ensure anything (*ibid.*); from a systems thinking perspective, one might observe that they represent information inputs that are actually not automatically associated with any flow of energy or resource.

In principle, the size of a *bolo* can be comparable throughout the world, i.e. hosting nearly 500 people and extending upon a land surface able to feed them all (with the aforesaid 10% more for their guests). Conversely, P.M. underlines that the territorial, cultural, architectural, organisational, and other values (if any) may be many (*ibid.*).

As mentioned above, the utopia at hand is not about designing new settlements. It is rather about re-using existing structures. An urban *bolo* in presently large cities may be composed of a couple of blocks, a small neighbourhood, or a series of adjoining buildings (*ibid.*):

> You just have to build connecting arcades, overpasses, using first floors as communal spaces, making openings in certain walls, etc. So, a typical older neighborhood could be transformed into a *bolo* like this: Larger and higher housing projects can be used as vertical *bolos*.

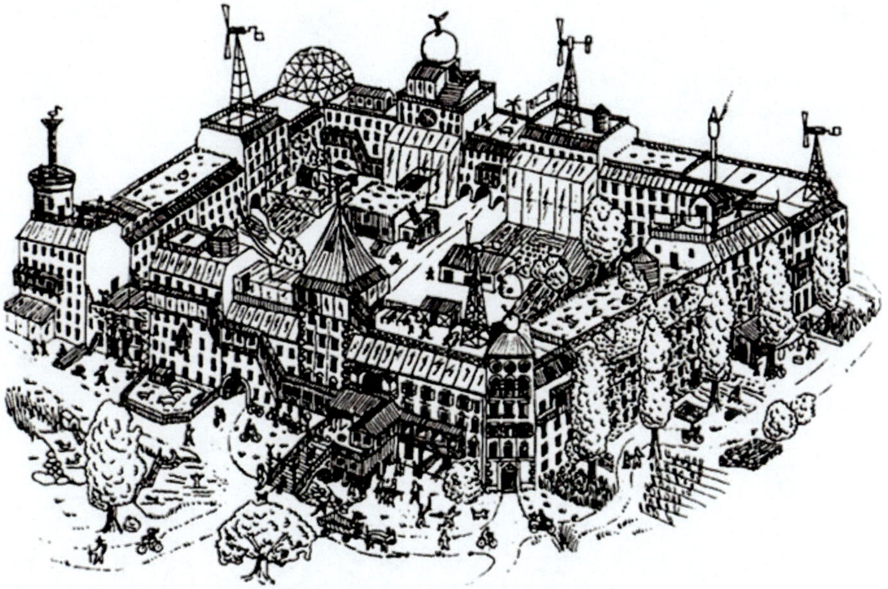

Fig. 2 An imaginary urban *bolo* resulting from the re-use of an existing block (P.M. 1983, anti-copyright)

An example of an urban *bolo*—later become iconic of the whole pamphlet—is offered in Fig. 2.

A rural *bolo* may instead correspond "to a small town, to a group of farmhouses, to a valley", and not necessarily be morphologically homogeneous (*ibid.*):

> In the South Pacific, a *bolo* is a coral island, or even a group of smaller atolls. In the desert, the *bolo* might not even have a precise location; rather, it is the route of the nomads who belong to it (maybe all members of the *bolo* meet only once or twice a year). On rivers or lakes, *bolos* can be formed with boats. There can be *bolos* in former factory buildings, palaces, caves, battleships, monasteries, under the ends of the Brooklyn Bridge, in museums, zoos, at Knotts Berry Farm or Fort Benning, in the Iowa Statehouse, shopping malls, the University of Michigan football stadium, Folsom Prison. The *bolos* will build their nests everywhere, the only general features are their size and functions.

In the pamphlet, such diversity in the possible shapes of a *bolo* is shown in Fig. 3. Based on the cultural features of a *bolo*, subdivisions may exist (*kanas*), i.e. groups comprising up to 15–20 members: these social bodies would more easily and immediately be able to live together, but would be not large enough to be independent for their food and goods requirements (*ibid.*).

It might be interestingly noted that the spatial and the societal organisation of P.M.'s utopian human settlements depend upon the local and regional features of the context in which these are inserted.

Although the idea of a *bolo* starts from the re-use of existing built environments, the intensity of the latter is imagined as thinned out, and the division between the rural and the urban is thought as less pronounced (*ibid.*):

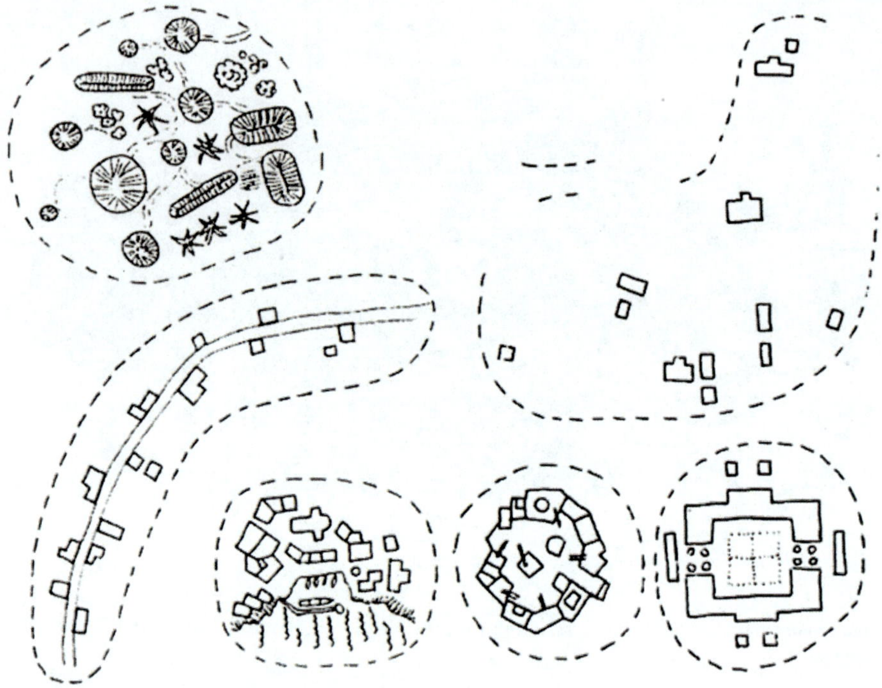

Fig. 3 Examples of morphologically varied *bolos*: urban, rural, compact, scattered (P.M. 1983, anti-copyright)

> The conflict of interest between farmers struggling for high prices and consumers demanding cheap food no longer exists. Moreover, nobody can be interested in waste, artificial shortages, deterioration, maldistribution, or planned obsolesence of agricultural products. Everybody is directly interested in the production of qualitatively good and healthy food, because they produce and eat it themselves.

This results in the protection from land consumption and fertile soil depletion, as it can be deduced from the following passages (*ibid.*):

> Careful treatment of the soil, the animals and themselves becomes self evident, for every *bolo* is interested in long-term fertility and the preservation of resources. The use of land or other resources and their distribution among *bolos* must be discussed and adapted carefully.

The way to reach land conservation varies. In pure countryside *bolos*, the surrounding land is likely used for farming activities. As to *bolos* in larger cities, small vegetable gardens are suggested "around the houses, on roofs, in courtyards, etc." (*ibid.*).

Concerning the shape and geographical organisation of an urban *bolo*, yet developing in a thinned out city, what we could mean as P.M.'s inspiring draft for general food-resilient planning (Fig. 4) builds upon the "three-zone model" by German urban ecologist Merete Mattern (*ibid.*):

Fig. 4 Mattern's "three-zone model" as applied to the spatial planning of a *bolo* (P.M. 1983, anti-copyright). The inner circle is dedicated to the built living spaces, with small food production such as herbs; the intermediate ring is farmed with vegetables for daily or weekly use; and the outer zone is dedicated to seasonal crops and to woods

> Around the city there would be a garden zone, where every *bolo* would have a larger plot for vegetables, fruits, fish ponds, etc., i.e., for produce that is needed fresh almost every day. These gardens could be reached by foot or bicycle within minutes, and the quantities needing special transport would be relatively low. The real agricultural zone, larger farms of up to 80 hectares (200 acres) or several farms of smaller size, could be about 15 kms or so from the city-*bolo*. (Particularly in the case of certain cultures using lakes, peaks, vineyards, hunting grounds, etc.) These *bolo*-farms would specialise in large-scale production of durable foods: cereals, potatoes, soya, diary products, meat, etc. Transportation would be on the scale of tons (by chariot, trucks, boat, etc.). For the *kodu* of larger cities, a system of three zones could be practical.

Based on such a model, P.M. reports[1] that (*ibid.*):

> A 15-km large agricultural zone could feed such a large city as Munich. For this purpose, she proposes two wood zones (for a good micro-climate) and an intensive compost system. This means that agricultural self-reliance is also possible in densely populated areas. But this would imply that every square foot is used, and that there be no space for waste, experimentation or parks. A more flexible system of three zones and additional farms would be more practical, as distance, required freshness and harvest-cycles could be optimally combined.

In other words, crops requiring seasonal tillage (e.g. cereals or legumes) would be placed further from the final recipients than small cultivation (e.g. daily or weekly plants such as herbs and other ingredients to be eaten fresh). The design of the food landscapes to supply the areas dedicated to other human activities follows the principle of transport minimisation, i.e. reducing both time and energy. Labour intensity, local varieties, and appropriate methods (e.g. soil protection, alternation, crop

[1] To the best of the author's knowledge, no English manuscript by Merete Mattern is available on this topic.

combinations, etc.) would not damage productivity compared to current agri-business (*ibid.*). Together, such wise and local cropping would result in increased sustainability when much less importation of oil-based fuels and petrochemical fertilisers is needed.

Although urban *bolos* are envisioned as plausible, all of these requirements would imply cities to be resized. In *bolo'bolo*, the above-mentioned thinning out would consist of "larger cities with more than 200,000 inhabitants" pursuing some degree of depopulation, which—in some cases—would let deserted or underpopulated villages to revive.

One last description of the utopia at issue regards the cuisine and dietary habits (*yalu*). As to this aspect, P.M. seems to implicitly invite to be realistic when it comes to imagine what food to desire and produce when trying to be local, sustainable, and resilient. Vegetable calories and proteins are recommended since animal production requires much larger amounts of the same crops that would provide humans with those calories and proteins (*ibid.*). However, the idea is not to necessarily go vegan: farmyard animals are imagined to be mostly running free in courtyards or alike, and fed with food scraps instead of virgin vegetables. Neither monotony is sought since fantasy in recipes might make up for the lack of (still less and less tasty) exotic ingredients that are instead familiar today.

4.3 Possible Learnings Towards Groundbreaking Planning and Resilient Food Supplies

In the light of all of the above, possible learnings can be drawn to enrich current debates around urban agriculture, resilient food systems, and the planning of resilient rural and/or urban systems. From utopia to practice, prospective connections and adaptations to geographically varied existing environments and culturally heterogeneous communities look at the very least recommendable when almost none of the first ecological warnings from the 1970s seems being seriously, systemically taken into account. A challenging century such as the twenty-first has been already showing by means of interconnected crises that abrupt changes are possible—and not always desirable—and that a dependency upon imported goods and services risks to mine even the basics of human requirements.

Starting from the aftermaths of another period of crises, another path is suggested in *bolo'bolo*, in some way addressing both the social and ecological dimensions of sustainability and resilience, and with a strong focus on one pivotal need: food. With stunning flexibility to local social and ecological contexts, P.M.'s utopia is seen here as able to sketch and inspire a possible path for a conscious and durable transformation towards food-resilient landscapes and human settlements in such a century. Therefore, contrary to other more theoretical and dreamy utopias, the one at issue is quite rooted in reality and in major global issues that are likely to become more evident in the future.

Almost forty years after its first edition, *bolo'bolo* seems still relevant. First of all, it deals with the "zero hunger" Sustainable Development Goal #2 (United Nations 2015) that is currently postponed to 2030, admittedly with poor chances to be met, as seen above. Moreover, it deals with the "zero soil consumption", not directly addressed by SDGs—but somehow compatible with goal #15, "life on earth"—and attracting an ever-increasing attention in ecological warnings, geographical studies, urban and regional planning, and public debates. The utopia at hand also pioneeringly deals with sustainable cities and communities (goal #11); it does this from an integrated ecological and social point of view, with a look at resilience too (as far as the provision of basic requirements is met within each settlement), and with an open differentiated path to be chosen locally.

The main reasons for long relegating the idea of *bolo* to the realm of outdated utopias—at least in a mainstream perception—may be found in the following. On the one side, even though when it comes to food in hard times this can be seen as a great goal and achievement, seeking sufficiency is mostly incompatible with a widespread global cultural flattening leading to desire abundance instead; on the other side, the idea of a world of *bolos* still stands as an alternative strategy to the "winning" one, made of overall globalised markets, economic demand, and industrial provision. If the ecological and social ambitions of P.M.'s pamphlet seem genuine and forward-looking, the utopia seems not compatible with the prevailing economic paradigm seeking—among other things—an endless expansion. Anyway, the limits to growth as defined in the 1970s have been confirmed earlier in the present century, and the intrinsic unsustainability of the current global agenda was and is still denounced by many, as partially recalled earlier in this chapter. Ecological and economic projections, matched with facts and current strategies, suggest that a need will come to reconsider the founding drivers of *bolo'bolo*. The economic incompatibility has so far acted as a barrier for not even minimally considering to elevate the *bolo* not necessarily to the dignity of a proper planning model, but at least to a source of inspiration. In the light of an uncertain close future, such a barrier might represent instead the greatest potential for change—in systems thinking words, a leverage point (Meadows 1997). As a matter of fact, systems teach us that their goal is one of their main drivers. If we seek profit, we are unlikely to find sustainability and resilience (Cristiano and Gonella 2020); so we are if we pursue endless technical progress, as in a 19th century's practical utopist like Robert Owen, as recalled by Benevolo (2019). Conversely, if *bolo'bolo* argues about how to find food resilience, ecological sustainability, and personal and social wellbeing, then the inspiration for change might be great. While this chapter is being written, the world is suffering from a pandemic, in which health protection is pursued as a compromise with economic stability, and human deaths are increasing day after day, thus providing some food for thought about what the currently prevailing systemic goal is.

Back to P.M.'s utopia, the concept of *kodu* (subsistence) can be seen as a source of inspiration for possible new and lighter societal metabolisms and for their consequences in rural and/or urban food landscapes. If our planet is finite and so is its fertile soil, this seems as a promising point to be addressed: in the light of the scarce arable land available per capita on planet Earth, human settlements ought not to consume further areas and, in parallel, human requirements should shift from an ever-growing demand, often not linked to any biophysical reality, to keep in mind what the actual possibility to get something is, granted that a minimum of 2,000 kcal a day per person ought to be ensured, as recalled by the very P.M., and—also very importantly—decommodified. Speaking of optimisation of the remaining requirements that are compatible with the challenges of a difficult century, more pioneering proposals can be found: one is represented by the idea of reducing animal proteins, replacing them with vegetal ones, in the light of the now renown massively larger impact of the former; again as a complementary action to a minor consumption of animal products (so not necessarily to go vegan), another proposal regards a really "circular economy" in animal breeding, i.e. out of intensive breeding plants but re-using everyday scraps. Ethical concerns now affect the production of biogas from crops; in closer and closer times of scarcity and more pronounced ecological impacts, similar ethical concerns may soon arise from wasting valuable food in order to less efficiently produce animal food. (All of this leaving apart the still highly respectable vegetarian and vegan ethical concerns—only because they are not part of P.M.' utopia.)

Animals and plants for human nutrition are central in *bolo'bolo* under multiple aspects, starting from the organisation of space. They are all intertwined with other functions of the cities and the villages. As to cities and villages, P.M. notes that resizing and thinning out cities would let villages revive. This might add some thoughts to the increasing discourse about internal and marginal areas (see e.g. De Rossi 2019), whose differences compared to urban systems would then become less pronounced: from marginality, a *bolo* might lead to polycentrism (Kloosterman and Musterd 2001). The organisation of space so as to allow for walking and riding bicycles also pioneeringly anticipates planning discourses on car-free districts and cities, or at least partial (e.g. in Barcelona's Example; Rueda 2019) or temporary measures (e.g. in New York City in pandemic times; NYC DOT, 2020). The urban design aimed at letting its inhabitants reach more easily (within minutes) what they need most frequently, and still easily—on their feet—what they need less frequently (up to a quarter of hour) seems already going much beyond the very recent proposals for 15-min cities (Meng 2017; Moreno et al. 2021; Pozoukidou and Chatziyiannaki 2021)—where most services can be reached in such a time interval—thus representing some possibly useful insights to enrich such current planning ideas.

The learnings suggested above confirm P.M.'s explicit statement that the utopia he proposes in his pamphlet is not just about spatial organisation. The allocation of scarce resources and the interactions between humans and territories pass through the rethinking of the relations among individuals, among communities, and between humans and nature. This passage marks a clear discontinuity in both social and ecological attitudes compared to the main policies and trends of the end of an era of growth and expansion: indeed, it is rather oriented towards a re-organisation able to be more compatible with those completely different scenarios that are far from being unlikely in the twenty-first century. Therefore, the matter can no longer be about how current habits are *willing* to change, but rather about how they *will* change when no other option is possible; Odum and Odum (2006; 2008) warn us that a collapse of civilisation is at stake, but Odum (2007) also maintains that humanity has always found some ways to develop cultures able to program its resource use, providing "appropriate morality to adapt people to new conditions and times of descent" (*ibid.*). Besides acknowledging the still relevant projections of the report "The Limits to Growth" (Meadows 1972, 2004), and compatible with the systems ecology concept of descent (Odum and Odum 2008) in the pulsing paradigm (Odum et al. 1995), *bolo'bolo* may stand as a model for inspiration while getting close to a brand-new era. Consciously or not, the utopia shows features that have been recognised as pertinent to such an epoch; if abundant resources generally push humans and other animals to exhibit some degrees of competition, scarcity is the time when collaboration is more frequent instead: as a matter of fact, it is systemically more efficient (Odum 1973; Odum and Odum 2006). Collaboration to deal with scarcity seems to be right the case in P.M.'s *bolo'bolo*, and the very concept of a *bolo* can be seen as a flexible, spontaneous, culturally-differentiated approach to organise resources in times of scarcity, as envisioned by Odum (2007).

In the light of such a diversity, complexity, and anticipatory attitude, the open and flexible utopia at hand might serve now as a source of inspiration, not involving just urban and regional planning, but also—among others—political philosophy, sociology, economic political geography, and political economy in order to more comprehensively and transdisciplinarily get prepared for a new era to come.

References

Benevolo L (2019) [1963] Le origini dell'urbanistica moderna. Edizioni Laterza

Bini E, Garavini G, Romero F (eds) (2016) Oil shock: The 1973 crisis and its economic legacy. Bloomsbury Publishing

Bonaiuti M (2001) La Teoria bioeconomica. La «nuova economia» di Nicholas Georgescu-Roegen. Carocci

Campbell K (2008) Women, Mother Earth and the business of living. In: Steyaert C, Hjorth D (eds) Entrepreneurship as social change: a third new movements in entrepreneurship book, vol 3. Edward Elgar Publishing

Cristiano S (2018a) L'approccio sistemico eMergetico. Prospettive per una valutazione integrata della sostenibilità di progetti civili e piani urbani. RIV Rassegna Italiana di Valutazione. https://doi.org/10.3280/RIV2018-071008

Cristiano S (2018b) Systemic thoughts on ecology, society, and labour. In: Through the working class ecology and society investigated through the lens of labour. Edizioni Ca'Foscari, Venice, pp 9–23. https://edizionicafoscari.unive.it/en/edizioni4/libri/978-88-6969-297-0

Cristiano S (2020a) Sviluppo urbano sostenibile? Di ecologia, economia politica e città post-crescita. Territorio 94:183–186. https://doi.org/10.3280/TR2020-094020

Cristiano S (2020b) Tavola vuota? Tabula rasa! Per un ripensamento urgente della forma e dei rifornimenti della città in una stagione di crisi e vulnerabilità: la priorità del cibo. Working papers, 1/2020 (Urban@it - Centro Nazionale di Studi per le Politiche Urbane) https://www.urbanit.it/wpcontent/uploads/2020/07/BP_Cristiano-02.pdf

Cristiano S, Gonella F (2020) 'Kill Venice': a systems thinking conceptualisation of urban life, economy, and resilience in tourist cities. Humanit Soc Sci Commun 7(1):1–13. https://doi.org/10.1057/s41599-020-00640-6

Cristiano S, Zucaro A, Liu G, Ulgiati S, & Gonella F (2020) On the systemic features of urban systems. A look at material flows and cultural dimensions to address post-growth resilience and sustainability. Front Sustain Cities 2:12. https://doi.org/10.3389/frsc.2020.00012

De Rossi A (2019) Riabitare l'Italia: le aree interne tra abbandoni e riconquiste. Donzelli editore

FAO—Food and Agriculture Organization of the United Nations (2020a) Sustainable development goals. http://www.fao.org/sustainable-development-goals/en/ (accessed November 2020).

FAO—Food and Agriculture Organization of the United Nations (2020b) Tracking Progress on Food and Agriculture-Related SDG Indicators 2020. A Report on the Indicators under FAO Custodianship.

FAO—Food and Agriculture Organization of the United Nations (2020c) Land use in agriculture by the numbers. Sustainable Food and Agriculture. http://www.fao.org/sustainability/success-stories/detail/en/c/1295695/ (accessed January 2021)

Forrester JW (1973) Principles of systems: text and workbook. MIT Press

Friedman Y (1982) Alternatives énergétiques. Editions Dangles

Georgescu-Roegen N (1977) Inequality, limits and growth from a bioeconomic viewpoint. Rev Soc Econ 35(3):361–375. https://doi.org/10.1080/00346767700000041

Hirsch F (1977) The social limits to growth. RKP

Holling CS (1973) Resilience and stability of ecological systems. Annu Rev Ecol Syst 4(1):1–23

Huber D (2020) New European Commission's green deal and geopolitical language: A critique from a decentring perspective. Istituto Affari Internazionali (IAI). https://www.jstor.org/stable/resrep25276

Issawi C (1978) The 1973 oil crisis and after. Journal of Post Keynesian Economics 1(2):3–26

Kloosterman RC, Musterd S (2001) The polycentric urban region: towards a research agenda. Urban Stud 38(4):623–633

Kraehmer K (2019) Geography matters: Ideas for a degrowth spatial planning paradigm–on Xue and Vansintjan II. In: Nelson A, Schneider F (eds) Housing for degrowth: Principles, models, challenges and opportunities. Routledge

Krausmann F, Schaffartzik A, Mayer A, Eisenmenger N, Gingrich S, Haberl H, Fischer-Kowalski M (2016) Long-term trends in global material and energy use. In: Haberl H, Fischer-Kowalski M, Krausmann F, Winiwarter V (eds) Social ecology. Springer, Cham, pp 199–216

Lappé FM, Collins J, Fowler C (1977) Food first. Beyond the myth of scarcity. Houghton Mifflin Co.

Latouche S (2004) Survivre au développement: De la décolonisation de l'imaginaire économique à la construction d'une société alternative. Fayard/Mille et une nuits

Leakey E, Lewin R (1979) People of the Lake: Mankind and its Beginnings. Avon

Meadows D (1997) Places to Intervene in a System. Whole Earth 91(1):78–84

Meadows DH (2008) Thinking in systems: A primer. Chelsea Green Publishing

Meadows DH, Meadows DL, Randers J, Behrens WW (1972) The Limits to Growth

Meadows D, Randers J, Meadows D (2004) Limits to growth: The 30-year update. Chelsea Green Publishing

Meng LI (2017) The planning strategies of a 15-minute community life circle based on behaviors of residents. In: Urban Planning Forum, vol 233, pp 111–118. https://en.cnki.com.cn/Article_en/CJFDTotal-CXGH201701016.htm

Miller C (2020) An Ideological Criticism of The Green New Deal and Media Responses. UNLV Theses, Dissertations, Professional Papers, and Capstones. 3929. https://doi.org/10.34917/19412132

More T (1971) [1516] Utopia. Washington Square Press

Moreno C, Allam Z, Chabaud D, Gall C, Pratlong F (2021) Introducing the "15-Minute City": Sustainability, Resilience and Place Identity in Future Post-Pandemic Cities. Smart Cities 4(1):93–111. https://doi.org/10.3390/smartcities4010006

NYC DOT—New York City Department of Transportation (2020). Pedestrians. Open Streets. Steets Opened for Social Distancing. https://www1.nyc.gov/html/dot/html/pedestrians/openstreets.shtml (retrieved January 2021).

Odum HT (1971) Environment, power, and society. John Wiley and Sons
Odum HT (1973) Energy, ecology, and economics. Ambio, 220–227
Odum HT (1983) Systems Ecology; an introduction
Odum HT (1994) Ecological and general systems: an introduction to systems ecology. University Press of Colorado
Odum HT (2007) Environment, power, and society for the twenty-first century: the hierarchy of energy. Columbia University Press
Odum HT, Odum EC (2006) The prosperous way down. Energy 31(1):21–32
Odum HT, Odum EC (2008) A prosperous way down: principles and policies. University Press of Colorado.
Odum WE, Odum EP, Odum HT (1995) Nature's pulsing paradigm. Estuaries 18(4):547–555. https://doi.org/10.2307/1352375
P.M. (1983) *bolo'bolo*. Paranoia City Verlag. Zurich. [here cited the edition by The Anarchist Library]
P.M. (1993) *bolo'bolo*, 2nd, English. Paranoia City Verlag, Zurich
P.M. (2011) *bolo'bolo*, 3rd, English. Paranoia City Verlag, Zurich
Parrique T, Barth J, Briens F, Kerschner C, Kraus-Polk A, Kuokkanen A, & Spangenberg JH (2019) Decoupling debunked: Evidence and arguments against green growth as a sole strategy for sustainability. European Environmental Bureau.
Pozoukidou G, Chatziyiannaki Z (2021) 15-Minute City: Decomposing the New Urban Planning Eutopia. Sustainability 13(2):928. https://doi.org/10.3390/su13020928
Rosen AM (2015) The wrong solution at the right time: The failure of the kyoto protocol on climate change. Polit Policy 43(1):30–58. https://doi.org/10.1111/polp.12105
Rueda S (2019) Superblocks for the design of new cities and renovation of existing ones: Barcelona's case. In: Integrating human health into urban and transport planning. Springer, Cham, pp 135–153
Schneider (2018) 15 November 2018. Zurich. https://housing.degrowth.net/zurich/
Seymour J (1976) The complete book of self-sufficiency. Dorling Kindersley.
Spash CL (2016) This changes nothing: The Paris Agreement to ignore reality. Globalizations 13(6):928–933. https://doi.org/10.1080/14747731.2016.1161119
Springett D, Redclift M (2019) Sustainable development. In: Davoudi S, Cowell R, White I, Blanco I (eds) The Routledge companion to environmental planning. Routledge.
Sterman JD (1994) Learning in and about complex systems. Syst Dyn Rev 10(2–3):291–330. https://doi.org/10.1002/sdr.4260100214
Sterman JD (2012) Sustaining sustainability: creating a systems science in a fragmented academy and polarized world. In: Weinstein M, Turner R (eds) Sustainability science. Springer, New York, NY, pp 21–58. https://doi.org/10.1007/978-1-4614-3188-6_2
The World Commission on Environment and Development (1987) Our common future. United Nations. https://archive.org/details/ourcommonfuture00worl/page/n3/mode/2up
Ulgiati S (2004) HT Odum and EC Odum, the prosperous way down. Ecol Model 178(1–2):247–250. https://doi.org/10.1016/j.ecolmodel.2003.12.033
United Nations (2015) Transforming our world: the 2030 agenda for sustainable development. https://sustainabledevelopment.un.org/post2015/
Xue J (2021) Urban planning and degrowth: a missing dialogue. Local Environ. https://doi.org/10.1080/13549839.2020.1867840
Viglia S, Civitillo DF, Cacciapuoti G, Ulgiati S (2018) Indicators of environmental loading and sustainability of urban systems. An Emergy-Based Environmental Footprint. Ecol Indic 94:82–99. https://doi.org/10.1016/j.ecolind.2017.03.060
von Bertalanffy L (1969) General system theory: Foundations, development, applications. George Braziller, New York, NY

Widmer H, Schneider F (2019) Neighbourhoods as the basic model of the global commons. In: Nelson A, Schneider F (eds) Housing for degrowth: Principles, models, challenges and opportunities. Routledge

Wiedmann T, Lenzen M, Keyßer LT, Steinberger JK (2020) Scientists' Warning on Affluence. Nat Commun 11(1):1–10

Thinking Utopia: A Resilient Approach to Productive Landscapes by Yona Friedman

Carla Brisotto

Abstract The Hungarian architect Yona Friedman (1923–2020) founded the collective Groupe d'études de Architecture Mobile (GEAM) with other European architects in 1956. The group's manifesto advocated for spatial mobility and a flexible city. Although literature describes Friedman as the theorist of a social and scientific approach to spatial design, this chapter unfolds a new perspective of his theories from the lens of their urban agriculture components. This chapter delivers an in-depth narrative about his projects through archival research. The author identified three aspects of urban agriculture in Friedman's philosophy: mobile agriculture as a means of non-urbanization, self-sufficiency as means of food security, and non-paternalism as means of pragmatic utopia. These findings suggest the design of a food system where strategic regional land use and tactical approaches to self-sustenance co-exist. The regional scale is a flexible model to operate land for a certain number of urban settlements. The individual scale employs a series of tactics to guarantee survival and contributes to the flexibility of the regional system. Strategies (long-term plans) and tactics (short-term plans) are part of the same vision, building a food production system that can be efficient and reliable.

Keywords Productive urban landscapes · Yona Friedman · Food self-sustenance · Resilient thinking · Pragmatic Utopia

1 The Inevitable Return of Utopia

French Hungarian architect and theorist Yona Friedman is among the most influential thinkers of the nineteenth century. His conceptual work on the redefinition of city design did not only foresee the struggles between "individual freedom and institutional constraints" (Busbea 2007: 67) but it also foresaw contemporary issues, such as the necessity to work from home (Friedman and Decavèle 2011: 1:19), the abuse of land development, and the urgency to stop agricultural sovereignty over individual

C. Brisotto (✉)
Florida Institute for Built Environment Resilience (FIBER), University of Florida, Gainesville, FL, USA
e-mail: c.brisotto@ufl.edu

C. Brisotto and F. Lemes de Oliveira (eds.), *Re-Imagining Resilient Productive Landscapes*, Cities and Nature, https://doi.org/10.1007/978-3-030-90445-6_15

self-sustenance. He accomplished this achievement by patiently drawing thousands of vignettes, plans, and illustrations over his long and prolific career until he passed away in February 2020. His ability to capture the problems of an era with the stroke of a few lines and to turn them into possibilities for more equitable living conditions made him a precursor of resilient thinking, the predisposition to adapt to changes (Walker and Salt 2006).

Architectural and planning historical literature has covered Friedman, mainly focusing on mobile architecture and technological principles. Scholars like Hans H. Obrist, Manuel Orazi, Sabine Lebesque, and Helene van Fenteber van Vlissingen collaborated with Friedman to publish monographs and oeuvre collections. Other researchers interpreted his technocratic participatory techniques in anticipation of software gaming (Pearson 2016: 76; Veloso 2014: 386). They viewed the current return to Friedman's philosophies as anarchist-scientism's exploitation of his theories (Roche 2020). They understood his flexible design as a strategy to undermine the figure of the architect/planner from the design process (Lockard 2017; Rouillard 2018).

Above all, the notion of utopia in relation to his projects is a recurrent theme across academic papers (Pearson 2016; Roche 2020; Ferri 2015; de Wit 2021; Rouillard 2018). As Italian architectural critic, Marco de Michelis states, Friedman's utopic projects have to be understood within the return of postwar utopia. At that time, the hegemonic modernist approach to architecture was challenged by figures like Giancarlo De Carlo in Italy and Alison and Peter Smithson in the United Kingdom. Ideas that developed when "a new synthesis between science, technology, and artistic practices [put at its center] humanity with its needs and its desires" brought everyday life with all its pluralities to the center of the architectural discourse (de Michelis 2008: 37 in Friedman et al. 2008). Friedman's approach to utopia is, therefore, a reaction to years of architectural autocracy. It is not a coincidence that he founded the Groupe d'Etudes d'Architecture Mobile (GEAM) after failing to align with the modernist group CIAM (de Wit 2021: 191). In this light, his utopia has to be considered a call to collectively imagine a possible different way to design the city as a vision in which everyone could contribute.

Among Friedman's numerous projects, what stands out is the recurrent topic of the productive landscape—how to grow food in the city and conceive a more accessible food system. Literature has not thoroughly and comprehensively investigated this aspect yet; however, it is vital to understand the role of this facet of Friedman's design. If utopia shows a possible world, it is essential to explore why the food question was part of his philosophy and approach to city design. This chapter aims to answer this question thereby shedding new light on several of his plans and concepts such as the *Spatial Settlements* (1959), the *Continental City* (1963–1993), the *Manuals* (1989) for vulnerable populations.

The chapter employs primary sources and microhistory analysis developed by the Italian School of historians Carlo Ginzburg and Edoardo Grendi.[1] This method

[1] Microhistory is typically recognized as an Italian method. The most cited study on microhistory is Il Formaggio e i Vermi: Il Cosmo di un Mugnaio del "500" by Italian historian Carlo Ginzburg,

acknowledges the importance of focusing on microscopic units of investigations—the family, the village, one single person, or one brief event—to see history from a more comprehensive or holistic perspective (Grendi 1977: 512). Clearly, the aim is to advance the understanding of complexity through the analysis of the particular (Magnússon and Szíjártó 2013; Trivellato 2015). Friedman himself was fond of this approach as signified by a statement in one of his several interviews: "Don't under-estimate the role of the very small, unknown person in history—and I'm speaking of everyday history, not of the great history of historians" (Obrist 2007: 53).

The method can be used in various disciplines, including the critique of architecture, which addresses Ginzburg's request to challenge his historical method with other perspectives (Ginzburg 2014). This study's unit of investigation is the design of a plan rather than a person. As such, the author gathered as archival sources the plans, drawings, and concepts that described Friedman's productive landscape approach and relied on published interviews he released throughout his life. Looking at these sources from the perspective of productive landscapes makes them "eccezionalmente normal[i]"[2] since they narrate a different story by enlightening unique content (Grendi 1977: 512).

Hence, the adoption of this method aims to find new evidence of Friedman's productive landscape theories to complement the accepted interpretations of his work, like the dismissal of the architect's role in the design process and the vernacular advocacy as the sole drive of his philosophy. Specifically, this chapter argues that Friedman's utopia can inspire a more sustainable productive landscape. More broadly, since grounded on bold possibilities, utopian visions can propel resilient practices for future changes.

The chapter is composed of three sections. The first section presents Friedman's historical and educational background giving an overview of the main principles of his theories. The second section describes three diverse approaches to a productive landscape. The third section and conclusion draw lessons for a more contemporary, resilient productive landscape.

2 The *Forma Mentis* of Emergency Times

Friedman looked at the city as a human habitat made of distinct ways of living. As he stated on the first page of his book *Pro Domo* (2006),

who is considered by many microhistorians to be the most influential scholar of the discipline. However, according to Ginzburg's investigation, the term microhistory was previously used by fictional writer Primo Levi (Ginzburg 1993). The word was applied to historical studies developed in the United States in 1958, in Mexico in 1968, and France in 1958 (Ginzburg 1993). Hence, microhistory appears to have a broader origin in terms of both time and geographical boundaries. Hence, microhistory has a broader application focusing on a broad range of topics and subjects (Muir 2003: 30–31; Brisotto 2019).

[2] A document is exceptionally normal because it is not usually studied. This kind of document reveals a representation of a culture, otherwise intelligible.

> To imagine one house is to imagine the whole world. Each individual imagines his own house. Thus, each one imagines his own world. Each house that is imagined by its inhabitant is different; each world imagined by an individual is different. A house and a world do not have to look like the house and the world that is imagined by others. At the same time, houses and worlds imagined by others are real for each individual. We live with the others, our "neighbours" and their imagined houses and worlds also belong to our world, our "environment." We live in a world made up of unique individuals. Our world is formed by our "neighbours," a common reality (Friedman 2006: 9)

Friedman spent his career trying to solve this intrinsic living contradiction by seeking the equilibrium between individual rights and a collective understanding of the different. In his view, the city's paradigm had to change to achieve this equilibrium and, therefore, a more democratic society. He investigated problems of overpopulation (how to size the ideal dimension of a social group and therefore of the spatial unit), urban density (the city in relation to the countryside), production of food (the issue of individual self-sustenance as well as of the city as an organism), and spatial freedom (the belief that individuals craft their space in independent and personalized ways making it difficult for the designer to predict how his/her project will be used). His unusual education coursework, attitude to positivist philosophy, and personal and professional experience in Israel highly influenced the *forma mentis* of Friedman.[3] This influence led him to develop resilient design approaches to the problems mentioned above, subverting the idea of a uniformed city with an idea of multiple connected habitats while safeguarding individual autonomy.

Born in Hungary to a Jewish family in 1923, Friedman was not allowed to enroll in a university due to the *Numerus Clausus* law that limited Jewish students' access to education in the interwar period (Nagy 2005; Friedman 2006: 9). He managed to study architecture as an auditor as deemed a promising talent under Ivan Kotsis, professor at Polytechnic School of Budapest (Orazi and Friedman 2012: 61; Orazi 2008: 88 in Friedman et al. 2008, Lebesque, Fentener van Vlissingen 1999: 116, Friedman 2006: 9). He also trained with architect Lajos Kozma who practiced an architectural style not included in the curriculum of Hungarian universities, the Bauhaus style. This training exposed Friedman to the latest architectural discourse of his time (Orazi and Friedman 2012: 62; Orazi 2008: 88 in Friedman et al. 2008). This alternative educational pathway was, in his own words, an opportunity rather than an obstacle as he "[…] was actually lucky: as [he] couldn't go down the main road, by necessity [he] got into a situation in which [he] was more informed than the main road allowed" (Orazi andFriedman 2012: 62).

The positivist thought that characterized Budapest's environment during Friedman's early years also played an essential role in his formation (Obrist 2007: 8; Orazi 2008: 88 in Friedman et al. 2012). This influence is essential to understanding Friedman's passion for technology and science to the extent that he titled one of his books *Toward a Scientific Architecture* (1975a). In this book, he tried to define his philosophy mathematically using notions of critical group size, the flatwriter, and the urban mechanism. All these concepts attempted to calculate human behaviors in relation to spatial design. Interestingly, he admitted the impossibility of predicting

[3] Mindset.

human behaviors when he was a child visiting Austria, something he saw as a failure of the positivist thought (Lebesque and Fentener van Vlissingen 1999: 116). To a more significant extent, this failure to predict how people will behave would be a restraint to planning the future (Obrist 2007: 41).

Nevertheless, he strived his whole life to define objective and mathematical design principles for achieving a more democratic and sustainable city system. It is in this belief that we witness the resilient significance of Friedman's work. In fact, Friedman tried to define the problems that concerned him through the methodological solutions he proposed across his career. His attempts shifted his definition of what a group in society should be in order to achieve a sustainable life, moving it from being synonymous with family in *L'Architecture Mobile* (1958) to the critical group size in *Pour une architecture scientifique* (1972). Analogously, using such evolutionary thinking, he approached and investigated a range of other issues he cared about by developing them throughout the years. We can see this approach as an embodiment of what Rittel and Webber described in 1973 as *wicked* problems of the urban condition, extremely complex issues (i.e.,: transportation, poverty, criminality) that a cause-effect approach could not solve. In these cases, solutions along with identification of the problems themselves emerge gradually during the planning process (Rittel and Webber 1973: 162). Hence, a *wicked* problem is constantly evolving and can be defined through an argumentative process. Rittel and Webber pointed to the lack of theories capable of addressing the *wicked* problem of the urban question. However, Friedman's visions demonstrate an attempt to develop such a theory—a theory that strives to be scientific and objective through a utopian and imaginative approach.

The Second World War taught him that access to primary resources, such as food and water, means survival (Orazi and Friedman 2012: 63), while his critique of the positivistic approach led him to elaborate on methods to design for the individual as well as the collective. His stay in Israel from 1946 to 1957 (Friedman 2006: 9–11) was essential to complete those views of architecture. Not only did he study and complete his education in architecture at Technion in Haifa but he also developed practical architectural skills. He saw first-hand how small groups of people could be self-sufficient when he worked in a kibbutz for six months (Friedman 2006: 10). He also learned building techniques when working as a construction worker and serving in the engineering corps during his one-year military duty. Both experiences probably led him to pursue the design of what he called *spatial infrastructures* (Friedman 2020) in one of the most pragmatic utopias, the *Ville Spatiale* in 1957 (Routilliard 2018: 7). A uniform framework of trusses that would let residents design their own space within it providing for their essential needs as sheltering, feeding, and socializing.

He experimented with these ideas when he designed cylindrical infrastructures for new settlements for the Israel Prime Minister in 1953 (Orazi and Friedman 2012; Obris 2007; Friedman 2006; Lebesque, Fentener van Vlissingen 1999). The proposal was not approved and Friedman had to design more traditional settlements, something that he saw as exploitation of space and its residence (Friedman 2006: 11). Although Friedman felt for the first time a "citizen equal to anybody else" in Israel (Orazi and Friedman 2012: 64; Lebesque and Fentener van Vlissingen 1999: 116), the failure to build his pragmatic utopia made it impossible for him to remain (Friedman and

Orazi 2015: 548). His ideals of a new society reflected in a new sense of space were disillusioned. This event along with a more general disappointment about his expectation from the Israel experience led him to leave for Europe.

He moved to France where he founded the incubator of all his concepts, the *Groupe d'Études d'Architecture Mobile* (GEAM) in 1958 with other young European architects. He worked and taught these ideas in numerous North American universities. To mention only two of these experiences, Percival Goodman, professor and theoretician of utopian communities invited him to Columbia (Ferri 2015: 337; Orazi 2011: 134 in Friedman and Rodriguez 2011), and in 1973, Massachusetts Institute of Technology (MIT) offered him a visiting position to develop his mathematical diagrams of self-design (the flatwriter) (Vardouli 2012). His career shifted from a planning utopia of the *Ville Spatiale* (1958) to a diagrammatic utopia of scientific exploration with *Toward a Scientific Architecture* (1971) and again to a graphics utopia represented by the vignettes of the *Manuals* for vulnerable populations in developing countries (1989). In this evolution, we can observe his strive to realize his utopias, make them real, and balance individual rights with collective living.

All his life experiences forged his mindset and design objectives. He turned the limitations imposed on the Hungarian Jewish population into opportunities. The experiences during and after the war taught him about survival. The participation in the construction of a new country shaped his pragmatic approach to utopias.[4] Friedman's work represents the idea that anything can become an opportunity, crisis leads to creativity, and the designer/planner/architect has to become a producer of methods for the users to employ, avoiding the dominance of the discipline. Emergency became his mode of thinking and the main drive beyond the design of infrastructures that allowed flexible uses (what he later defined, respectively, as hardware and software of the city) to be resilient, ready to adapt to emergent situations in this way mitigating the development of future emergencies.

[4] Although the design of the built environment of the new country of Israel might be seen as controversial in more contemporary views, we must focus here on the genuine interest that Friedman had to build a more democratic society. A sentiment that pushed him to pursue this utopia elsewhere when he saw the impossibility to realize it in Israel.

3 The Emergent Patterns of Productive Landscapes in Friedman's Oeuvre

Looking at Friedman's theories from the lens of their urban agriculture components, we will see concerns for survival, adaptability, and opportunities. Works like *Spatial Settlements* (1958), *Ou Commence la Ville?* (1977), and the *Manuals* (1989) point to the utopian flexible design of productive landscapes and the empowerment of the citizens for a self-resilient life, including how to grow food and maintain a garden. Specifically, there are three angles of how urban agriculture contributes to the discourse of city design in Friedman's philosophy: mobile agriculture as a means of non-urbanization, self-sufficiency as means of food security, and non-paternalism as means of pragmatic utopia. We will analyze these three views in the following sections.

3.1 Mobile Agriculture as a Means of Non-Urbanization

> All these ideas of self-planning–the continent city, urban agriculture, and mobile architecture are interrelated. They are not separate projects (Obrist 2007: 21).

Starting in 1958, Friedman elaborated on his theory of *L'Architecture Mobile,* which was visually embodied by the *Ville Spatiale*. This work was an overarching project that he kept reviewing until the end of his career, with the latest edition in 2020. Starting from the assumption that cities are fast-changing organisms, he concluded that they should be able to adapt and adjust to the societal transformations every five years (Friedman 2020: 65). This conclusion led to two postulates, (1) the city's changes are created by its inhabitants and, therefore, they—the inhabitants—are entitled to design these changes, and (2) the architect can only design the general framework in which such societal changes can occur. Mobility was his proposed axiom through which these posits exist. He admitted that the word *mobility* did not represent wholly the meaning he was trying to convey (Friedman, 2020: 19).

Nonetheless, he attempted to describe *mobility* as the transformations of society and the adaptive system of its tectonics. Hence, it is not the city that moves, but its transformations. It follows that its residents are not nomadic, contradicting Henri Lefebvre's main critique on Friedman's work (as cited in Bubea 2007: 67).[5]

The result of this introspection was a flexible city according to the individual needs while respecting a certain level of organization and coexistence between different groups, either the family (biological groups) or cultural groups (non-biological groups) (Friedman 2020: 19–22). Friedman saw infrastructure—a system of modular grids fluctuating over existing cities or landscapes—as the possible conduit of this

[5] Henri Lefebvre, "Avant-garde de la présence," Internationale Situationniste, no. 8 (January 1963), 17. Facsimile edition (Paris: Librairie Arthème Fayard, 1997).

flexibility. In Friedman's city, residents could turn and adapt their habitats at their measure. In his words,

> [...] Ville Spatiale project [...] presents a structure within which everything can change. By contrast, with the space-frame structure, even the structure itself can be changed. I always use this expression: architecture without building. This doesn't mean without super-structures, but rather that the super-structures can be changed. They are not eternal; they are not imposing. (Decavèle 2012: 24)

Hence, the city is not about its shape—an end-product—but about how it works (IAAC 3:33–5:10). In this vision, the hardware of the city (its physicality) has become software (its processes) to create a highly adaptable city (Obrist 2007: 14).

Friedman envisioned the occurrence of life inside this space-frame infrastructure in what he called the *Spatial Settlements.* The settlements were places for the inhabitants' nourishments. They had to contain houses (sheltering), public spaces (socializing), transportation (moving), industries (supplying), and agricultural spaces (eating) (Friedman 2020: 98). The infrastructure shifted from being an engineering object "to one touching upon the question of living" (Rouillard 2018: 8), intended to fulfill the basic needs of life (Banham 1976: 60).

Specifically, Friedman saw at least two advantages from bringing agricultural production inside the city:

- Cultivating legumes, fruits, and dairy products would eliminate costs of transportation and storage (Friedman 1982: 40; Friedman 2020: 99; Friedman 1978: 128–130)
- Bringing peasants into the city would enhance their quality of life (Friedman 2020: 99)

Although the infrastructures of the settlements were thought to be flexible and changeable, Friedman calculated a very specific agricultural surface of 40m^2 per inhabitant (Friedman 2020: 220). We see this approach within urban agriculture discourses as more in line with farming and industrial production than gardening for self-sustenance. In fact, the third principle of his theory on a "new urbanism" stated, "large cities should incorporate agriculture together with industrial activity. The urban farmer is a social necessity"[6] (Friedman 2020: 236; Friedman 2018: 66). Thus, the agricultural components of the *Spatial Settlements* were places of production in line with Le Corbusier's sensitivity toward this city activity. The Swiss architect envisioned the *City for Tomorrow* (1920) with blocks of apartments surrounded by fruit orchards and gardens, determining their surfaces at 150 m^2 per patch. The vegetable gardens would be clustered into a single field so that a hired farmer could cultivate it on behalf of all the apartment's residents. Le Corbusier introduced the professional figure of the urban farmer, thinking that the urban residents were not interested in farming or gardening:

> It is complicated and difficult to keep up and involves endless pains [...] for the householder and his wife to keep things tidy, to weed it, water it, kill the slugs and the rest; long after

[6] Author's translation.

> twilight the watering-can is still on the go. Some people may call all this a form of healthy exercise. On the contrary, it is a stupid, ineffective and sometimes dangerous thing. The children cannot play there, for they have no room to run about in, nor can the parents indulge in games or sports there. And the result of all this is a few pears and apples, a few carrots, a little parsley and so on. The whole thing is ridiculous (Le Corbusier 1971: 214–215).

Friedman did not explain the role of the urban farmer at the service of the city dwellers as Le Corbusier did. However, the alignment with the industrial activity suggests that Friedman saw urban agriculture as a functional distribution of space as Le Corbusier advocated. A superficial analysis would classify this approach as dangerously inclined to promote the already occurring migration flow from the countryside to the city. This phenomenon involved all the major cities of the Western World during the Industrialization era, depopulating the rural villages and leading to their neglect. A more in-depth view demonstrates that Friedman intended to protect the countryside and its landscape by limiting the expansion of existing cities through flexible infrastructures. This recommendation is evident in the sixth principle of the already mentioned "new urbanism" manifesto: "a new city in the desert is not doable. Big cities will develop from the small ancient ones: the city must be the intensification of the existing cities (Friedman 2020: 236)."[7]

The vertical and floating structures of the *Ville Spatiale* that Friedman imagined over Paris (Friedman 2020: 21–27), Tunis, New York (Friedman 2011: 42), Tel Aviv, and London (Friedman and Orazi 2015: 150–153) are all representations of this principle. In this light, his vision for urban agriculture is a meta-analysis of what he believed the relationship between countryside and city should have been: an act that advocates for the proximity between the production of food and consumers, replicating the rural life in the city. Rural villages were set close to one another and thus more efficient in terms of food production. In *Où Commence la Ville?* (1977), Friedman stated, "the city starts in the countryside" which manifests not a geographical origin of the city, but the imitation of that rural model (Friedman 1980). The city dwellers should "learn to live in small cities to make the big cities livable again" (Friedman 1977: H3-F12).[8] That is, Friedman's *Spatial Settlements* are an assemblage of small cities surrounded by suspended agricultural fields (Fig. 1) since the city must be self-sufficient by producing its food (Friedman 1977: F12).[9]

This understanding of Friedman's infrastructural city completely reverses the endless expansions of other mega-structures such as Archizoom's No-Stop City (1970) or the Superstudio metaprojects. The *Continent City* (1961–1994) is the ultimate manifestation of this attempt to limit urbanization in favor of a more balanced interrelation between countryside and city. The concept is based on Friedman's observation of Europe's territory where high-speed train networks virtually eliminated distances between cities. This continental fabric became the "new city," a system made of connected European existing municipalities such as Lyon, Paris,

[7] Author's translation.

[8] Author's translation.

[9] Author's translation.

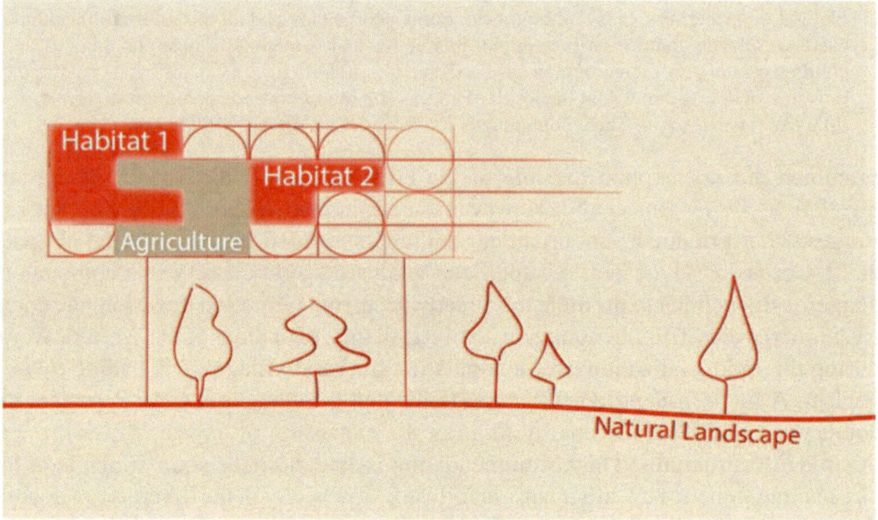

Fig. 1 Illustration inspired by one of Yona Friedman's sketches. The suspended infrastructure hosts small habitats separated by agricultural areas. The complete overview of Friedman's Spatial Settlements drawings can be found in his official website (http://www.yonafriedman.nl/?page_id=1053). Illustration by Carla Brisotto

and London (Obrist 2007: 21; Friedman 2018: 77).[10] The city-states (nodal points of the system) are self-sufficient and self-contained social organizations becoming "the building block for sustainable development" (Friedman 2018: 75). It is sustainable for the economy creating shared values that can be exchanged as also noted by German sociologist Ulrich Beck in his analysis of post-global cities development (Beck and Cronin 2012: xi). It is also sustainable in terms of consumption of resources, especially food production.

Agriculture is essential for the sustainability of the *Continental City*. Friedman noted that "[the continental city] has certain additional advantages: agriculture is included in the city. That means it can be a self-sustaining city system" (IAAC 11:04–11:17). The system does not exploit the countryside at the hand of urbanization since there are no city expansions. In fact, each of these cities is bound to a specific area (Friedman 2018: 77). Agriculture is that connective tissue between the cities that guarantees proximity between the food it produces and the consumers it serves. That is, the *Continental City* is a *Spatial Settlement* on a regional scale. In this light, urban agriculture inside the city is not different from the agricultural fields in-between cities. They are both mobile productive landscapes since they are adaptive to the needs of the city's inhabitants. Mobile agriculture becomes a means of non-urbanization.

[10] Paper written for UNESCO in 1994 and published in *Pro-Domo*.

3.2 Self-sufficiency as Means of Food Security

> Earth is overbuilt, Earth is overplanned, Earth is overfarmed. This does not mean that we don't need architects, planners, and farmers (Friedman and Decavèle 2011: 00:24–00:50).

The purpose of the infrastructure imagined by Friedman was to empower citizens to design their own space and place, "an early representation of so-called 'participation'" (Rouillard 2018: 9). However, this empowerment is not an attempt to completely cede the design of space to the dwellers of a city. On the contrary, it is an effort to advance the disciplines of architecture and urbanism toward new roles. As Italian architect Giancarlo De Carlo said, reflecting on his life-long interest in design participation, "the problem was no longer how to make people participate in architecture, but how to make architecture that can be appropriated and participated in by people" (Serrazanetti and Schubert 2011: 214; Brisotto 2018).

In 1968, elaborating on the nature of architecture and urbanism, Friedman identified three possible professional subcategories. The laboratory architect/urbanist seeks theories and methods that the other professionals will follow. The contact architect/urbanist applies such theories and methods in collaboration with the users. The architect/artist designs monuments in concert with the public in a participative decision-making process (this last option resonates with the archistars of current times, although often deprived of the public participation facet) (Friedman 2020: 232–233).

The *Spatial Settlements* and the *Continental City* are the results of Friedman's work as a laboratory architect. In the *Manuals* for the Communication Centre of Scientific Knowledge for Self-Resilience, an institute of the United Nations, he directed his attention to particular users, the vulnerable population of developing countries, especially India (Friedman 1986: 333; Friedman 1987: 43; Orazi 2008: 89). With this project, he became a contact architect, a facilitator of applied methods. The *Manuals* cover topics on food growth and preparation (gardening on shelves, minimal kitchen gardens, food storage and conservation, as well as plant nourishment), but also water collection, accessible energy sources, health care, environment, and enterprise (Friedman 1986: 335; Friedman 1987: 44). The *Manuals* teach how to be self-sufficient by providing simple instructions about basic necessities, the same necessities he imagined in the *Spatial Settlements.*

The main principle behind this series of manuals is a very simple equation that Friedman published in Leonardo in 1986 (333; Friedman 1987: 43). The equation represents Friedman's belief that the most appropriate indicator for the economic capacity of a nation is the gross national resource per capita (GNR p.c.) rather than the usual gross national product per capita (GNP). For resources, he considered both natural and human, the latter either in terms of labor and intellect. The GNR p.c. consists of the sum of all natural and human resources (raw materials, land, energy sources, human labor force, and human intelligence) divided by the total number of citizens (Eq. 1).

$$\text{GNR p.c.} = (\text{natural resources} + \text{human labor} + \text{human intelligence})/\text{number of citizens} \quad (1)$$

If natural resources are finite and citizens and human labor are proportionally related, it derives that the only factor that can be manipulated to increase the GNR p.c. is human intelligence (Friedman 1987: 42–43; Friedman 1986: 333–336). Of course, Friedman did not consider that demographic growth could be unbalanced in regard to human labor. For instance, an increase in senior citizens does not equate to an increase in human labor. Nevertheless, the equation shows the architect's aim to achieve an equally distributed profit through education and, therefore, disenfranchisement of a nation's vulnerable population (Brisotto 2019: 79). The idea that human intelligence was the only factor that could yield to the enhancement of the wealth of a nation generated the teaching purpose behind the *Manuals*. Friedman designed these manuals so that they were mighty accessible to the general public. In fact,

> The manuals are highly simplified communication supports, with many drawings simple enough to be copied easily by unskilled people and with brief text captions. The presentation has to appeal to knowledge already possessed by the target public and must emphasize the effective benefit the recipients can expect from the innovation (Friedman 1986: 334)

This approach to self-sustenance is what Friedman called l'*Architecture de Survive* (Friedman 1978: 79), an architecture of survival. This architecture is possible when architects design it in a way that it is easy to be self-realized. This kind of architecture favors food production and all the other primordial resources such as water, privacy, climatic protection, social organizations, and aesthetic satisfaction (Friedman 1978: 79,122). Food production is deemed essential not only for survival but also for creating a human habitat that satisfies life quality, which is viewed as even more important than shelter (Obrist as cited in Friedman and Rodríguez 2011: 147). Therefore, the productive landscape is a domain that the individual generates to reconnect with the land and achieve food security.

Friedman's mobility encouraged people to participate in architecture by providing a framework (Lockard 2017: 123) to guide humans' return to the production of space—and we would also say, of food—either inside the city, as in the *Spatial Settlements* or the network system of the *Continental City*, or as survival techniques as in the *Manuals*. The latter provides practical and explicit instruction on how to build simple technologies to become self-sufficient. This approach is an expression of democratic access to well-being, leveraging individuals' potentiality.

3.3 Non-paternalism as Means of Pragmatic Utopia

> Man has long tried to formulate the conditions for achieving an ideal environment. We call such a model a utopia. [...] Utopian proposals are provoked by dissatisfactions within a society that has not found remedies for them (Friedman 1975a: 37).

The etymology of the word "utopia" originated from ancient Greek with the combination of the words "ou" and "topos," meaning, respectively, "no" and "place." Hence, utopia is the negation of place. The word was first used as the title of a book by English writer Thomas More in 1516. More narrated the story of a fictional island called precisely Utopia, the site of an ideal society. Utopia residents lived in separated districts. Mobility between the districts was impeded by authority. In Utopia, wine bars and ale-houses were strictly forbidden as anything related to what was considered an immoral lifestyle according to More's time standards. Ironically, what was supposed to be an ideal world was neither idyllic nor just. Instead, it was an overregulated "no place" that operated tyrannically where anyone making autonomous decisions was severely punished (More 2005: 58). Thus, initially and contrary to the current adopted meaning, Utopia represented a standardized world achieved by controlling any possible controversial events, behaviors, and values. Utopia was unattainable precisely because it aimed to homogenize the behaviors of everyday life (Brisotto 2019: 149). Using one of Friedman's definitions of utopia, More's Utopia failed because it was a paternalistic dream (Friedman 1975a: 160; Friedman 1975b: 27–34).

Friedman dedicated a book, *Utopies réalisables* (2008), to investigating utopias and their attainability. In Friedman's thinking, utopias occur in history when society is balanced any longer. Hence, utopias are models to pursue an ideal environment where people and the physical world interact harmoniously and democratically. As in More's island, utopias are not realizable when they become a mirror of one single solution applied using control over people. In opposition to this approach, Friedman suggested a non-paternalistic utopia. This kind of utopia would be a model that would allow several, adaptable living conditions.

Furthermore, he suggested creating a communication system that would allow individuals to verify how their decisions might impact the lives of others while informing them of the potential impacts to themselves due to someone else's decision (Friedman 1972: 39). To make this model feasible, the community's dimension is of critical importance. Friedman understood that utopias pursue changes for a problem that involves a vast number of individuals (Obrist 2007: 43). Therefore, finding the ideal environment that achieves the well-being of all its inhabitants is the real challenge.

Friedman believed that utopias are realizable where there is consensus, and such consensus can occur only by controlling the size of the community that uses the resources of its territory. He named this concept "critical group size" (Friedman 1975a: 158; Friedman 2018: 175–176), the maximum size of a group not only to be able to be sustainable but also to be capable of collectively solving problems that might arise. The size is not a standardized dimension. Instead, it must be calculated according to the specific case, territory, resources at stake, and urgency of the problem that needs to be solved. If a decision has to be made in a short amount of time, then the group size must be small. Conversely, if a decision can take a longer time, the group can be larger (Obrist 2007: 35–41). The notion of critical group size is foundational for the sizing of the *Spatial Settlements* and the principle of non-expansion of the *Continental City*.

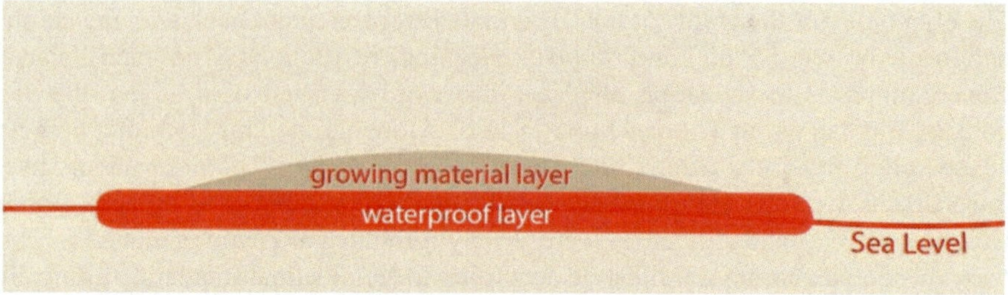

Fig. 2 Reproduction of Yona Friedman's sketch. The island was made of artificial materials allowing the cultivation of food on the water. The original drawing was published in Friedman (1972) On Models of Utopias and Social Ecology. Leonardo 5 (1): 39. Illustration by Carla Brisotto

De Michelis points out that utopias identify "something [that] is missing" in the present condition of humanity perfection—a status that people have always strived to achieve (de Michelis 2008: 32 in Cerizza, Daneri). Hence, utopias show imperfections to induce improvements and changes. Friedman did not imagine a perfect society in which everyone consented to the greatest good. The imperfection was what interested him as characterized by the idea that in a non-paternalistic utopia the individual still had the opportunity to choose differently. If these individuals found it impossible to pursue their well-being without affecting others, they could choose "elsewhere," creating another group.

Interestingly, his proposal to exemplify the elsewhere was an artificial island cultivated with hydroponic gardens for the food provision of the new residents (Fig. 2) (Friedman 1972: 39).

Although contradictory with the notion of a non-expanding built environment, this concept was a very innovative idea for the year it was formulated. The floating garden is an example of what utopia might signify, a vision for a possible future that "technology makes feasible" (Friedman 1972: 39).

Friedman's utopia is an emblem of "the topological utopias of spatial urbanism" that characterized the French discourse of the sixties (Busbea 2007: 191). Those utopias were designed and invented to solve the social struggles between over-structures of powers and the individuals. The environmental and ecological crisis are the struggles of today's scenario, substituting the social concerns for lost freedom in living, experiencing, and producing space. Nevertheless, Friedman's utopias are inspirational in understanding how to implement resilient solutions often obstructed by public opinion due to the general inability to envision future long-term scenarios. This aspect is mainly due to the individual nature of contemporary society that favors immediate benefits rather than "required behavioral changes, even when doing so is clearly contrary to our long-term interests" (Levin et al. 2012: 125; Rittel and Webber 1963). Furthermore, the increased diversity of the population has broadened the spectrum of individuals with diverse social and cultural needs that exponentially increase the difficulty to obtain consensus. Ideas such as the critical group size and non-paternalistic utopias can inspire and indicate directions for the acceptance of innovative urban agriculture solutions that encounter society's skepticism (Lappe and Bailey 2002; Miller 2018).

4 Conclusions

In 2050, most of the world population (68%) will live in urban areas (United Nations 2018). We are familiar with this datum and its implication for the future of cities. One outcome of this phenomenon is the conversion of rural areas into urban areas in an unstoppable process of urbanization of the countryside (Sun et al. 2020; Seto et al. 2011). Hence, scholarship must concentrate on the process of rural area transformations and not exclusively on the urban core changes. In this light, the productive landscape design is key to understanding this transformative process: How do we protect the countryside from urbanization? How do we integrate it into the urban fabric?

Friedman's work on the productive landscape informs recent architectural attention on productive landscapes providing insights about their design. Friedman approached these questions from the agricultural regional system scale and at the individual scale. These two conditions, as we will see, are tightly connected.

The Continental City represents *the first approach.* With this vision, Friedman foresaw the development of networks of cities that now characterize most of the world. The megaregions of the United States—connected municipalities, suburban areas, and rural lands that expand over multiple states—are examples of such urban mechanisms (Barnett 2020; Florida et al. 2007). These regions are rapidly expanding without an overarching strategic plan, a scenario that would lead to unavoidable, uncontrolled sprawling, and consequent disruption of agricultural land. Friedman's proposal suggests that agriculture is nested in-between existing urban settlements belonging to different geographical and political contexts. This productive landscape relies on the proximity between food production and consumers to achieve sustainability while becoming a blockage system to urbanization. This strategy leads us to think about the productive landscape more in terms of foodsheds than localized production (Horst and Gaolach 2015; Zasada 2019). By establishing foodsheds, we can overcome the political limitation that geographical boundaries would apply to extensive food systems.

The second approach is the design of a productive landscape in terms of individual self-sustenance. Friedman suggests that the production of food at the individual level is necessary to create sufficiency when the food system fails to provide nourishment due to emergencies. Literature observed how easily disruptive the food system could be after a disaster or sanitary emergency. The individual resilience suggested by Friedman can address and prepare for such unexpected conditions. It is a tactical methodology established through educational toolkits that must be in place as mitigation plans or as adaptive infrastructures used by inhabitants of a city according to needs and personal habits.

At first glance, these two levels of resilience, regional and individual, seem in contrast. On the contrary, they are complementary. The productive landscape serving the region is a strategic infrastructure, a flexible model to operate land for a certain number of urban settlements. The infrastructure is also in place at the individual level to guarantee survival tactics and contribute to the flexibility of the regional system. Strategies (long-term plans) and tactics (short-term plans) are part of the same vision, building a production system that can be efficient and reliable.

Friedman was skeptical about predicting the future as it "[...] cannot be planned. It is a no-plan time" (Obrist 2007: 41). While we cannot completely predict the future, which is a possibility that is embedded in resilient thinking ideas, we can design a productive landscape on a large scale while addressing unpredictable events on a smaller scale. This goal can be reached by designing a resilient, productive landscape with two simultaneous levels: one that serves the majority and one that is flexible and self-regulatory for those excluded by the system due to choices, accidents, or events. We must pursue the sustainable goal for the United Nations number fourteen (feeding), knowing that our plans and designs might fail. Therefore, we also need to nurture, embrace, and sustain individual resilience to overcome those failures.

This chapter is not arguing that Friedman's work has no limitations. His ideas looked into the social sphere as the ecological solution for the problems of his time. In doing so, his designs risk to frame an ecology that is still anthropocentric. Furthermore, his theory of self-substance addresses basic needs that might lead to a bare life hindering the possibility of achieving fulfilling civic rights. However, we are proposing to look at his work acknowledging the powerful message of his pragmatic utopia. In order to face the future's unpredictability, architects/planners need to create a flexible food system engaging different stakeholders.

Acknowledgments The author expresses her deepest gratitude to Mrs. Helene Fentener van Vlissingen. This chapter would not have been possible without her precious feedback, enthusiasm, and support.

References

Banham R (1976) Megastructure: urban futures of the recent past. Harper and Row, New York

Barnett J (2020) Designing the megaregion: meeting urban challenges at a new scale. Island Press, Washington, DC

Beck U, Cronin C (2012) Twenty Observations on a World in Turmoil. Polity, Cambridge, Malden

Brisotto C (2019) City of agro-fragments: how ecological thinking influences architectural processes in the era of post-globalization. Dissertation, University of Florida

Brisotto C (2018) Sense of place in the era of Anthropocene: how architectural drawings imagines the urban ecological shift. In: Bean J, Dickinson S, Ida A (eds) Critical Practice in An Age of Complexity - An Interdisciplinary Critique of The Built Environment. Amps/University of Arizona conference 12, Tucson, February 2018, vol. 11, Amps, N/A, p 303

Busbea L (2007) Topologies: the urban utopia in France 1960–1970. The MIT Press, Cambridge, USA

Decavèle J-B (2012) Dear Yona. Frieze. 151:23–24

de Wit W (2009) The Papers of Yona Friedman. Getty Research Journal 1:191–196
Etymonline (2019) Utopia. Available via https://www.etymonline.com/word/utopia. Accessed March 1, 2019
Ferri P (2015) Utopie Realizzabili. Scienze Del Territorio 3:336–339
Florida R, Gulden T, Mellander C (2007) The Rise of the Mega-Region. Camb J Reg Econ Soc 1(3):459–476
Friedman Y (1972) On Models of Utopias and Social Ecology. Leonardo 5(1):37–41
Friedman Y (1975) Toward a scientific architecture. MIT Press, Cambridge
Friedman Y (1975) Utopies réalisables. Union Générale d' Éditions, Paris
Friedman Y (1977) Où Commence la Ville? UNESCO, Paris
Friedman Y (1978) L'Architecture de Survie: Ou Invente Aujourd'hui le Monde de Demain. Casterman, Paris
Friedman Y (1980) Que peuvent apprendre les architectes des villages urbains et que peuvent apprendre les villages urbains des architectes? Espaces Et Sociétés 32:43–54
Friedman Y (1982) Alternatives énergétiques: Ou la civilisation paysanne modernisée. Editions Dangles, St.-Jean-de-Bray, France
Friedman Y (1986) Communication Centre of Scientific Knowledge for Self-Reliance. Leonardo 18(4):333–336
Friedman Y (1987) Lezioni di Tecnologie Semplici. Spazio e Società 10(37):42–43
Friedman Y (1989) Food: Manuals for Sample. CNEI, Paris
Friedman Y (2006) Continental City. Domus 896
Friedman Y (2018) Pro domo / Yona Friedman. Actar, Barcelona
Friedman Y (2020) L'architecture mobile: vers une cité conçue par ses habitants (1958–2020). Éditions de l'éclat, Paris-Tel-Aviv
Friedman Y, Rodríguez VMI (2011) Arquitectura con la gente, por la gente, para la gente = Architecture with the people, by the people, for the people. Musac - Actar, Barcelona
Friedman Y, Obrist H U (2007) Yona Friedman. Walther König, Köln
Friedman Y, Cerizza L, Daneri A (2008) Yona Friedman. Charta, Milan
Friedman Y, Orazi M (2015) Yona Friedman: the dilution of architecture. Park Books, Zurich
Horst M, Gaolach B (2015) The potential of local food systems in North America: a review of foodshed analyses. Renewable Agric Food Syst 30(5):399–407
Institute for Advanced architecture of Catalonia (IAAC) (2012) Closing Lecture July 10, 2012 - Yona Friedman. Available via https://www.youtube.com/watch?v=BEX1dk0cJxw. Accessed 15 January 2021
Yona Friedman Y, Decavèle J B (2011) Yona Friedman architecture without a building, Balkis Productions. Available via https://vimeo.com/25050690. Accessed 19 December 2020
Ginzburg C (1976) Il formaggio e i vermi: il cosmo di un mugnaio del '500. Einaudi, Torino
Ginzburg C, Tedeschi J, Tedeschi AC (1993) Microhistory: two or three things that I know about it. Crit Inq 20(1):10–35
Grendi E (1977) Microanalisi e Storia Sociale. Quad Stor 12(35/2):506–520
Lappe M, Britt B (2002) Engineering the farm: the social and ethical aspects of agricultural biotechnology. Island Press, Washington, DC
Lebesque S, Fentener van Vlissingen H (1999) Yona Friedman: structures serving the unpredictable. NAi Publishers, Rotterdam
Le Corbusier (1971) The city of to-morrow and its planning. M.I.T. Press, Cambridge, MA
Levin K, Cashore B, Bernstein S, Auld G (2012) Overcoming the tragedy of super wicked problems: constraining our future selves to ameliorate global climate change. Policy Sci 45(2):123–152
Lockard J (2017) Yona Friedman: architecture mobile = architecture vivante. J Soc Archit Hist 76(1):123–125
Magnússon S, Szíjártó I (2013) What is Microhistory?: theory and practice. Routledge, Milton Park, Abingdon
Miller HI (2018) Genetic engineering applied to agriculture has a long row to hoe. GM Crops & Food 9(1):45–48

More T (2005) Utopia. Cambridge University Press, Cambridge
Muir E W (2002) Modi diversi di leggere Il formaggio e i vermi. In: Colonnello A, Del Col A (eds) Uno storico, un mugnaio, un libro: Carlo Ginzburg, il formaggio e i vermi, 1976–2002, Edizioni Università' di Trieste, Trieste, p 29–33
Nagy PT (2005) The numerus clausus in inter-war Hungary. East European Jewish Affairs 35(1):13–22
Orazi M, Friedman Y (2012) A Conversation with Yona Friedman. Log 26:58–68
Pearson LC (2019) System cities: building a 'quantitative utopia.' Archit Des 89(4):70–77
Rittel HWJ, Webber MM (1973) Dilemmas in a general theory of planning. Policy Sci 4(2):155–169
Roche F (2020) Yona Friedman®: A case study from radicalism to idiocracy. Flash Art International 53(332):146–159
Rouillard D (2018) Megaspace structure Yona Friedman and Eckhard Schulze – Fielitz. Histories of Postwar Architecture 1(3):3–18
Serrazanetti F, Schubert M, De Carlo G (2011) Giancarlo de Carlo: inspiration and process in architecture. Moleskine, N/A
Seto KC, Fragkias M, Güneralp B, Reilly MK (2011) A Meta-Analysis of Global Urban Land Expansion. PLoS ONE 6(8): e23777. https://doi.org/10.1371/journal.pone.0023777
Sun L, Chen J, Li Q, Huang D (2020) Dramatic uneven urbanization of large cities throughout the world in recent decades. Nat Commun 11(1):1–9. https://doi.org/10.1038/s41467-020-19158-1
Trivellato F (2015) Microstoria/Microhistoire/Microhistory. Fr Polit Cult Soc 33(1):122–134
Vardouli T (2020) How architect Yona Friedman used math to design utopian cities of the future. In: The Conversation. Available via https://theconversation.com/how-architect-yona-friedman-used-math-to-design-utopian-cities-of-the-future-132474. Accessed 10 April 2021
Veloso PLA (2014) Cybernetic diagrams: design strategies for an open game. Int J Archit Comput 12(4):379–398
Walker BH, Salt D (2006) Resilience thinking: sustaining ecosystems and people in a changing world. Island Press, Washington, DC
United Nation. World Urbanization Prospects. Department of Economic and Social Affairs Population Division. 2018. https://population.un.org/wup/. Accessed 18 April 2021
Zasada I, Schmutz U, Wascher D, Kneafsey M, Corsi S, Mazzocchi C, Monaco F, Boyce P, Doernberg A, Sali G, Piorr A (2019) Food beyond the city – Analysing foodsheds and self-sufficiency for different food system scenarios in European metropolitan regions. City Cult Soc 16:25–35